Fundamental Questions of Practical Cosmology

Astrophysics and Space Science Library

For further volumes:
www.springer.com/series/5664

Yurij Baryshev • Pekka Teerikorpi

Fundamental Questions of Practical Cosmology

Exploring the Realm of Galaxies

Springer

Yurij Baryshev
Institute of Astronomy
St.Petersburg State University
Staryj Peterhoff, 198504
St.Petersburg
Russia
yubaryshev@mail.ru

Pekka Teerikorpi
Tuorla Observatory
Department of Physics and Astronomy
University of Turku
21500 Piikkiö
Finland
pekkatee@utu.fi

ISSN 0067-0057 Astrophysics and Space Science Library
ISBN 978-94-007-2378-8 e-ISBN 978-94-007-2379-5
DOI 10.1007/978-94-007-2379-5
Springer Dordrecht Heidelberg London New York

Library of Congress Control Number: 2011939769

Cover illustration: R. Williams (STScI), the Hubble Deep Field Team and NASA
Cover design: eStudio Calamar, Berlin/Figueres

Printed on acid-free paper

Springer is part of Springer Science+Business Media (www.springer.com)

Preface

Allan Sandage used the term "Practical cosmology" to denote the study of the large-scale universe and the search for the world model which best describes it. In our book we, as students of practical cosmology, guide the reader through modern cosmology, with emphasis on cosmological physics within our gradually deepening sample of the universe. We have restricted this treatise to the realm of galaxies where we can measure distances to individual objects, which in itself is a classical and fundamental problem in astronomy. Therefore, our discussion of the problems of the early universe and the cosmic background radiation and its angular fluctuations, so central in modern cosmology in general, is relatively limited.

This book presents cosmology as a physical science based on observations, experiments and theoretical interpretations. We remind the reader that fundamental physics as understanding of natural phenomena, including the whole universe, is the modern *Philosophiae Naturalis* started by Galileo and Newton. We do not presuppose advanced knowledge of astronomy and do not go into detailed descriptions of observing techniques. Basic mathematical concepts used in modern cosmological models are presented in a simple way. We hope that all this will make the book useful for both astronomers and general physicists, and also for university students of physical sciences. If needed, suitable background reading on astronomy may be found in the book by Karttunen et al.: "Fundamental Astronomy" (Springer 2006), and on history of astronomy and cosmology in Teerikorpi et al.: "The Evolving Universe and the Origin of Life—The Search for Our Cosmic Roots" (Springer 2009). The development of the Cosmological Principle has been discussed by Baryshev and Teerikorpi: "Discovery of Cosmic Fractals" (World Scientific 2002).

The following subjects are close to our own experience in the study of the galaxy universe and form the main contents of the book: The cosmic distance scale and the Malmquist bias. Gravitation and world models. Observational tests of cosmological models. The large-scale structure of the universe. Conceptual problems of cosmological physics.

We emphasize those aspects which give practical cosmology its special character, due to a subtle interplay between observations, data analysis, and fundamental physics. We hope that our book will help the reader to achieve useful conceptual understanding of some central but not so often emphasized problems in modern cosmology (such as selection effects and methods of analysis of the large-scale structure).

We underline throughout the book the essential fact that to obtain the correct picture of cosmological physics one should be aware that cosmology is—in a sense—"a science of cosmic selection effects". The interpretations from directly observed astronomical quantities and from astronomical data in general, may be in a subtle manner distorted by physical and technical limitations of observations and by inadequate methods of data analysis.

Cosmology has not been finished yet, and many fundamental questions are still open problems. The late Edward Harrison wrote to us in a letter: "Cosmology is always in the state of triumph and crisis. It is the natural state of the art." And this is because the initial assumptions and physical consequences are often difficult or impossible to test directly. There is the need for continuous testing of the accepted cosmological framework, which is, in the terminology of Thomas Kuhn, the working paradigm which rules and inspires contemporary cosmologists. Assumptions made by cosmologists determine the theory, its predictions and the inferred properties of deep space phenomena. At the same time the assumptions themselves are open for investigation—a constructive duality. Therefore we also wish to point out the positive role of alternative cosmological ideas which serve as a test-bench on the way towards a true world model.[1]

In principle, one would like to see practical cosmology as wider based than any specific cosmological model. This is because its methods are especially aimed at testing the initial assumptions and basic predictions of different world models. It should guide cosmologists between the Scyllas of empirism and the Charybdises of pure thinking ever tempting us away from the correct route towards progressively closer approximations to the true world model and to a deeper understanding of Reality.

We wish to mention with gratitude several people whose collaboration with us or whose own work have made this book possible. Some of them have also read parts of the manuscript.

Andrej Berdyugin, Lucette Bottinelli, Alexander Butkevich, Gene Byrd, Arthur Chernin, Timo Ekholm, Chris Flynn, Andrea Gabrielli, Lucienne Gouguenheim, Alik Gromov (†), Mikko Hanski, Toivo Jaakkola (†), Michael Joyce, Igor Karachentsev, Boris Komberg, Francesco Sylos Labini, Ari Lehto, Benoit Mandelbrot (†), Georges Paturel, Luciano Pietronero, Fred Rost, Allan Sandage (†), Vladimir Sokolov, Gilles Theureau, Mauri Valtonen.

We dedicate our book to the memories of Allan Sandage (1926–2010), Geoffrey Burbidge (1925–2010), and Benoit Mandelbrot (1924–2010). In the 20th century, Sandage represented the classical approach to observational cosmology, Burbidge

[1]During the preparation of the book we organized the conference *Problems of Practical Cosmology* held in St.Petersburg in June 2008. The meeting offered examples of mainstream and critical views, both of which are fruitful for the advancement of cosmological physics (Proceedings at http://ppc08.astro.spbu.ru).

defended alternative cosmological views, and Mandelbrot introduced novel mathematical concepts for describing cosmologically distributed matter. We hope that we have been able to convey, not only their specific contributions to cosmology, but also some of their spirit of approach to science.

St.Petersburg, Russia — Yurij Baryshev
Paimio, Finland — Pekka Teerikorpi

About the Authors and the Book

Dr. Yurij Baryshev (b. 1948) works as Senior Research Associate at the Astronomical Institute of St.Petersburg University, Russia. He has done research in various fields of extragalactic astronomy and cosmology, where his main research interest is relativistic astrophysics including cosmology and gravity physics. He has investigated Einstein's and Feynman's approaches to gravitation, active galactic nuclei, quasars, gravitational radiation, gravitational lensing by dark matter, and crucial cosmological tests. He has also studied the implications for cosmology of the recent discovery of the dark energy component and the fractal structure of the luminous matter distribution. He has been working as invited professor at the Rome University "La Sapienza" (Italy) and at the Lyon University (France). He has co-authored with P. Teerikorpi the book on the history of the Cosmological Principle *Discovery of Cosmic Fractals* (World Scientific 2002).

Dr. Pekka Teerikorpi (b. 1948) works as Senior Research Associate at Tuorla Observatory of the Department of Physics and Astronomy of Turku University. He has done research in various fields of Galactic and extragalactic astronomy and practical cosmology, with special interest in the selection effects and biases influencing the determination of the cosmic distance scale and the value of the Hubble constant. He has also made excursions to the Milky Way (the rotation curve, interstellar dust), the realm of active galactic nuclei and the problem of the local detection of dark energy. He has co-authored with Yu. Baryshev the book on the history of the Cosmological Principle *Discovery of Cosmic Fractals* (World Scientific 2002) and has recently written, together with a few other astronomers and exobiologists the book *The Evolving Universe and the Origin of Life—The Search for our Cosmic Roots* (Springer 2009), telling about the history and spirit of our science.

The book guides the reader (astronomer, physicist, university student) through central questions of Practical Cosmology, a term used by the late Allan Sandage to denote the modern scientific enterprise to find out the cosmological model best describing the universe of galaxies, its geometry, size, age, and material contents. The authors draw from their personal experience in astrophysics and cosmology to explain key concepts of cosmology, both observational and theoretical, and to highlight several items which give cosmology its special character:

- idiosyncratic features of the "cosmic laboratory"
- Malmquist bias in determination of cosmic distances
- theory of gravitation as a cornerstone of cosmological models
- crucial tests checking the reality of space expansion
- methods of analyzing the structures of the universe as mapped by galaxies
- usefulness of fractal as a model to describe the large-scale structure
- new cosmological physics inherent in the Friedmann world model

Contents

Chapter 1
The Golden Age of Cosmological Physics

The tremendous growth of observational efforts devoted to cosmological questions shows that cosmology is becoming a mature physical science with its own subject and methods. This is a novel situation for a field which up to recent times has been characterized by a respectable collection of theoretical ideas, but a small number of crucial observations to constrain them. Only six decades ago Hermann Bondi (1952) expressed the state of cosmology at the time so that "... the checking of a prediction, which usually forms such a vital link in the formulation of physical theories, does not occur in this field, and we have to rely on less objective and certain criteria, such as how satisfying and how simple a theory is".

In 1914 Vesto Slipher had a total of 15 nebulae in his list of measured spectra. Today the surveys of galaxy redshifts contain millions of objects and allow one to study the large-scale structure of the universe. The deep pencil-beam surveys in many colours extend up to redshifts z of 6, and one can study how galaxies and their clustering evolve. The gamma-ray bursts (z up to 9), the distant Ia supernovae, the gravitational lensing by galaxies and clusters, the observations at radio to gamma frequencies by space observatories—all these characterize the new era where "the checking of a prediction" is an intimate part of cosmology.

1.1 Our Sample of the Universe

Albert Einstein introduced the term practical geometry for the empirical determination of the geometry of physical space. Practical cosmology (the term used by Allan Sandage) is a science on large-scale physics, where the scales of distances, times, and masses are the largest ones available for us. It deals with world models and observations performed for testing them, and is, in certain sense, the modern inheritor of natural philosophy at such a deep level that allows one to regard the whole observable universe as an immense test ground of fundamental physics. It has an ambitious goal to build a trustworthy world model, in itself a major goal of science. The model is also a necessary instrument for interpreting deep-space phenomena and celestial bodies whose distances, sizes and luminosities we otherwise

Y. Baryshev, P. Teerikorpi, *Fundamental Questions of Practical Cosmology*,
Astrophysics and Space Science Library 383,
DOI 10.1007/978-94-007-2379-5_1, © Springer Science+Business Media B.V. 2012

cannot infer—in fact, nothing, it has been said, is more practical than a good theory. Obviously the problem is that the theory must be good.

Cosmology is based on theory (cosmological principles and theories of physical interactions) and observation (detection of photons and other cosmic particles). Thus its advancement is linked to the development of fundamental physical theories and the growth of observational data.

1.1.1 The New Building Blocks

During the 1920s, in a fine phase of the history of science, it became clear that a majority of the faint nebulae in the sky are distant galaxies far beyond the borders of our Milky Way system. This gave impetus for new classification systems after the older ones had mixed objects from inside and outside of the Milky Way galaxy. The basic pattern among the different kinds of galaxies was identified remarkably quickly and became iconized by Hubble's "tuning fork".

Hubble & Lundmark In 1926 two systems were proposed by Edwin Hubble and Knut Lundmark. Hubble's system remained in use and has almost unchanged stood the test of time. His "extragalactic nebulae" included

A. Regular
1. Elliptical (En) ($n = 0, 1, \ldots, 7$ indicates the ellipticity of the image)
2. Spirals: Normal spirals (Sa – Sb – Sc); Barred spirals (SBa – SBb – SBc)

B. Irregular (Irr)

Lundmark's "anagalactic nebulae" were
1. Elliptical, elongated or lenticular nebulae (Ae) (different degrees of concentration)
2. Spiral nebulae a. spiral structure barely seen (As0); b. different degrees of concentration (As1-As5); c.-f. certain special types of spirals
3. Magellanic clouds (Am) (different degrees of concentration)

Both systems contain elliptical, spiral and irregular (Magellanic) types. Finer divisions differ. Lundmark (1926) used the concentration of light as a basic index.[1] Hubble (1926) grouped ellipticals in terms of the ellipticity of their images. The division into a, b, c spirals was based on three criteria: (1) relative size of the central bulge, (2) extent to which the spiral arms are unwound, and (3) resolution in the arms, from Sa "with closely coiled arms of unresolved nebulosity" to Sc with its arms "wide open and the nuclei inconspicuous".

Hubble's way of arranging the spiral galaxies has proven successful in its simplicity and physical relevance: many properties not seen "by eye" have been found to change systematically along the spiral sequence. Instead of the gradual shift from

---

[1] Over thirty years later, in Morgan's system the classes (a, af, f, fg, g, gk, k) change in the sense of increasing domination of the nuclear component, together with a simultaneous change in the spectral type. Lundmark's and Morgan's classes are rather well correlated.

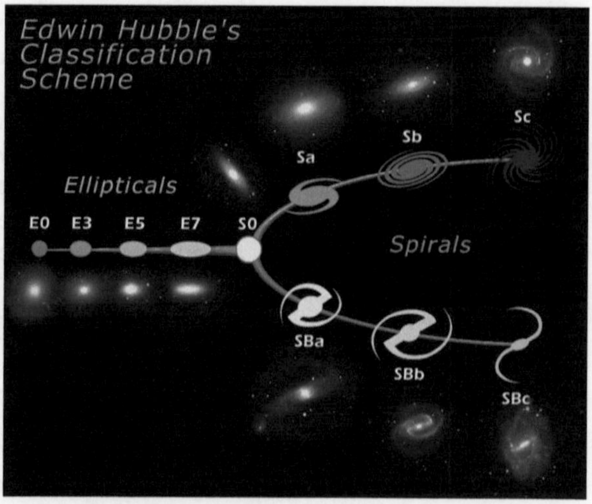

Fig. 1.1 Hubble's tuning fork diagram. A modern presentation (NASA)

"early" to "later" types, Lundmark's eyes were caught by physically interesting subclasses now studied by numerical N-body simulations of spiral arms and interacting galaxies. His spiral classes c.–f. included descriptions: *one-branched spirals, spiral arms form a bright ring, doubtful connection of the ring with the centre, ring or arms connected with center through a bar*. Later Lundmark added the category g. "spiral arms have an appendix" (interacting companion galaxies—an example: M51).

Already in 1922 Lundmark, during his visit to the Lick Observatory, had worked on nebula classes and used e.g. the term Magellanic type. His catalog and other evidence suggest that Hubble's concern of plagiarism (Berendzen et al. 1976; Smith 1982) was unfounded (Teerikorpi 1989). In fact, the two systems are rather different, in spite of similarities in the outline. Furthermore, Lundmark's Magellanic nebulae are not fully irregular, but may contain rudimentary spirals. In the modern classification the galaxies called Magellanic by Lundmark, range from Sd to I0.

Modern Galaxy Classes Hubble (1936a) put the S0 or lenticular galaxies between ellipticals and spirals in his "tuning fork" diagram (Fig. 1.1). These hypothetical entities were later found in nature, as systems flatter than E7, but without any spiral structure, though some show dust features. It is of interest to note that Lundmark's "lenticulars" and "As0" may be related to S0 (zero 0 in both refer to the absence of spiral structure).

Sandage's Carnegie Atlas (Sandage and Bedke 1994) describes well developments in galaxy classification. De Vaucouleurs extended Hubble's system in several ways, making a transition from Sc to irregulars via the classes Sc, Scd, Sd, Sdm, Sm, and Im (m means "Magellanic"). He emphasized the presence of a ring, out of which spirals start tangentially at the points where the bar ends, by the symbol (r), while in (s) the spirals start from the ends of a bar. Though the r, s distinction in its purest refers to barred galaxies in this system (with intermediate cases SAB between ordinary SA and barred SB spirals) it is smoothly extended to all lenticulars and spirals.

A T-index denotes the main types of de Vaucouleurs's system, used, e.g., in the extragalactic LEDA database: Ellipticals: −6 to −4. Lenticulars: −3 to −1. Spirals: 0 to 9 (0/a a ab b bc c cd d dm m). Irregulars (Im): 10.

A way to realize the value of galaxy classes is to consider similar galaxies widely separated in space. If galaxies may be divided into a small number of classes, we can reason that there are not so many distinct processes producing galaxies and these occur everywhere in the universe. Also, the appearance of a galaxy may tell about its luminosity or size, providing a (very) rough estimate of its distance (Chap. 2).

From Aristotle to Hubble Classification has been a part of natural sciences since Aristotle who looked at Nature as a field naturalist, and recognized as a philosopher that *essential* properties of individuals define classes which they belong to (Grene 1963). We might see only a small finite part of the huge, possibly infinite-sized class, but this may be sufficient to characterize it, if we are lucky to have a representative sample and if we are perceptive enough. For Aristotelian method of science it was important to get things under the right headings and grouping them in the right way. Clearly Hubble succeeded very well in grouping the nebulae as he did.

Hubble himself told that his system was "descriptive and entirely independent of any theory", even though "the results are almost identical with the path of development derived by Jeans". Evidently Hubble "saw" in the sky what he thought to be a sequence of evolution. In a new physical realm and just from inspection of a "snapshot", it is difficult to decide between an evolutionary sequence (with specimen of old and young) and a sequence of grown-ups whose appearances reflect different pristine states ("initial conditions"). But a perceptive eye may discern a pattern, nevertheless, which carries a clue to the origins of those strange bodies, even though its correct interpretation is left for later students of the subject.[2]

1.1.2 Observational Cosmology in Our Sample of the World

In his Rhodes Memorial Lectures at Oxford (where he studied in 1910–13), published as the book *The Observational Approach to Cosmology*, Edwin Hubble (1937) stated the gist of practical cosmology: "The observable region of space, the region that can be explored with existing instruments, is a sample of the universe. If the sample is fair, its observed characteristics should furnish important information concerning the universe at large."

[2]Sandage devoted much thought to the birth and meaning of Hubble's classification (Sandage and Bedke 1994; Sandage 1995b, 2005). He suggested that this faculty helping one to go directly to the useful, physically relevant classification system is "imagination" or "intuition". For Aristotle, scientific knowledge was essentially demonstrations from better known premises. Then how to find the very first axioms, the starting-points of demonstration? His solution was to argue, in *Posterior Analytics*, that all begins from the intuition of the thinker observing and contemplating Nature. This seems to fit well with what Allan Sandage proposed to be the secret of Hubble and his classification.

"Our sample of the universe" within about 1000 Mpc, where observational data are the most accurate, gives the bulk of information on the galaxy universe (in comparison, the distance of the nearby Andromeda galaxy is about 0.77 Mpc). This region, where reality is in contact with theory, is the starting point for building world models that attempt to extend our cosmic picture far beyond the observable limits.

In the classical paper *The ability of the 200-inch telescope to discriminate between selected world models* Sandage (1961) formulated several cosmological tests (among them the magnitude-redshift $m(z)$ relation), which, together with a large telescope, hopefully might be used to decide between different Friedmann models (which Alexander Friedmann had introduced in the 1920s and which are used to describe the expanding universe). At that time the first task of cosmology was to derive two numbers: the present expansion rate (the Hubble constant) H_0 and the cosmic matter density ρ_0.

It was gradually realized that severe problems complicate the use of real celestial bodies in the tests. Selection effects and poorly known cosmic evolution easily hide from view true model parameters and may even deceive the analyst of the observations into interesting, but erroneous conclusions.

Only on the verge of the 21st century it became possible to detect stellar standard candles at large distances and to make a new step in the Hubble-Sandage programme. In 1979 Gustav Tammann had proposed type Ia supernovae occurring in very distant galaxies as a way to test whether Einstein's cosmological constant Λ is non-zero. Finally the redshift-distance relation for high-redshift SNe Ia led to the conclusion that the Friedmann model should include $\Lambda > 0$, or exotic dark energy causing the observed acceleration of the universe (Riess et al. 1998; Perlmutter et al. 1999). "The search for two numbers" in a simple universe containing ordinary matter and radiation, had to be extended to include the densities of dark matter and dark energy, and the equation of state. About 15 parameters characterize the different components in the modern version of the standard cosmological model (Spergel et al. 2007).

1.1.3 Empirical Cosmological Laws

The major empirical steps in modern cosmology are connected with advances in instrumentation. Thanks to astrophotography and large optical telescopes, the realm of galaxies was accessed during the first decades of the 20th century. Three cosmological key laws were then unveiled:

- the cosmological redshift-distance law
- the thermal law of isotropic cosmic background radiation
- the power-law correlation of galaxy clustering

Astronomical spectroscopy made possible the discovery of the cosmological redshift in the 1910s and then the redshift-distance (Hubble) law. The developing radio techniques led to the serendipitous detection of the thermal ocean of 3 K photons.

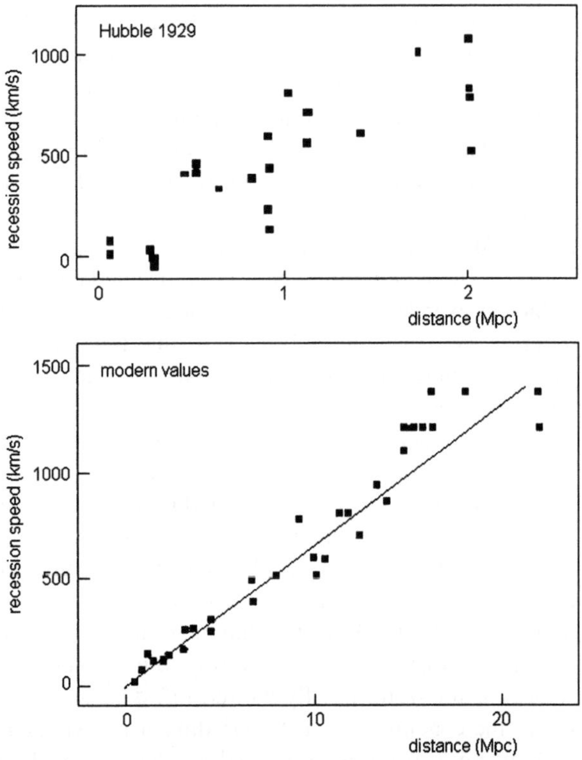

Fig. 1.2 Hubble's redshift law for local galaxies expressed as apparent recession velocity *cz* vs. distance. The *upper panel* is based on Hubble's 1929 work. Below we give an updated version of the Hubble diagram for about the same distance interval. The modern distances are almost ten times longer than those used by Hubble; his measurements had a large systematic error. The new Hubble relation is also less scattered than the original one. The *straigth line* represents the Hubble law for the Hubble constant $H_0 = 65$. Actually one determines the value of H_0 from a distance interval at least ten times longer

Finally, thousands of galaxy spectra obtained with telescopes using state-of-the-art multi-fibre spectrographs have revealed the filamentary galaxy distribution with fractal power-law behaviour.

The Hubble Law The cosmological redshift was the first major discovery of extragalactic astronomy (Slipher 1915), a genuine new phenomenon of cosmological physics. Then Hubble (1929) found in the local universe, mainly using brightest stars in a galaxy as standard candles (and Hubble and Humason in 1931 using brightest cluster galaxies) a relation between apparent magnitude and line shift which, using the inverse square law for light flux, can be written as a linear law between the redshift and the distance:

$$cz = H_0 r \tag{1.1}$$

In this empirical law $z = (\lambda_{obs} - \lambda_0)/\lambda_0$ is the redshift (λ_{obs} and λ_0 are the observed and emitted photon wavelengths), c is the speed of light, and r is the distance to a galaxy. Hubble's estimate for H_0 was 526 km s^{-1}/Mpc. To-day the value of this important cosmological quantity is measured to be in the range 60–75 (cf. Fig. 1.2).

The Thermal 3 K Cosmic Radiation The cosmic thermal background radiation was the next step. Working at 7.35 cm, Arno Penzias and Robert Wilson (1965) made the discovery. This brought them the Nobel prize in 1978 and is of great importance for cosmological physics, as recognized again in 2006, when the Nobel prize was handed to Mather and Smoot for their high-accuracy measurements of the spectrum of the CMBR and the tiny angular fluctuations of its intensity (10^{-5} K).[3] A viable cosmological model must be able to explain its thermal character and isotropy.

The COBE space observatory made the thus far most accurate measurement in cosmology, that of the temperature of the cosmic radiation $T = 2.725 \pm 0.002$ K (Mather et al. 1999). Its observed intensity as a function of the frequency ν was almost perfect blackbody radiation B_ν, containing the temperature of radiation T as the only parameter:

$$B_\nu(T) = \frac{2h\nu^3}{c^2} \frac{1}{\exp(h\nu/kT) - 1} \tag{1.2}$$

Small anisotropies of the smooth background have been measured by balloon and satellite experiments (de Bernardis et al. 2000; Spergel et al. 2007). Besides a dipole component due to our motion, tiny $\Delta T/T(\theta)$ fluctuations at the micro-Kelvin level were detected. From these, estimates of the main parameters of the standard cosmological model have been derived, especially that the spatial geometry of the Friedmann universe is close to Euclidean.

The Power-Law of Galaxy Clustering The third unexpected discovery was the very inhomogeneous, even fractal-like, spatial distribution of galaxies, revealed by means of the first discovery. The small-dispersion Hubble law is a distance indicator and gave the possibility to study directly 3D maps of the galaxy universe.

Indications for a continuous hierarchy of galaxy clustering were found, e.g., by Kiang (1967), Totsuji and Kihara (1969), and de Vaucouleurs (1970). But only after the work of the Rome University team (Sylos Labini et al. 1998) who analysed all major catalogues of galaxy redshifts using adequate statistical methods (Pietronero 1987), it was accepted that the galaxy distribution is "essentially fractal" (Wu et al. 1999). This history is described by Baryshev and Teerikorpi (2002, 2006).

The strong large-scale non-uniformities have gathered storms over the simple picture where smoothly scattered matter fills the universe according to the Cosmological Principle. The galaxies are clustered on a wide range of scales, and the lumps reveal self-similarity, which calls for the mathematics of fractal geometry (Mandelbrot 1977, 1982). The fundamental relation between mass density ρ and scale r is given by Mandelbrot's law:

$$\rho(r) \propto r^{D-3} \tag{1.3}$$

where D is the fractal dimension of the structure.

---

[3] Already in 1941 a residual radiation corresponding to a black body temperature of 2.3 K was found by A. McKellar (1941) who investigated spectral lines of interstellar CN molecules. T. Shmaonov (1957) detected with the Large Pulkovo Radio Telescope at a wavelength of 3.2 cm an isotropic radiation with an effective temperature of 4 ± 3 K. These results were published, but they did not yet attract the attention of cosmologists.

When discussing the fractal model for the galaxy distribution, two numbers are in focus: the fractal dimension D and the maximum scale of fractality R_{hom} where one may expect the crossover to homogeneity (from 100 Mpc up to the size of the Sloan Great Wall, 500 Mpc; Gott et al. 2005).

There is also the problem of dark substances. What about their spatial distribution? Observations unveal a clumpy dark matter, which seems to repeat the distribution of visible galaxies (Massey et al. 2007a). Dark energy might serve as a truly homogeneous background for the Friedmann universe.

Important Astrophysics In addition to the cosmological redshift, the background radiation and the complex structure of the galaxy universe, many other phenomena have cosmological relevance. Neutron stars, X-ray binaries, gamma-ray bursts and active galactic nuclei are extraordinary objects with strong gravitational fields, where accretion phenomena lead to a liberation of super-high energies. Especially quasars and gamma-ray bursts are very luminous lighthouses, visible from large distances.

Among other subjects of astrophysics relevant for cosmology are the chemical composition of stars and interstellar medium, the morphology and contents of galaxies, the dependence of galaxy properties on redshift, collisions and merging of galaxies, and the properties and evolution of active galactic nuclei. At a first sight secondary or local observations can become crucial for cosmology, such as calculations of the ages of globular clusters and studies of formation processes of dust and molecules.

1.1.4 The Standard Cosmological Model: Modern Paradigm

The standard cosmological model is based on Friedmann's (1922) model of expanding space—a solution of Einstein's equations of general relativity, presupposing homogeneity. It explains the linear Hubble law of redshifts as due to space expansion and the Planck law of the cosmic background radiation as the relict from the recombination epoch in the hot early universe.

The standard model as a whole includes the hot big bang scenario and in particular the process of growth of local inhomogeneities as the origin of galaxies and the large-scale structure. Here the cold dark matter has played the major role, while at the present epoch the global dynamics of the expansion of the universe is ruled by the antigravity of the dark energy. While the first variants of the big bang model were concerned with the "ordinary" matter (stars, gas, dust) only, now this luminous matter makes mere 0.5 percent of the mass of the universe.

Dark energy is postulated to explain an accelerating universe. It is a kind of exotic substance with negative pressure. A particular case is Einstein's cosmological constant Λ, which may be interpreted as cosmological vacuum.

1.2 Idiosyncratic Features of the Cosmological Laboratory

Galileo Galilei started modern physics with laboratory experiments. Remarkably enough, this same man, a mathematics professor at Padua, four centuries ago also opened the gate to the "cosmological laboratory", by making a telescope and observing the structures of the universe much better than was possible by the naked eyes of all previous philosophers.

1.2.1 Fundamental Limitations

Modern observers with their 10 metre telescopes see much deeper in space than Galileo could with his 2 cm magnifying tube. However, fundamental astronomical and physical limitations will always cause difficulties, many of which are typical for astronomy in general. Two main features are that

- the observer is attached to one dot in space–time, and
- measures processes within cosmologically large distances and times.

We have learned to believe that the Earth, the Sun, the Milky Way ... occupy an ordinary place in the universe. If so, then observations here are representative of what one would see at other places, too—for example, every observer would measure the linear Hubble law and saw a rather isotropic cosmic landscape. This Copernican principle, if valid, allows us to a certain extent as if to free ourselves from our permanent position.

Non-locality of Observations In museums "you may look, but not touch!". This is also a handicap for the Earth-bound cosmologist: our restricted location in the universe. We cannot exert influence on the studied objects. Furthermore, when we compare a photon from an atomic process in a laboratory and another photon generated by the same phenomenon but in a very distant galaxy, the time that a photon is travelling in space is an inevitable part of a cosmological experiment.

Practical Necessity of a Cosmological Model Our inability to move freely in the universe means that an adequate cosmological model is needed for the study of celestial bodies at high redshifts, where derived sizes, luminosities, energies and time scales much depend on the model used. At the same time, cosmologists use these objects to test the model itself; hence there is a complex interplay between the theory and the practical work.

The Past Light Cone and Evolution Astronomical observations are time-limited and confined to radially moving photons in our past light cone. Thus observations made during, say, one century can even in principle detect only a small fraction of the events within 10^{10} light years ("Hubble radius") producing isotropically moving photons. On the other hand, one should note the remarkable transparency of space along the long lines of sight between us and distant galaxies.[4]

Because of the light travel time, astronomers, like archeologists, probe different past time layers at different distances, and thus can study evolution and its time scale in some cases. Among the various types of evolution (for individual celestial bodies, their classes, and the universe), the evolution of the classes of cosmic test objects severely complicates cosmological tests.

Making the Distance Ladder The usual methods of measuring distances on the Earth fail on cosmological scales. Astronomers have built an elaborate cosmic distance ladder, which starts from the Solar System and is extended by different methods deep into extragalactic space. In this enterpise an important role is played by *standard candles*, classes of stars and galaxies having a known average luminosity with a small dispersion. One may also use objects with constant sizes, *standard rods*. It is presupposed that the standards are the same all over the universe. This may seem plausible in principle, but is hard to check from the observations themselves.

Selection Effects Today there are tremendous amounts of data obtained at different wavelength bands. One might think that the more there are objects, the easier it is to test world models. Unfortunately this is not exactly so, because collected data are influenced by various selection, distortion, and evolution effects. They distort the original physical relations between observed quantities, and may make observed relations imitate laws, which actually are not true and just originate from the observing procedure. To perform cosmological tests properly, we should know how to detect and treat systematic errors and biases which are hidden in astronomical data on distant objects and may affect seemingly secure methods of data analysis.

A classical case is the *K-effect*. Because of the cosmological redshift and fixed wavelength bands of detectors, different parts of the spectra are seen from different distances (redshifts). Another kind of classic is *the Malmquist bias*. When probing the deep universe, one cannot measure arbitrarily faint fluxes, hence one necessarily observes exceptionally powerful objects not representative of the typical population at such large distances. This "iceberg effect" is one aspect of selection effects often collectively termed "Malmquist bias".

The One-Sided View Sometimes our one-point position leads to especially bizarre problems. For instance, the unification scheme tells that an active galaxy is seen as a radio galaxy, or as a quasar, or as a blazar, depending on the viewing direction. If one has a "face-on" view of the accretion disc in the centre of the galaxy, one sees the active nucleus in all its lustre—this is a quasar or, in the extreme case, a blazar. If one looks at the same galaxy more from the side, the active nucleus is shadowed behind thick dust and there is a radio galaxy before one's eyes. One

[4]For instance, if there were obscuring dust in the extragalactic space between us and the nearest galaxy M31, just one fifth of the density typical within the Milky Way, M31 (faintly visible by plain eye) were blocked out of the sight of even the largest optical telescopes.

cannot move around a quasar, so one has to device indirect tests of the unification theory.

Astronomical data contain, along with the searched-for phenomenon, a blend of selection effects, effects of intervening space, and evolution. We would like to extract the true cosmological signal from this mixture.

1.2.2 Physical Laws on the Largest Scales

Astronomy pushes the frontiers of our cosmic sample deeper in space, and cosmology tries to extend the picture even over the visible borders. On this road one expects to encounter new global physical laws.

Empirical and Theoretical Cosmological Laws The subject of cosmology is exceptional, the entire physical universe that we can never see as a whole. Nevertheless, it begins modestly—as any physical science—with a number of empirical facts among which one hopes to find fundamental laws. This process is complicated by great limitations and even under the paradigmatic grip of any current standard cosmology one should be careful to distinguish between two kinds of cosmological laws:

- experimentally measured *empirical laws*,
- logically inferred *theoretical laws*.

The empirical laws, being based on repeatable observations, are independent of existing or future cosmological models. The theoretical laws are valid only in the frame of a specific model. Good examples are the empirical magnitude-redshift (m–z) Hubble law and the corresponding theoretical linear velocity-distance (V–R) law within the Friedmann model.

Local and Global Laws In our laboratories physical laws are studied at the very position of the experimenter. In astronomy, new laws specific for very large scales may appear. When one deals with spatial scales of $\sim$ 1000 Mpc, times $\sim 10^{10}$ yrs, and masses $\sim 10^{56}$ g, the validity of the local physical laws cannot be taken for granted, but should be tested by observations.

A local observation may be cosmologically relevant, if one may assume that it is valid for every observer and then use it as an ingredient of a (reasonable) world model. What locally may seem very important is not necessarily so globally. A historical example is the rotation of the sky! On the other hand, the local gravitation has proven to be relevant.

Theoretically Inferred Laws The theoretical laws require a jump from observation to interpretation. One may introduce a new essence into a cosmological model for explaining some observed effect, but it may take years to test its reality and physics by independent ways. Fortunately, theoretical inferences may be so unusual that they stimulate intensive study (like dark energy). An inferred law emerges via the machinery of a cosmological model. Different models may have different views of empirical laws, depending on their initial principles, fundamental theories, and inferred laws.

Every world model has its beginnings in cosmological principles which are regarded as valid for the whole, perhaps infinite universe, even though observations can be made only from a finite part of it. For instance, the statement that matter has a homogeneous and isotropic distribution (the cosmological principle of Einstein) is at the basis of the standard cosmological model.

All four fundamental interactions—the strong, the weak, the electromagnetic, and the gravitational—are used for modelling and understanding cosmological phenomena and for predicting observed astrophysical effects. A special role in cosmology is played by the theory of gravitation.

Examples of Inferred Cosmological Laws A cosmological model should have consequencies that may be compared with observations. Some of these explain the key empirical cosmological laws, giving the initial feasibility for the model. Other inferred laws gives predictive power to the model, relationships between quantities that may be observed.

For example, the Friedmann model predicts that the temperature of the thermal background radiation increases proportionally to $1 + z$ (where z is the redshift), whereas in the Steady State model it remains constant. Hence, the measurement of the temperature at different redshifts (distances) is an important, albeit difficult task.

Together with a cosmological model appear theoretical concepts, laws, processes, and substances. The standard model concludes from the Hubble law and the background radiation that the universe is expanding and cooling, hence has begun from a hot state. The laws of expanding space and decreasing density and temperature are fundamental for this model.

From the lumpy distribution of galaxies and the smooth background radiation one may infer another process: the large-scale structure has its origin in the gravitational growth of tiny initial density fluctuations of dark matter. New dominating substances, dark matter and dark energy, were deduced from the observations of distant supernovae and the background radiation.

1.3 Important Questions of Cosmology

Modern cosmologists share with their old predecessors the passion to find out fundamental properties of our physical universe, such as its size and geometry, age, material contents, and evolution.

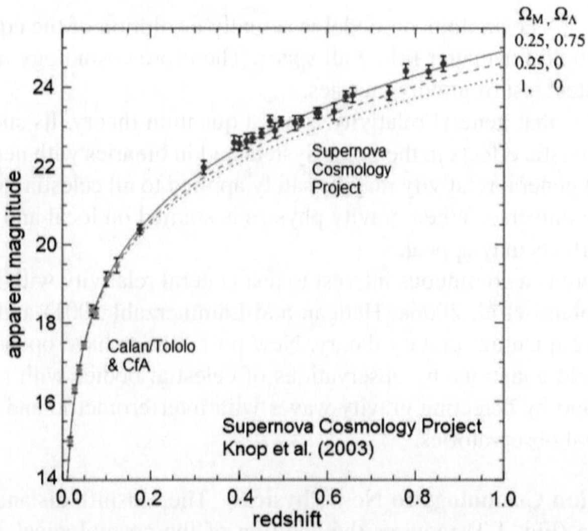

Fig. 1.3 The observed Hubble relation for high-redshift Ia supernovae. The three curves from top to bottom correspond to the Friedmann models with different density parameters (Ω_m, Ω_Λ) = (0.25, 0.75), (0.25, 0.0), (1.0, 0.0). The observations show a deflection upwards from the prediction of the model with a critical matter density (the *bottom curve*) leading to the view of dominating dark energy. A typical task of practical cosmology is to find if this observed deflection in measured magnitudes is really cosmological in origin and not caused by, e.g., subtle observational selection effects or imperfect standard candles (credit: Supernova Cosmology Project)

1.3.1 Astronomical Tests of Fundamental Physics

Today's cosmology is deeply influenced by the close interaction between astrophysical research and theoretical physics at the level of the study of fundamental laws of Nature. This includes questions about the universality of local physical laws and about new fundamental laws on large scales.

Fundamental Constants and Laws One test of cosmological principles is the search for possible changes in physical constants (Uzan 2003). Here astronomical observations take the role of experiments attempting to decide between concurrent fundamental theories of physics.

Different extensions of the standard model and their possible cosmological consequences have been intensively discussed (e.g., Peacock 1999; Uzan 2003). Multidimensional theories predict changes of fundamental constants over cosmological time scales (e.g., the electron charge e, the speed of light c, and the fine structure constant α; Magueijo et al. 2002).

Special relativity and quantum mechanics are both being tested astronomically (Bertolami et al. 2006b). String theory predicts that the Lorentz invariance may be violated inside "quantum foam"; this might be observed on long paths of photons and other particles (Amelino-Camelia et al. 2005).

Gravitation A cosmological model is actually a solution of the equations of the gravitational field for matter filling all space. Therefore cosmology may be said to form the greatest test of gravity theories.

One point is that general relativity is not a quantum theory. Its successes in explaining relativistic effects in the Solar System and in binaries with neutron stars, do not mean that general relativity may be safely applied to all celestial bodies, including the whole universe. When gravity physics is studied on local and cosmological scales, new effects may appear.

In fact, there is a continuous interest to test general relativity with increasing accuracy (Bertolami et al. 2006a; Haugan and Lämmerzahl 2001) and to develop a more adequate quantum gravity theory. New possibilities have opened for tests of the gravity field equations by observations of celestial bodies with strong gravitational fields and by detecting gravity waves with interferometric and bar detectors, or gravitational observatories.

From Precision Cosmology to New Physics The redshift–distance relation for Ia supernovae (Fig. 1.3) requires the addition of the cosmological Λ term to the equations of general relativity. The anisotropies of the thermal background radiation detected by the WMAP satellite confirmed that a dominating dark energy is needed for Friedmann models. The anisotropy measurements have also led to so-called precision cosmology: the data can give precise values of the main cosmological parameters within the Big Bang model.

These findings shattered the hope that the universe consists of some ordinary (gravitating) matter having the critical density ρ_{crit}. Now we have a world where the negative-pressure dark energy is the major component (about 70%) and together with dark matter (about 30%) makes the density critical and space Euclidean. Thus the dream of precision cosmology has shifted to a vision of new cosmological physics (Peebles 2002) meaning that cosmologists now think they know the values of the main cosmological parameters of the standard model, but they do not know the physical sense of the exotic substances which the parameters refer to (Turner 2002).[5]

1.3.2 Central Subjects of Cosmological Research

The breakthrough into the dark has brought into light new puzzling aspects of modern cosmology (Turner 2002; Baryshev 2006) and makes one ask: Is this strange world true, or perhaps rather a construction necessiated by the pressure of new observations and made possible by the relative flexibility of the Friedmann model? Both possibilities are quite inspiring.

[5] According to Webster's dictionary "exotic" means "striking or unusual in effect or appearance; strange; exciting; glamorous". All this applies to the new cosmological physics which differs from what is known from laboratory studies. It is even as if "introduced from abroad, but not fully naturalized or acclimatized".

The Distance Scale Accurate extragalactic distances, the classical goal of astronomy, have become increasingly important in the era when the fluctuations of the cosmic background radiation give (model-dependent) information on global cosmological parameters, including the Hubble constant (i.e. the distance scale). The directly measured distance–redshift relation of the galaxy universe at large distances is an important constraint on global models and cosmological physics involved. Precise local distances are also needed for new tests of dark matter and dark energy.

The Physics of Dark Matter Ever since Zwicky (1933, 1937) reported a surprisingly high mass-to-luminosity ratio for the Coma cluster of galaxies, dark matter has been with us. Its reality has often been cast in doubt, but its presence is now demonstrated by surveys of gravitational lensing, independently of a particular cosmological model. For instance, observations of weak gravitational lensing in deep space have revealed large amounts of filamentarily distributed dark matter (Massey et al. 2007a).

The dark substance is not expected to be ordinary, baryonic matter. This follows from the primordial nucleosynthesis theory that yields for the baryonic component a low density of about 5% of the critical density ρ_{crit}. Furthermore, models of structure formation, together with the observed smoothness of the background radiation, suggest that the non-baryonic dark matter is cold, i.e. its thermal energy and pressure are negligible in comparison with the rest mass energy.

The problem of dark matter is chiefly the problem of its carriers (Feng 2010). The list of well-motivated candidate particles is rather long. For instance, axions are predicted by the quantum chromodynamics, while the much heavier neutralinos appear in supersymmetry theories which unite fermions and bosons and are extensions of the Standard Model of elementary particles. Facilities exist and new projects are planned with the aim to detect particles of dark matter. Up to now their existence has not been directly confirmed.

Existence and Nature of Dark Energy In his first cosmology paper Einstein (1917) showed that his gravitational field equations allow an extra term Λg^{ik}. Currently the cosmological constant Λ is put into the right side of the equation, into the energy-momentum tensor. Interpreted as an additional source of gravitation, it may represent some vacuum-like substance (as reviewed by Chernin 2001). Dark energy generalizes the cosmological constant (Caldwell et al. 1998); the term, coined by Turner, reminds us that this sort of thing emerges "from the dark" in expanding space and the pressure of such a fluid is comparable (in absolute value) to its energy density. In Einstein's equations, the negative pressure leads to antigravitation.

It would be desirable to have more observational evidence for dark energy (and its possible evolution) from the Hubble relation at all redshifts up to $z \sim 1$ and beyond. At the same time, studies of the local Hubble law promise independent evidence on dark energy possibly influencing the properties of the outflow which is strangely quiescent (Sandage 1999) even in the lumpy environment close to the Local Galaxy Group (Chap. 12).

Power-Law Correlations on Large Scales Studies of the spatial distribution of galaxies, using large redshift surveys, probe the space within about 1000 Mpc and may essentially constrain world models. But it is not easy to extract reliable information on the distribution of galaxies from the 3D maps, and one needs appropriate methods of data analysis (Chap. 10).

The data from redshift surveys, combined with the anisotropies of the cosmic background radiation, can be used to check the crucial prediction about the existence of baryon acoustic oscillations having a spatial scale of about 100 Mpc. Observations also allow a check of the Cosmological Principle for the luminous matter. On how large scales is the distribution still inhomogeneous? We discuss these questions in Chaps. 11 and 12.

Numerical simulations of gravitating particle systems are the basic way to study how large-scale structures are formed within the cold dark matter models. Among the major problems in N-body simulations are the choice of the initial conditions and the representation of the cosmological fluid by a discrete particle set (Baertschiger et al. 2002; Gabrielli et al. 2005). On small scales, if there is an additional self-interaction of the CDM particles, simulations lead to a different density profile for the forming condensations (Spergel and Steinhardt 2000). On large scales, it is still an open question how the observed power-law correlations ("megafractality") have emerged from the tiny initial fluctuations.

The hypothetical dark matter with unknown properties makes it difficult to build a reliable model of the large-scale structure formation. Peebles (2002) made here the illuminating comment that the main unknown element in the standard model is the physics in the dark sector and therefore tests of fundamental physics have a high priority, as they may clear up the nature of the dark substance. Until this is done structure formation is a hazardous basis for testing cosmological models.

Physics of High-Redshift Phenomena At high redshifts one tries to observe a very remote past in the history of the universe. According to the standard cosmology, around the redshift $z \approx 1000$ the cooling temperature reached 3000 K and electrons and protons could unite ("recombine") and form neutral atoms. The radiation decoupled from matter. After this, though before the epoch when we already can see galaxies ($z \approx 9$), denser regions of the primordial gas started condensing into young galaxies forming stars. This long period (about 10^9 years) before the "first light" appeared is the "Dark Age", where the origin and dynamical and chemical evolution of the first generations of stars, galaxies, and quasars are still poorly known.

The cosmic background radiation is usually regarded as a relic from the recombination epoch. In view of possible other explanations (e.g., Hoyle et al. 2000), this crucial element of standard cosmology deserves to be tested as carefully as the reality of expansion has been tested by Sandage.

Geometry and Expansion of Space The analysis of the anisotropies of the background radiation gives zero curvature of space. From the times of Gauss, there has been a dream of seeing differences from "Euclidean predictions" that could be without doubt ascribed to intrinsic non-Euclidean properties of our space. Starting from the opposite points of view on the geometry of physical space as expressed by Poincaré (1902) and Einstein (1921), the traditional question, whether space is finite or infinite and what its curvature is (Sandage 1992), has been extended into inquiries about the topological structure of space and the number of possible extra dimensions.

The scale factor $S(t)$ gives a mathematical description of space expansion in Friedmann models. All distances between uniformly scattered particles are changing with time as $r(t) = S(t)\chi$. As the real world is highly inhomogeneous, one has to ask what expands and what does not. One useful way to view the expansion of space is as creation of space together with the cosmological vacuum. In fact, the physics of space expansion is a subtle subject of few observational facts and somewhat nebulous theoretical concepts (Chap. 12).

The standard cosmology suggests that behind the observed Hubble law there is a not-directly-observable linear velocity–distance law. "Velocity" refers to space expansion which causes the redshift by the Lemaître effect. As space expansion is not yet directly measurable, it is important to make tests confirming its reality as Sandage has done using the surface brightness. In future, it might be possible to perform Sandage's changing redshift (dz/dt) test using the very large new generation telescopes.

Chapter 2
Distance Measurement and Cosmography

Combining observation with mathematics can result in unexpected ways of probing the world—small and large. Examples start in Antiquity when Thales measured the height of a pyramid, using the length of its shadow at the very moment when the shadow cast by a vertical rod was as long as the rod itself. It is also told that he inferred the distance of a boat from the shore, without stretching any tape measure between them. Astronomers have made analogous things throughout history when constructing the cosmic distance ladder.

2.1 Cosmic Distance Indicators

It is common knowledge in our neighbourhood that of two similar shining objects the more distant one looks fainter and smaller, and has a smaller parallax shift. In Euclidean space, an object's *luminosity distance*, *angular size distance* and *parallax distance* have the same value when the same unit of length is used. These familiar kinds of distances are also used to construct the distance ladder beyond the Solar System, inside and even outside of our Galaxy, finally reaching the realm where a novel kind of distance measure appears: the *redshift* of light. The empirical Hubble law indicates that a higher redshift corresponds to a larger distance. Here "distance" is still a somewhat misty concept of remoteness.

With a cosmological model, the redshift and other distances are linked to the *metric distance* contained in the theoretical framework of Riemannian space. It can be said to correspond to the distance obtained with metersticks put one after the other between the observer and an object. In non-Euclidean spaces, as considered in large-scale physics, different distances have different values, giving a way to find out the geometry of space.

The Distance Ladder This concept refers to the steps making possible the measurement of progressively larger distances (see, e.g., Webb 1999).

One may say that *a distance indicator is a method where an astronomical object is placed in 3D space so that its properties observed through space agree with what*

Y. Baryshev, P. Teerikorpi, *Fundamental Questions of Practical Cosmology*, Astrophysics and Space Science Library 383,
DOI 10.1007/978-94-007-2379-5_2, © Springer Science+Business Media B.V. 2012

we know about such objects, their constituents, and the propagation of light. A good distance indicator would restrict the object's position in a narrow range around the true distance. For finding good distance indicators the diversity of cosmic phenomena and the knowledge and imagination of the cosmographer should meet. Thus numerous methods for "placing an object in space" have been invented, each requiring different kinds of measurements at the telescope. The most basic is to measure angles. Then there is the flux of light, and one may also measure its time variation and spectrum at different wavelengths.

2.1.1 Geometry and Photometry in Euclidean Space

In Antiquity, the term "parallax" meant the shift in the direction of a celestial body as looked from two places on Earth. In his *Almagest* Ptolemy (ca AD 150) stated that "none of the stars has a noticeable parallax (which is the only phenomenon from which distances can be derived)". One sees a sound method, but the observational means are insufficient. In fact, only after Copernicus was there strong motivation to search for stellar parallaxes, as a crucial test for proving the Earth's motion and as a distance indicator.

The Triangulation If the length of the baseline d is known in needed units and the measured peak angle in the triangle is θ, then the distance r is

$$r = \frac{d}{2\tan(\theta/2)} \approx \frac{d}{\theta}. \tag{2.1}$$

One may use two kinds of baselines. The baseline may be either "here" like the astronomical unit AU (the Earth–Sun distance) in the annual parallax method or "out there", like the radius of a galaxy or some other object.

The local baseline (2×AU) is obtained in physical units using Kepler's Third Law together with distances to nearby planets and asteroids, measured with radar techniques nowadays. So a stellar parallax angle can be expressed as a distance in cm (see Appendix A for useful numerical data).

In the stellar realm in our Galaxy, the triangulation appears in various forms. The Solar System moves relative to the local standard of rest (about 13 km/s or 2.8 AU per year), so one can expand the local baseline much beyond the 2 AU of the annual parallax method and derive "secular parallaxes" (leading to a mean distance to a class of stars). Another related method derives "statistical parallaxes".

However, when the baseline is "out there", we have a problem: how to know its actual length? This critical question arises often in astronomy. Ideally, one should have a *standard rod*, a class of objects whose linear size has a small scatter around an average value. From the measured angular size of a standard rod, one can derive its angular size distance using Eq. (2.1).

The expanding nebula method, a special kind of triangulation, was suggested and first applied by Lundmark in the 1920s for the Crab nebula (Trimble 1973). Knowing the expansion velocity of the nebula in km/s (from the Doppler shift), one can calculate its distance from its transversal proper motion μ in arc seconds per year derived from the angular size and the date of explosion:

$$V_r = 4.74\mu r, \tag{2.2}$$

where the expansion velocity V_r is in km/s and the distance r is in parsecs.

For the Milky Way, our distance from the Galactic centre, R_0, is fundamental. It is also important for extragalactic distances through its impact on the calibration of stellar standard candles. Earlier, the round number $R_0 \approx 10$ kpc was often used, but for years it has been known that the distance is shorter, around 8 kpc. In a summary of the results, Eisenhauer et al. (2003) reported a geometric measurement to the centre with an uncertainty of about 5%. They combined astrometric and spectroscopic measurements of the star S2 orbiting the massive black hole candidate in the Galactic centre. The solution for the best-fit Kepler orbit gave the needed parameters to calculate the distance $R_0 = 7.94 \pm 0.38$ kpc (the black hole mass turned out to be about 3.6×10^6 solar masses).

Triangulation in Nearby Extragalactic Space Water masers, observed in interstellar clouds, are strong and compact radio sources. Their positions can be measured accurately with Very Long Baseline Interferometry (VLBI) techniques. A few nearby galaxies have known water maser sources.

Brunthaler et al. (2005) made true the dream of van Maanen in the days of the Great Debate: to detect the angular rotation of M33, in order to infer its distance. Comparing the angular rotation rate, as measured with VLBI from water masers on opposite sides of the galaxy, with the rotation speed and inclination, they derived a distance of 730 ± 168 kpc. This comes from the measured angular motion of 30.8 ± 4 μarcsec/year (in RA) and the velocity of 106 ± 20 km/s from the rotation model, together with Eq. (2.2).

Miyoshi et al. (1995) observed megamasers close to the centre of the spiral galaxy NGC 4258. They could deduce the radius (0.13 pc) of the disk rotating around the massive black hole candidate from the rotational speed (1080 km/s) and centripetal acceleration (9.5 km/s/yr) of its edge. Then the angular radius (4.1 mas) gave the distance of 6.4 Mpc (revised to 7.2 ± 0.3 Mpc by Herrnstein et al. 1999).[1]

The Photometric Method: Standard Candles Before annual parallaxes were measured by Bessel and others in the 1830s, brightness had been recognized as a possible distance indicator. James Gregory (1638–1675) assumed that stars are other suns observed through transparent space where the inverse square law of light flux from a point source works (introduced in Kepler's *Optics*). Newton applied it to Sirius, using the Sun as the calibrator. He inferred a distance of one million solar distances (twice the true one) and was thus well aware of the remoteness of stars.

[1] A variant of this method applied to M33 (Argon et al. 2004) resulted in the distance 800 ± 180 kpc.

Consider an object radiating isotropically with the power output (luminosity in erg/sec) L. Then at the distance r it produces a flux f (erg/sec/cm^2) through the surface of the sphere having the total area $4\pi r^2$:

$$f = \frac{L}{4\pi r_{\mathrm{L}}^2}. \tag{2.3}$$

If one knows the luminosity L and measures the flux f, one can thus derive the *luminosity distance* r_{L}. It is good to underline that the photometric method is also geometric in nature. One essentially measures—in a special way—the solid angle made by the cone whose 1 cm^2 bottom is at the Earth and the peak is at the centre of the object in question.

Though we indicated above the units erg, sec, and cm, the method of standard candles is usually relative, i.e. one obtains the distance to an object relative a nearby calibrator. The distance modulus $m - M$ often appears in such calculations. It is related to the luminosity distance r_{L} as

$$m - M = 5 \log r_{\mathrm{L}}/10\ \mathrm{pc}. \tag{2.4}$$

This formula has its origin in the $1/r^2$ law in Euclidean space. We remind the non-astronomer reader that the apparent magnitude is connected to the observed flux f as $m = -2.5 \log f + \mathrm{const.}$ (e.g., Karttunen et al. 2006).

Some history: The familiar unit light-year was used before parsec which was suggested by Herbert H. Turner (1861–1930) and apparently first mentioned in text by Frank W. Dyson in 1913. The concept of absolute magnitude M was defined by the Dutch astronomer Jacobus C. Kapteyn (1851–1922) in 1902, as the apparent magnitude m which a star would have if moved to a distance corresponding to a parallax of $0.1''$, i.e. the distance of 10 pc (Hughes 2006).

2.1.2 Relative and Absolute/Physical Methods

When using a "standard rod" or a "standard candle" as a distance indicator, in both cases one must ascribe a value to the object whose distance is measured (length or luminosity). For this there are two possibilities:

- Calibration based on nearby objects of the same kind, whose distances are known using other distance indicators.
- Derivation of the linear size or luminosity for the object directly from observation and physical theory, with no information about its distance.

The first variant requiring nearby counterparts is "relative". The latter type of methods bypasses the local distance ladder, and may be described as "physical" (e.g., Sandage et al. 2006) or "absolute" (Nikiforov 2004) or "one-step distance method" (Jackson 2007).

Relative Methods The very first application of the standard candle method used the calibrator "here", the Sun. Hence, Gregory's method had 1 AU as a natural distance unit. For several distance indicators of this kind, the size or the luminosity are inferred from a relation $X = ap + b$, where p is an observable parameter whose value can be measured without knowing the distance. Examples are the period–luminosity relation of Cepheid stars, the rotational velocity—luminosity (TF) relation for spiral galaxies, and the decay time—maximum luminosity relation for Supernovae Ia.[2]

The parameter p can refer to quite different observations and time-scales. A spiral galaxy remains in its position in the TF-relation for uncounted generations of astronomers, and observations of its maximum rotation velocity and magnitude may be repeated at will. Cepheid observations are limited to relatively nearby galaxies and require a sufficient time interval, so that different pulsational periods can be observed. Supernova explosions cannot be predicted, and when they occur, one should start following the process before the peak luminosity is reached.

Absolute/Physical Methods Relative methods rely on the calibrator distances. Physical methods result in distances directly expressed in physical units (cm). Such methods are rare at large extragalactic distances. One example is the Sunyaev-Zeldovich method for distant rich galaxy clusters which utilizes the interaction of the hot cluster gas and the cosmic background radiation (Sect. 4.3.1).

Redshifts and Distances Thanks to the regular universe, the measured redshift provides an estimate of relative distances (the Hubble law). It is quite different from the indicators mentioned above. In the assumed Friedmann model, (or any other sufficiently developed cosmological model) the redshift links other kinds of distances to the fundamental metric distance. For instance, locally calibrated standard candles lead to luminosity distances d_{L}, which can be measured independent of the cosmological model. Then, if one assumes that a flat Friedmann model describes the universe, one gets from the luminosity distance (in Mpc) the metric distance (in Mpc) just by dividing it by the redshift factor $1 + z$.

The procedure of distance measurement may explicitly involve a cosmological model. For instance, if one needs a time interval occurring at the object, one transforms it to the rest-frame by dividing the observed time by $1 + z$, when the Friedmann model is assumed. This happens, e.g., when the decline rate is used to correct the peak luminosity of a supernova of type Ia.

[2] Also the Faber-Jackson and Fundamental Plane (Davis–Djorkovski) methods for elliptical galaxies, utilizing relations between the luminosity and stellar velocity dispersion (FJ), and the effective size, surface brightness and velocity dispersion (DD) and a modification by Dressler et al. (1987) belong to this category.

Table 2.1 Some stellar distance indicators

Type of star	Pop	Absolute V mag	Range for V = 27
RR Lyrae variables	II	0.8	2 Mpc
TRGB (Red Giants)	II	−2.5	5 Mpc
Cepheid variables	I	−5 (for $P \approx 20$ d)	25 Mpc
Supernovae type Ia	I, II	≈ -19.5	$z < \sim 2$, (~ 10 Gpc)

2.2 Lower Rungs in the Distance Ladder

Lundmark (1919, 1920) made an early determination of the distance to the Andromeda galaxy using novae (the result was 175 000 pc). He also suggested the use of the brightest stars in galaxies: "If it were possible to determine the apparent magnitude at which the main body of giants in a spiral nebula begin to appear separated, it would give us an additional means of estimating its distance".

In his 1929 classic Hubble introduced the brightest blue stars as distance indicators. He identified these from the photographs of galaxies up to about half distance to the Virgo cluster. They usefully indicated relative distances, though Sandage (1958) showed that these objects actually were HII regions, bright ionized hydrogen clouds around blue stars. Calibrated by Hubble using true stars in nearby galaxies, those indicators gave too short distances.

2.2.1 *The Brightest Giant Stars in Galaxies*

But is there a sharp upper limit in the luminosities of normal stars? From the HR diagram for the solar neighbourhood, Sandage and Tammann (1974) concluded that the brightest blue stars surpass $M_B = -9$ mag. In other galaxies $M(1)$ depends on the galaxy luminosity (the number of stars), roughly as expected if the stellar luminosity function is exponential at the bright end. Indeed, if the luminosity function of a class of stars does not go abruptly to zero at a high luminosity, it is clear that in a large sample of those stars the extreme one is usually more luminous than in a small sample. This makes blue supergiants problematic as distance indicators.

The trend between $M(1)$ and the host galaxy magnitude M_{gal} depends on the steepness of the bright end of the luminosity function. In some cases the apparent magnitude $m(1)$ of the brightest star and the host magnitude m_{gal} can be used to make a distance estimate. If we know the dependency $M(1) = aM_{gal} + b$, then $(1-a)\mu = -(am_{gal} + b)$, where $\mu = m(1) - M(1)$ is the unknown distance modulus. If the slope $a \approx 1$, one cannot solve for the distance.

The brightest red supergiants ($\langle M(1) \rangle \approx -8.0$ mag) also show a trend with the host galaxy luminosity, though weaker than the blue stars (Sandage 1984). They have been used to derive distances to a number of nearby galaxies with an accuracy of ≈ 0.3 mag in the distance modulus.

Instead of extremal stars within a broad class, one may consider inherent features in the luminosity function, such as a maximum or a discontinuity. These maintain their location (absolute magnitude) in samples of different sizes, when there is no selection depending on the absolute magnitude.

The TRGB Method The Hubble Space Telescope could measure the luminosity distribution of bright red stars in many galaxies (earlier studied, e.g., by Sandage 1971), making a new way of using red giants possible: the Tip of the Red Giant Branch. The modern TRGB method was developed especially by Madore, Freedman and associates (e.g., Lee et al. 1993; Sakai et al. 1997) and is now an important way to measure distances in the local volume within about 10 Mpc (e.g., Karachentsev et al. 2007), where the Population I Cepheid method is not applicable to elliptical galaxies (cf. Table 2.1). The TRGB method is also efficient in using the HST telescope time, as one can measure distances out to 7 Mpc with a single orbit (a single-epoch observation at two wavelengths is enough).

The "tip" refers to a sudden discontinuity in the luminosity function of the stars in the red giant branch of the HR diagram and it is observed at the absolute magnitude $M_I \approx -4$ in the I photometric band (around 8200 Å). This feature is understood as marking the core helium flash of old, low-mass stars (less than ~ 1 solar mass) which evolve up the red giant branch, but very quickly change their physical characteristics upon ignition of helium.

The TRGB I-magnitude has been shown to be quite stable, only slightly sensitive (~ 0.1 mag) to age 2–15 Gyr and metallicity between $-2.2 < [Fe/H] < -0.7$ dex (the range for the Galactic globular clusters). The statistical problem affecting the use of extremes in a population is reduced and it is regarded that with the TRGB method one reaches an accuracy of 0.1–0.2 mag for the distance modulus, requiring more than ~ 100 red giants detected in the one-magnitude bin below the tip.

The calibration in the I-band is based on Galactic globular clusters whose distances have been determined using the RR Lyrae variable stars. Tabur et al. (2009) have made a direct geometrical calibration of the TRGB K-band magnitude in the solar neighbourhood using parallaxes measured by the HIPPARCOS satellite.

2.2.2 Cepheid Pulsating Stars

On the cosmic distance ladder the classical Cepheid variables continue to have a special role. They are young objects of Population I found in spiral and irregular galaxies. Their predecessors in the Main Sequence are stars two times heavier than the Sun. With their (V) absolute magnitudes in the range centered around -4 (for $P = 10$ d), they can be used up to $m_{\mathrm{lim}} - M \approx 28.5$ ($r \approx 5$ Mpc) with earth-bound telescopes ($m_{\mathrm{lim}} \approx 24.5$) and $m_{\mathrm{lim}} - M \approx 31$ (≈ 25 Mpc) from space ($m_{\mathrm{lim}} \approx 27$).

Some Physics A simple argument concerning the physics of the PL relation was outlined by Sandage (1958; also Madore and Freedman 1991)—such understanding helps one to formulate the law, to add relevant parameters, and perhaps to recognize a reason for possible discrepant results.

The Stefan's law connects the luminosity, radius, and effective surface temperature of a star: $L = 4\pi R^2 \sigma T_{\rm eff}^4$, thus mapping all stars on the magnitude-temperature diagram, with stars of a constant radius defining a straight line. Such a line is also one of constant volume, so moving along it the average density of stars changes, as there is a luminosity–mass relation. A result from classical mechanics states that the radial pulsational period of a star $P \propto 1/\sqrt{\rho}$. Here ρ is the average density of the star, hence *the pulsational period of a star tells us about its density, and, at the end, about its luminosity*. For the bolometric (total luminosity) absolute magnitude, we may write the PL relation as:

$$\langle M_{\rm bol} \rangle = A \log P + B \langle \log T_{\rm eff} \rangle + \text{const.} \tag{2.5}$$

In practice, instead of the total flux one observes a Cepheid through different wavelenght bands. Each has its own PL slope and zero-point. A major advance has been the near-infrared photometry, complementing the measurements in the B and V systems. In longer wavelengths (1) the extinction and reddening of the light by the dust are reduced, (2) the scatter of the PL relation is smaller, and (3) the amplitude of magnitude variation is smaller.

The PL Relation As the colour is the natural third parameter in the PL relation, reflecting the surface temperature, the precept behind the use of a simple PL relation is as follows: If one assumes that the calibrating sample of Cepheids has the same properties as the distant sample on which one applies the relation, the colour term may be considered as the same for all galaxies. The PLC relation is thus transformed into a simpler PL relation.

Soon after the discovery of the PL relation by Henrietta S. Leavitt (Leavitt and Pickering 1912), Hertzsprung applied it to distance determination. He and then Shapley used statistical parallaxes for stars in our Galaxy to derive the zero-point of the relation. In the modern situation there are several routes to the calibration. Assuming a universal PL relation, the slope of the PL relation has been usually taken from observations in the Large Magellanic Cloud, while the zero-point comes from our Galaxy, together with independent distance measurements of the LMC (see e.g. Fouqué et al. 2007). For example, the PL relation for the LMC adopted by the HST Key Project (for the distance modulus $\mu_{\rm LMC} = 18.5$) was $M_V = -2.76 \log P - 1.46$ (Freedman et al. 2001).

The HIPPARCOS astrometric satellite measured many parallaxes for Cepheids in our Galaxy. However, the large distances cause large parallax errors, which make it difficult to derive the PL zero-point in a proper statistical manner. The future GAIA astrometric space observatory will open here new possibilities.

The PL relations may be systematically different in different galaxies. It is known that later galaxy types (like the Large Magellanic Cloud) have a lower metallicity than earlier types (like our Galaxy) (Paturel et al. 2002), and the relations in the Galaxy and in the LMC may be different (Tammann et al. 2003). Hence, the Cepheid properties could be different due to metallicity of the interstellar gas from which

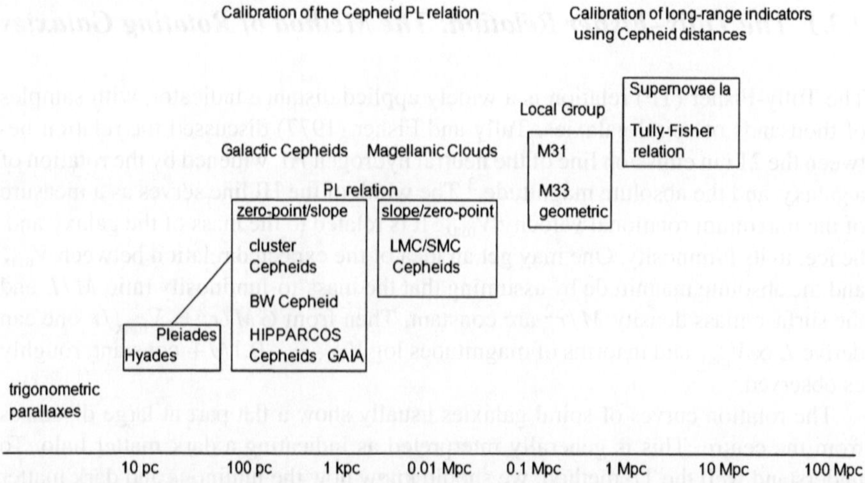

Fig. 2.1 Parts of the cosmic distance ladder involving Cepheids, the TF method, and supernovae Ia. The PL relation is calibrated in the Galaxy and in the Magellanic Clouds and checked in nearby galaxies. It is used to measure distances to galaxies and groups within about 20 Mpc, where the TF-relation and the SNe Ia luminosities are calibrated

these stars were initially formed. Such differences could influence the determination of the distance scale and the value of the Hubble constant (we return to this point in Chap. 4).

The absolute magnitude being predicted by the PL relation, the distance modulus can be derived from the mean apparent magnitude. It should be corrected for the sum of the extinctions in our own Galaxy and in the host galaxy. This is possible by using two different photometric bands (V and I; Madore and Freedman 1991). Then the corrected distance modulus μ_0 is derived from the separate moduli μ_V and μ_I as

$$\mu_0 = \frac{R_V}{R_V - R_I}\mu_I - \frac{R_I}{R_V - R_I}\mu_V. \tag{2.6}$$

The factors R_V and R_I give the ratio of the total extinction to the reddening E_{B-V} for the photometric band (V or I). They depend on the extinction law. For instance, if in round numbers $R_V \approx 3$ and $R_I \approx 2$, one has $\mu_0 \approx 3\mu_I - 2\mu_V$.

2.3 Long-Range Indicators

A common route to large extragalactic distances (as outlined in Fig. 2.1) is to use the PL relation for Cepheids to derive distances for a set of galaxies that serve to calibrate the luminosity of a long-range distance indicator. Such indicators can extend the distance ladder from the local scales of ~ 10 Mpc to scales of ~ 100 Mpc (e.g., the TF relation) or even 1000 Mpc or more (Type Ia supernovae).

2.3.1 The Tully–Fisher Relation: The Method of Rotating Galaxies

The Tully-Fisher (TF) relation is a widely applied distance indicator, with samples of thousands of spiral galaxies. Tully and Fisher (1977) discussed the relation between the 21 cm emission line of the neutral hydrogen HI, widened by the rotation of a galaxy, and the absolute magnitude.[3] The width of the HI line serves as a measure of the maximum rotational velocity $V_{\rm max}$. It is related to the mass of the galaxy and, hence, to its luminosity. One may get an idea of the expected relation between $V_{\rm max}$ and the absolute magnitude by assuming that the mass-to-luminosity ratio M/L and the surface mass density M/r^2 are constant. Then from $GM/r^2 = V_{\rm max}^2/r$ one can derive $L \propto V_{\rm max}^4$ and in terms of magnitudes $\log V_{\rm max} = -0.1M+$ constant, roughly as observed.

The rotation curves of spiral galaxies usually show a flat part at large distances from the centre. This is generally interpreted as indicating a dark matter halo. To understand well the TF method, we should know how the luminous and dark matter are distributed inside galaxies (Theureau et al. 1997a) and how far away from the centre the horizontal rotation is reached for different galaxy types etc. The result about the universality of galactic surface densities (Donato et al. 2009; Gentile et al. 2009), both baryonic and dark matter, within one dark halo scale-length may bring more order to this complex field.

The Slope Different TF relations appear in the literature, depending on the photometric band, but also on how they were derived and from what data. The zero-point b poses a separate problem, requiring a sample of calibrating galaxies with known distances (the zero-point may also be normalized to a value of the Hubble constant, if derived from field galaxies or galaxy clusters with known (cosmological) recession velocity). We give a few examples of the slope a in some bands ($M = a \log V_{\rm max} + b$). The B-band: -5.82, -7.97 (Theureau et al. 1997b; Sakai et al. 2000); the I-band: -7.6, -9.24 (Giovanelli et al. 1997; Sakai et al. 2000); the H-band: -11.03 (Sakai et al. 2000).

In the first B-magnitude slope field galaxies (with distances from the Hubble law) were used and the linear fit is made with keeping all errors in the calculated absolute magnitude M. In the second one (the HST Key Project), the fit was made on cluster galaxies and takes into account errors in both M and $\log V_{\rm max}$. Because of the fairly large scatter in the B relation (≈ 0.5 mag), the two kinds of fits result in rather different slopes (cf. Fig. 2.2).

The slope of the inverse relation ($\log V_{\rm max} = a'M + b'$) is derived assuming that all errors are in $\log V_{\rm max}$. For the B magnitude Ekholm et al. (1999b) and Tully and Pierce (2000) derive the slopes $-1/10$ and $-1/7.3$, respectively, while for the I magnitude Tully and Pierce give $-1/8.14$.

When one uses the TF relation for distance determination, different kinds of systematic errors appear depending on the chosen slope (direct, inverse), the nature of

[3]Even earlier Gouguenheim (1969) had inferred distances to six galaxies, using angular sizes and 21-cm line observations made with the Nancay radiotelescope.

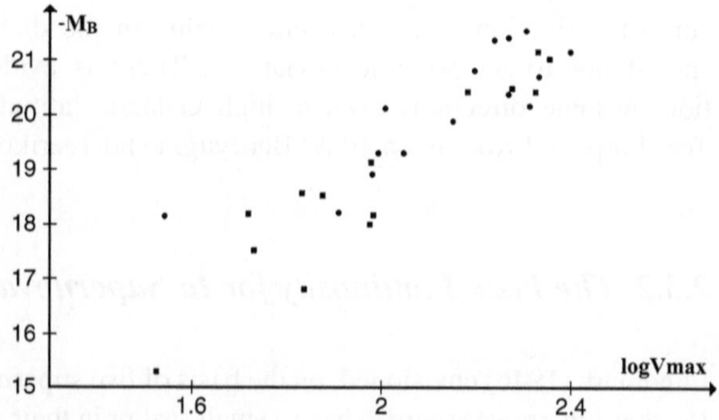

Fig. 2.2 Tully-Fisher (B magnitude) relation for nearby ($cz < 1000$ km/s) galaxies. The scatter is partly caused by distance errors. Based on calibration data in Theureau et al. (1997b)

the sample, and the situation (Chaps. 3 and 4). The systematic errors depend especially on the scatter in the relation. When one goes from shorter to longer wavelengths, the scatter decreases and may be quite small in the mid-IR range at 3.6 μm (±0.12 mag as tentatively given by Freedman and Madore 2010).

Corrections When working with the TF relation, one has to apply several corrections, most importantly the inclination correction, the Hubble type dependence of the zero-point, and the galactic extinction correction.

Disc galaxies are viewed from different directions. One needs the inclination angle i (between the line of sight and the polar axis of the galaxy) to correct two quantities: (1) the width of the HI line (as it would be seen for $i = 90°$), and (2) the magnitude (or diameter) of the galaxy (as seen for $i = 0°$ or "face-on"). The angle i is inferred from the axis ratio $R = D/d$ of the galaxy image, where D and d are the lengths of the major and minor axes, respectively, defined up to a fixed surface brightness (isophotal sizes). The influence of inclination on the magnitude is a topic with decades of history. It is interestingly connected to the degree of transparency of galaxy discs (the greater the opacity, the larger is the change in the magnitude when the disk is viewed from different directions; e.g., Kankare et al. 2009).

Bottinelli et al. (1995) derived the following correction for B-magnitude:

$$m_i = m_0 + a_{\text{incl}}(R) = -2.5 \log(k + (1 - k) R^{2C(1+0.2/K)-1}). \tag{2.7}$$

Here R is the axis ratio and k may be expressed as a function of the Hubble type t as $k = 0.754 \cdot 10^{-0.2t}$. The parameter K expresses how the apparent diameter changes with the surface brightness: $K = \partial \log D / \partial \log \mu$. The constant value $K = 0.094$ is used for disk galaxies. C is defined as $C = \partial \log D / \partial \log R$. The value $C = 0.04$ has been adopted in the LEDA extragalactic data base.

The *type dependence* is such that for a fixed V_{max} early type galaxies (Sa) are fainter than late type galaxies (or for a fixed absolute magnitude early types rotate more quickly than later types). This may be understood as due to the higher M/L ratio of the latter, as was modelled by Theureau et al. (1997a). The effect may be put into the zero-point of the TF relation.

The galactic extinction correction a_{gal} is usually taken (e.g., in the LEDA database) from Schlegel et al. (1998), based on infrared dust emission. This measure of extinction gives an average value in the direction of a galaxy and one should not forget possible deviations. There is evidence for high dust extinction in some directions even at high Galactic latitudes (e.g., Teerikorpi 1981a; Teerikorpi and Kotilainen 1989; Berdyugin and Teerikorpi 2002).

2.3.2 *The Peak Luminosity for Ia Supernovae*

Lundmark (1946) envisioned, on the basis of five supernovae observed in five galaxies, that supernovae seem to have a small scatter in their maximum luminosity so that they "seem to furnish an excellent distance indicator" and "will enable us to reach rather far out in the Metagalaxy". In fact, supernovae as a whole form a heterogenous blend of different origins and absolute magnitudes at their maxima. However, on the basis of their spectra one may group them into a few classes. Zwicky and Minkowski classified supernovae into Types I and II. Later Type I was divided into Ia, Ib, and Ic (all these originating from quite different processes).

Type Ia Luminosity The Ia supernovae are the most luminous of all supernovae, and they can be identified from the lack of hydrogen lines in their spectra. They form a rather uniform class of stellar explosions, in that more luminous objects have slower decline-rates (as was noticed by Pskovskii 1977). This simple behaviour allows them to be calibrated as standard candles. The peak luminosity is about -19.5 in both B and V magnitude bands, depending on the decline rate of the light curve.

The SNIa events that occur both in spiral and elliptical galaxies, have provided the sharpest Hubble relation along a wide range of distances (cf. Fig. 1.3). A drawback in supernova explosions is their unpredictability and even rareness in galaxies. Systematic search programmes are necessary so that one may catch a supernova before it reaches the maximum luminosity. We will discuss their use for cosmological purposes in later chapters (the Hubble constant in Chap. 4, the magnitude–redshift test in Chap. 8).

Lundmark (1919, 1920) had been the first to realize that there are novae and much more luminous supernovae in connection with his study of the distance of M31, with an "incredible foresight and imagination" as praised by Fritz Zwicky (Johnson 1961), who in 1934 together with Walter Baade connected supernovae with the death of massive stars. In 1925 Lundmark termed the giants and dwarfs among novae as "upper and lower class", with an upper class nova reaching a luminosity comparable to that of the whole host nebula).

2.3.3 *Morphological Galaxy Classes*

It would be fine if just by looking at the photo of a galaxy one could tell how big it is. Is the luminosity of a galaxy written on its appearance? This was asked by Van den Bergh (1960), after he noted that galaxies may differ in luminosity by a factor of about 10 000. He indeed found a correlation: the stronger its spiral structure is, the more luminous is a spiral galaxy. His luminosity classes range from I (supergiant) to V (dwarf galaxies), with intermediate cases (such as II–III). In the program "Steps toward the Hubble constant" by Sandage and Tammann in the 1970s, one goal was to calibrate the luminosity of the giant ScI galaxies, then use it as a standard candle. They found that the scatter within a luminosity class leads to rather inaccurate individual distances, and easily leads to a systematic error in distance determinations for a galaxy sample.

In connection with the morphology of galaxies it is interesting to mention that a half of S0 and S galaxies possess an inner ring structure (r), which is easy to observe. This finding led Buta and de Vaucouleurs (1983) to propose a distance indicator based on the size of the ring. The potentials of this method may not yet have been fully exploited (Teerikorpi 1986).

The Sosies Georges Paturel (1984) introduced the method of *sosies* into the distance ladder, whereby distant "clones" of nearby calibrators are searched for.[4] The idea is: if the absolute magnitude M depends on a measurable parameter p, but one does not know reliably the relation, one may restrict the study to such distant galaxies that have their values of p close to those of some nearby calibrators. The number of parameters p is in practice larger than one, including the morphological type and other more objectively measurable parameters. Van den Bergh's luminosity classes also define a kind of sosies, involving the Hubble type and the luminosity class index (labeled *beauty index* by Sandage). De Vaucouleurs's luminosity index Λ_c was calculated as $(T + L_c)/10$, where T is the morphological type code and L_c is van den Bergh's luminosity class number.

2.4 The Concept of Metric Distance in Curved Space

We described above some examples of practical ways to estimate cosmic distances, leaving a few other ones to later chapters. The physical concept of distance is deep-seated in the steps farther and farther in space, where one is compelled to take into account geometries more general than Euclidean. In cosmology, world models describe how distance and its measurement appear in the physical geometry of the large-scale universe. In order to facilitate understanding the meanings of different coordinates, forms of metric, and distances, we take advantage of analogies with more easily visualized spaces.

2.4.1 Universes of Constant Curvature Within E^3

We introduce main geometrical concepts in curved space so that the reader will see how Riemannian geometry enters cosmology and permits one to interpret astro-

[4] From *Larousse Classique*: Sosie—Personne avant une ressemblance parfaite avec une autre. Or—"Sosie—A person with a perfect resemblance to another."

nomical observations in terms of the geometry of the universe. For this purpose we first study the mathematics of two-dimensional homogeneous and isotropic spaces embedded in 3D Euclidean space E^3. Usually this subject is presented by differential geometry without referring to any external higher-dimensional embedding space. However, besides helping one to imagine and learn geometrical concepts, the embedding space may actually be essential for the physical concepts of distance, surface, and volume.

Systems of Coordinates for a 2D Sphere in E^3 Consider a spherical surface S^2 in the 3D Euclidean space E^3 and suppose that a 2D inhabitant, surveyor by profession, lives in this space. He cannot feel the third dimension, similarly as the fourth spatial dimension is not present for us. The surveyor is able to study his 2D universe by measuring distances with sticks and ropes or observing light sources scattered all around space. The light travels along the geodesics ("straight lines") of this spherical universe. The external 3D observer in E^3 sees that the trajectory of light is a great circle representing the shortest distance between two points as measured on the sphere.

To see what it means to measure lengths, angles, and fluxes in the 2D spherical space, we introduce four coordinate systems (Fig. 2.3). Three of these are defined in the embedding space E^3 and one in the space S^2 itself. The first three directly relate to the embedding space:

$$K_{\text{Cart}} = \{x, y, z\}, \qquad K_{\text{cyl}} = \{\rho, \phi, z\}, \qquad K_{\text{sph}} = \{R, \theta, \phi\}. \tag{2.8}$$

K_{Cart} is the Cartesian coordinate system, and x, y, z are its ordinary coordinates. These can be employed only in flat Euclidean space where it is possible to extend the local orthogonality over the whole space (for instance, any two parallel lines have everywhere the same distance between them).

K_{cyl} is the cylindrical coordinate system, where ρ is the polar radius on the XY plane, ϕ is the azimuthal angle, and z is the Cartesian coordinate orthogonal to the XY plane. K_{sph} is the familiar spherical system in E^3, so that R is the radial distance, θ is the polar angle, and ϕ is the azimuthal angle. These *external* coordinate systems are of course related by

$$x = R\sin\theta\cos\phi = \rho\cos\phi; \qquad y = R\sin\theta\sin\phi = \rho\sin\phi; \qquad z = R\cos\theta. \tag{2.9}$$

Let us suppose that in the spherical space S^2 the 2D observer at the point P can measure the angle between the light rays coming from the points A and B and also the distances to these light sources. Such a being would use a rope to measure the metric distance u from the stake pounded into the "ground" at point P. To measure the angle Φ he fixes a "zero-meridian" and the direction of increasing angle. In this manner the fourth coordinate system is generated, *internal* for the 2D spherical space:

$$K_{\text{int}} = \{u, \Phi\}. \tag{2.10}$$

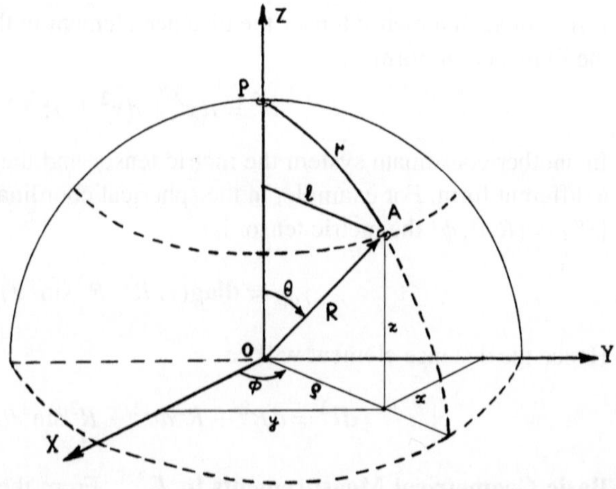

Fig. 2.3 Coordinates for 2D spherical space embedded in 3D Euclidean space

The coordinate u is the length of the arc PA of the great circle of the sphere and Φ is the angle which has the same value as the azimuthal angle ϕ in the embedding 3D Euclidean space. In his world the surveyor might, we guess, term u the polar radius and Φ the azimuthal angle.

Our one goal is to derive expressions that link the internal measurements by the 2D surveyor to the measurements made in the embedding E^3 space.

Metric Tensor and Metric Distance in E^3 The element of spatial distance dl in n-dimensional Riemannian space has the form

$$dl^2 = \gamma_{\alpha\beta} dx^\alpha dx^\beta, \tag{2.11}$$

where $\gamma_{\alpha\beta}$ is the metric tensor of the space and (the vector) dx^α gives the differentials of the coordinates $\{x^\alpha\} = \{x^1, x^2, \ldots, x^n\}$.

What is the practical importance of the metric tensor? The n-dimensional observer, who knows the mathematical expression for this tensor (i.e. how it depends on the chosen coordinate system), immediately knows all geometrical properties of his space. In particular, the metric tensor $\gamma_{\alpha\beta}$ defines the fundamental *metric distance* between any two points in that space: it is the integral of the distance element along the geodesic line connecting these points. One may identify the distance element dl with a local, very small standard rod, and the metric distance essentially means the length measured with such an unchangable "cm". When one applies this geometry to the real world, one must assume the existence of rigid rods.

In Euclidean 3D space the metric tensor assumes the simplest form when the Cartesian coordinates $\{x^\alpha\} = \{x^1, x^2, x^3\} = \{x, y, z\}$ are used:

$$\gamma_{\alpha\beta} = \text{diag}(1, 1, 1). \tag{2.12}$$

Here we employ the convenient expression diag(1, 1, 1) for the 3×3 matrice where the diagonal components $\gamma_{11}, \gamma_{22}, \gamma_{33}$ are equal to 1 and all other components are zero. For such a metric tensor the distance element in the embedding space E^3 has the Pythagorean form

$$dl^2 = dx^2 + dy^2 + dz^2. \tag{2.13}$$

In another coordinate system the metric tensor and the distance element will have a different form. For example, in the spherical coordinates of the system K_{sph} with $\{x^\alpha\} = \{R, \theta, \phi\}$ the metric tensor is

$$\gamma_{\alpha\beta} = \mathbf{diag}(1,\, R^2,\, R^2 \sin^2\theta). \tag{2.14}$$

Hence the distance element will be

$$dl^2 = dR^2 + R^2 d\theta^2 + R^2 \sin^2\theta d\phi^2. \tag{2.15}$$

Basic Geometrical Measurements in E^3 From the metric tensor one can calculate the metric distance between two fixed points in 3D space, and the result is *independent* of the choice of the coordinate system. For instance, the distance between the origin O and the point A is derived to be

$$l_{OA} = \int_O^A dl = \int_0^{R_A} dR = R_A = \sqrt{x_A^2 + y_A^2 + z_A^2}. \tag{2.16}$$

Though the end result is trivial, this calculation illustrates how in the step from the integral of dl (valid in any coordinate system) to the integral of dR one has chosen the spherical system and hence the distance element (2.15). This coordinate system is convenient, because R varies along the straight line connecting the points O and A, and the other coordinates θ and ϕ remain constant, hence the differentials $d\theta = d\phi = 0$. The metric distance l_{OA} thus calculated is indeed the shortest distance between O and A, because the geodesic curve in Euclidean space is a straight line.

We intentionally describe applications of the metric tensor for the simple Euclidean space. This helps one to grasp the role that the metric tensor has in curved spaces. Also, Euclidean geometry is the basis for the study of the local universe and the standard with which to compare observations from the deep universe where curved space may become measurable.

Several other familiar geometrical properties of E^3 can be rigorously derived from the fundamental metric tensor $\gamma_{\alpha\beta}$. The sum Σ of the angles a, b, c of a triangle, the length l_{circ} of the circumference of a circle of radius R, and the surface area A_{sph} of a sphere of radius R are, respectively,

$$\Sigma = a + b + c = \pi, \qquad l_{\text{circ}} = 2\pi R, \qquad A_{\text{sph}} = 4\pi R^2. \tag{2.17}$$

The volume element for the metric tensor $\gamma_{\alpha\beta}$ having the determinant γ is

$$dV = \sqrt{\gamma} dx^1 dx^2 dx^3 = dx dy dz = R^2 \sin\theta dR d\theta d\phi. \tag{2.18}$$

It gives the volume V_{sph} of a sphere of radius R as $V_{\text{sph}} = (4\pi/3) R^3$.

2.4.2 Metric Tensor and Distance Element in $\mathcal{S}^2$

Assume that the surveyor within his 2D spherical space is located at the "North Pole" P of the sphere that also belongs to the 3D embedding Euclidean space. P is the point where the Z-axis intersects the sphere, and it is the origin of the internal coordinate system K_{int}.

Distance in Spherical (External) Coordinates A sphere is defined in the spherical coordinates K_{sph} simply by the equation $R = \text{constant}$. In Cartesian coordinates K_{cart} this becomes

$$x^2 + y^2 + z^2 = R^2, \tag{2.19}$$

where x, y, z are the orthogonal coordinates, or the lengths of the projections of the radius vector $\vec{R} = \{x^\alpha\} = \{x, y, z\}$.

Because the sphere is a 2D manifold, any position P on it is given by two independent variables. The metric tensor and the distance element of this sphere can be expressed by these very coordinates.

By differentiating (2.19) we express the differential of z as $dz = -(xdx + ydy)/\sqrt{R^2 - x^2 - y^2}$. This allows us to eliminate z from the distance element of the embedding space $dl^2 = dx^2 + dy^2 + dz^2$, and we may write the distance element on the sphere in the Cartesian coordinates of E^3 as

$$dl^2 = dx^2 + dy^2 + \frac{(xdx + ydy)^2}{R^2 - (x^2 + y^2)}. \tag{2.20}$$

This form utilizes external Cartesian coordinates only. The following step is to express dl in the other two coordinate systems in E^3, and then to link these "external forms" to the internal coordinate system.

Distance in Cylindrical (External) Coordinates Recognizing that $x^2 + y^2 = \rho^2$ and using the azimuthal angle ϕ one may transform the distance element in Cartesian coordinates (2.13) into another form in cylindrical coordinates K_{cyl}:

$$dl^2 = \frac{d\rho^2}{1 - \frac{\rho^2}{R^2}} + \rho^2 d\phi^2. \tag{2.21}$$

Here R is the radius of the sphere in the embedding space E^3. For these coordinates $x^\alpha = (\rho, \phi)$ the metric tensor is

$$\gamma_{\alpha\beta} = \text{diag}\left(\frac{1}{1 - \frac{\rho^2}{R^2}}, \rho^2\right). \tag{2.22}$$

In the spherical coordinate system K_{sph} one may express ρ using R and θ as $\rho = R\sin\theta$. Then the distance element is

$$dl^2 = R^2 d\theta^2 + R^2 \sin^2\theta d\phi^2, \tag{2.23}$$

where the range of θ is $[0, \pi]$ and that of ϕ is $[0, 2\pi]$. In this case

$$\gamma_{\alpha\beta} = \operatorname{diag}(R^2, R^2 \sin^2\theta). \tag{2.24}$$

Distance in Internal Coordinates These expressions for the distance element dl (2.20, (2.21), (2.23) and the metric tensor $\gamma_{\alpha\beta}$ (2.22), (2.24) employ the embedding Euclidean space and are quite inconvenient for the practical 2D surveyor. Therefore, it is time to show how these geometrical quantities are related to the internal coordinate system K_{int}. The internal system is simply related to the spherical and cylindrical systems:

$$u = R\theta = R \arcsin\frac{\rho}{R} \Phi = \phi. \tag{2.25}$$

The surveyor, who naturally uses the internal (u, Φ) system, will find that the distance element has the following form:

$$dl^2 = du^2 + R^2 \sin^2\frac{u}{R} d\Phi^2 \tag{2.26}$$

hence the fundamental metric tensor of the 2D spherical space is expressed in the surveyor's (u, Φ) coordinates as

$$\gamma_{\alpha\beta} = \operatorname{diag}\left(1, R^2 \sin^2\frac{u}{R}\right). \tag{2.27}$$

These expressions written in *internal* coordinates contain the *external* radius R. This is interesting: in principle, the surveyor can recognize that the world has a non-Euclidean geometry and a finite area (2D volume). The radius R can be found from measurements of a circle (see below).

From our external viewpoint the rope that the surveyor uses for distance measurements is an inseparable part of his universe and cannot be straightened through the external space. In mathematical language this means that in order to find the metric distance on the sphere we must use the distance element in some of its "external" forms (Eqs. (2.20), (2.21), (2.23)).

A simple example is the metric distance u_A between the pole P, which is the origin of the internal coordinate system, and the point A:

$$u_A = \int_P^A dl = \underbrace{\int_0^{\rho_A} \frac{d\rho}{\sqrt{1-(\rho/R)^2}}}_{K_{\text{cyl}}} = R \arcsin\frac{\rho_A}{R} = \underbrace{\int_0^{\theta_A} R\,d\theta}_{K_{\text{sph}}} = R\Theta_A. \tag{2.28}$$

This calculation shows that the metric distance does not depend on the chosen coordinate system (K_{cyl} or K_{sph}). The internal surveyor uses his rope to determine the metric distance u_A, the length along a great circle of the sphere.

2.4.3 Basic Geometrical Properties of $\mathcal{S}^2$

The Gaussian curvature K of the spherical space is determined by the radius R of the sphere in E^3. A celebrated formula derived by Carl Friedrich Gauss expresses the (scalar) curvature K via the fundamental metric tensor. It yields in the present case the constant value

$$K = 1/R^2. \tag{2.29}$$

Hence, $\mathcal{S}^2$ is a Riemannian space of constant positive curvature.

Geometrical Measurements in $\mathcal{S}^2$ In a triangle ABC, whose sides are parts of three great circles, the sum of the angles a, b, c was found by Gauss to be

$$\Sigma = a + b + c = \pi + K \times \sigma, \tag{2.30}$$

where K is the curvature of the sphere and σ is the *area* of the triangle.

The length $l_{\rm circ}$ of the circumference of a circle of radius u is

$$l_{\rm circ} = \int_0^{2\pi} R \sin\theta d\phi = 2\pi R \sin\frac{u}{R}. \tag{2.31}$$

In this calculation we use external spherical coordinates and consider the circle with its centre in the polar point P. The needed distance element is given by (2.23). For the chosen circle the coordinate θ is constant, i.e. $d\theta = 0$, and the integration over ϕ yields 2π. For the last step we recall that $u = \theta R$. Note that here we compare the internal length u in cm with the external length R in the same centimetres.

It is useful to write out the first two terms of the Taylor series of sinus, in order to see how the length $l_{\rm circ}$ differs from the Euclidean circumference:

$$l_{\rm circ} = 2\pi u\left[1 - \frac{1}{6}\left(\frac{u}{R}\right)^2 + \cdots\right]. \tag{2.32}$$

When $u \ll R$, the small relative deflection from Euclidean is $\frac{\Delta l}{l} \approx -\frac{1}{6}\frac{u^2}{R^2}$.

The 2D volume element dA of the spherical space $\mathcal{S}^2$ with the determinant γ of the metric tensor (2.27) is

$$dA = \sqrt{\gamma} dx^1 dx^2 = R \sin\frac{u}{R} du d\phi. \tag{2.33}$$

Then the 2D volume $A_{\rm circ}$ of the circle of radius u is obtained as

$$A_{\rm circ} = \int_0^{2\pi}\int_0^u dA = 2\pi R^2\left(1 - \cos\frac{u}{R}\right) = \pi u^2\left[1 - \frac{1}{12}\left(\frac{u}{R}\right)^2 + \cdots\right]. \tag{2.34}$$

The 2D volume as measured by the internal surveyor is the area of the polar cap as seen by the external observer. When the surveyor makes only local measurements ($u \ll R$), the volume is close to the area of a circle of the same radius in Euclidean plane. The total volume of space is finite, in fact (2.34) tells that it equals $4\pi R^2$, simply the area of a sphere in E^3.

2.5 Physical Measurements in Spherical and Lobachevskij Spaces

Before going to the 3D curved spaces on which the expanding Friedmann models are based in Chap. 7, we inspect how the imagined surveyor will make observations from his fixed position at the point P.

2.5.1 *Angular Sizes, Fluxes, and Number Counts in $\mathcal{S}^2$*

Consider a small rigid rod at the metric distance u from the surveyor. During its motion in $\mathcal{S}^2$, the length of the rod does not change ("rigidity").

Angular Size and Metric Distance Then what is the angular size of the rod, if its length is $dl = d$ and it lies perpendicularly to the line of sight?

The expression (2.26) for the distance element dl with $du = 0$, gives $dl = R \sin \frac{u}{R} d\Phi$. Note that $d\Phi$ is the difference of the Φ-coordinates of the ends of the rod of length $l = dl$. It is the desired angular size $\alpha = d\Phi$:

$$\alpha = \frac{d}{R \sin \frac{u}{R}} = \frac{d}{u_{\mathrm{ang}}}, \tag{2.35}$$

where the *angular size distance* is denoted by u_{ang} and equals

$$u_{\mathrm{ang}} = R \sin \frac{u}{R}. \tag{2.36}$$

When the metric distance u grows from 0 to its maximum value πR, the angular size distance u_{ang} first increases from 0 to R, and then starts to decrease and becomes 0 when the metric distance is at its maximum (Fig. 2.4). This fictive construction is made in order to have an analogy of the ordinary relation between angle and distance. Simply, the angular size distance is by definition equal to the Euclidean distance at which a rod in Euclidean plane *would* have the angular size that it actually is observed to have in $\mathcal{S}^2$.

Flux and Distance Suppose a distant "galaxy" at the point A in the spherical universe $\mathcal{S}^2$ (Fig. 2.3) emits "light" isotropically around it with the rate L [energy/time] (its luminosity). In this 2D world the circle centered about A plays the role of a sphere in E^3. Define the flux of the light flowing through the circle with radius u as the energy per time unit per length unit (cf. unit of area in E^3). Then the flux observed at the point A is

$$F = \frac{L}{l_{\mathrm{circ}}} = \frac{L}{2\pi R \sin \frac{u}{R}} = \frac{L}{2\pi u_{\mathrm{lum}}}. \tag{2.37}$$

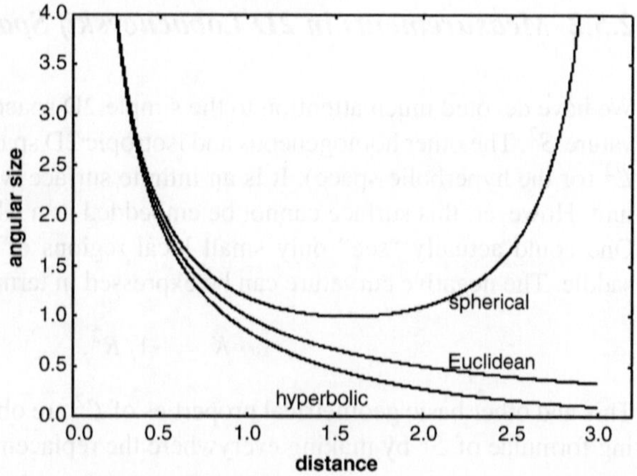

Fig. 2.4 Angular size–metric distance relation in different 2D spaces

Here appears the *luminosity distance* u_{lum}. It is such Euclidean metric distance at which the galaxy in Euclidean plane would produce the flux that is actually measured at P in S^2. Note that in the 2D spherical universe the angular size distance = the luminosity distance. This is valid also in other regular *static* spaces, both 2D and 3D, but not in expanding spaces such as used in the Friedmann model (Chap. 7).[5]

Counts of Galaxies in S^2 Suppose that S^2 is filled by some objects and their number density around the observer is $n(u)$. Then the total number $N(u)$ of these objects within the radius u (metric distance) grows according to a definite number-radius relation. To calculate $N(u)$, one must integrate the number density over the volume element dA of S^2 (Eq. (2.33)):

$$N(u) = \int_0^{2\pi} \int_0^u n(u) dA. \tag{2.38}$$

If the objects are uniformly distributed, i.e. $n(u) = n_0$, then:

$$N(u) = 2\pi R^2 \left(1 - \cos \frac{u}{R}\right) n_o.$$

As the spherical space has a finite "volume", the total number of these objects is finite, too, and is equal to $4\pi R^2 n_0$.

[5]Traditionally cosmologists speak about "luminosity distance". It might be more logical to term it "flux distance", emphasizing the observed quantity flux instead of the not directly observed luminosity. On the other hand, "luminosity" refers to the essential assumption that the object radiates isotropically with the rate L. With the angular size distance the analogous assumption is that the rod is always viewed perpendicularly.

2.5.2 *Measurements in 2D Lobachevskij Space*

We have devoted much attention to the simple 2D space with positive constant curvature $\mathcal{S}^2$. The other homogeneous and isotropic 2D space is the Lobachevskij space $\mathcal{L}^2$ (or the hyperbolic space). It is an infinite surface with negative constant curvature. However, this surface cannot be embedded as a whole in 3D Euclidean space. One could actually "see" only small local regions of $\mathcal{L}^2$ which remind one of a saddle. The negative curvature can be expressed in terms of the curvature radius as

$$K = -1/R^2.$$

This and other basic geometrical properties of $\mathcal{L}^2$ are obtained from the corresponding formulae of $\mathcal{S}^2$ by making everywhere the replacements[6]

$$R \Longrightarrow iR, \qquad \sin\theta \Longrightarrow \sinh\chi.$$

That is why it is sometimes said that the Lobachevskij space is a pseudospherical space with imaginary radius. Certainly, it is difficult to imagine. Fortunately, its mathematics is as simple as in $\mathcal{S}^2$. For example, the distance element for $\mathcal{L}^2$ in cylindrical coordinates is

$$dl^2 = \frac{d\rho^2}{1+\frac{\rho^2}{R^2}} + \rho^2 d\phi^2, \tag{2.39}$$

where the polar radius ρ can have values from zero to infinity. Another form of dl is obtained by substituting $\rho = R\sinh\chi$ into Eq. (2.39). Then

$$dl^2 = R^2 d\chi^2 + R^2\sinh^2\chi\, d\phi^2. \tag{2.40}$$

In the internal coordinates (u,Φ) of a surveyer in $\mathcal{L}^2$ the line element is

$$dl^2 = du^2 + R^2\sinh^2\frac{u}{R}d\Phi^2, \tag{2.41}$$

where $u = R\chi$, hence the fundamental metric tensor of the Lobachevskij 2D space is expressed in internal polar coordinates as

$$\gamma_{\alpha\beta} = \mathrm{diag}\left(1,\, R^2\sinh^2\frac{u}{R}\right). \tag{2.42}$$

The radius R refers to the pseudosphere in the external Euclidean space.

The sum of the angles of a triangle in $\mathcal{L}^2$ always fall shy of the Euclidean value π and the larger the triangle, the smaller the sum:

$$\Sigma = a + b + c = \pi - K \times \text{area}. \tag{2.43}$$

[6]We recall the hyperbolic functions: $\sinh x = (e^x - e^{-x})/2$, $\cosh x = (e^x + e^{-x})/2$.

The length of the circumsphere of a circle with radius u is given by

$$l_{\mathrm{circ}} = 2\pi R \sinh \frac{u}{R}.$$

The angular size–distance relation for the observer in $\mathcal{L}^2$ is

$$\alpha = \frac{d}{R \sinh \frac{u}{R}} = \frac{d}{u_{\mathrm{ang}}}.$$

This formula shows that the angular size of a "galaxy" in $\mathcal{L}^2$ monotonously decreases with increasing metric distance. Hence, different spaces E^2, $\mathcal{S}^2$, and $\mathcal{L}^2$ are characterized by clear differences in the behaviour of the angular size of a standard rod viewed at different metric distances (Fig. 2.3).

The volume of $\mathcal{L}^2$ is infinite, as may be seen from the volume up to u:

$$V(u) = 2\pi R^2 \left(\cosh \frac{u}{R} \right) - 1.$$

The fact that the volume (area) increases quicker than in the Euclidean plane may help one to understand why $\mathcal{L}^2$ cannot be embedded as a whole in $\mathcal{E}^3$—there is not enough space!

2.5.3 The Step to Three-Dimensional Curved Spaces

Before going to 3D spaces, we fix a convenient unified notation for the 2D spaces with constant curvature. This notation is easily transferred to the regular 3D spaces of modern cosmology.

Unified Notation for E^2, $\mathcal{S}^2$, and $\mathcal{L}^2$ First, the curvature K is simply

$$K = \frac{k}{R^2}. \tag{2.44}$$

The curvature constant k is 1, 0, or -1 (for $\mathcal{S}^2$, E^2, and $\mathcal{L}^2$, respectively). R is the radius of curvature of the space in question. Now consider the dimensionless variable χ expressing the internal coordinate u relative to R:

$$\chi = \frac{u}{R}.$$

Then the polar radius is

$$\rho = Ra(\chi),$$

where $a(\chi)$ is a new variable:

$$a(\chi) = \begin{cases} \sin\chi, & \text{for } k = +1, \chi \in [0, \pi], \\ \chi, & \text{for } k = 0, \chi \in [0, \infty], \\ \sinh\chi, & \text{for } k = -1, \chi \in [0, \infty]. \end{cases} \tag{2.45}$$

The differential of $a(\chi)$ is then given by the simple formula $da(\chi) = \sqrt{1 - ka^2(\chi)}d\chi$. The distance element for all three spaces now becomes

$$dl^2 = R^2(d\chi^2 + a^2(\chi)d\phi^2) \tag{2.46}$$

and the fundamental metric tensor and the volume element are

$$\gamma_{\alpha\beta} = \mathrm{diag}(R^2, R^2a^2(\chi)), \qquad dA = Ra(\chi)d\chi d\phi.$$

Spaces E^3, $\mathcal{S}^3$, and $\mathcal{L}^3$ The line element of homogeneous and isotropic 3D spaces is often written as follows:

$$ds^2 = R^2(d\chi^2 + I_k(\chi)^2(d\theta^2 + \sin^2\theta d\phi^2)). \tag{2.47}$$

χ, θ, ϕ are the spherical coordinates and $I_k(\chi) = \sin(\chi), \chi, \sinh(\chi)$ correspond to the curvature constants $k = +1, 0, -1$, respectively. The fundamental metric tensor is

$$\gamma_{\alpha\beta} = \mathrm{diag}(R^2, R^2 I_k^2(\chi), R^2 I_k^2(\chi)\sin^2\theta). \tag{2.48}$$

Note the similarity of this 3D line element to the internal line element of the 2D regular spaces, Eq. (2.46). It is also internal in the sense that all three coordinates (χ, θ, ϕ) are measured entirely within the 3D space.

Another form of the metric can be written in "cylindrical coordinates":

$$ds^2 = R^2\left(\frac{dr^2}{1 - kr^2} + r^2(d\theta^2 + \sin^2\theta d\phi^2)\right). \tag{2.49}$$

Now the coordinate r is not internal—Rr may be interpreted as the angular size distance. Integrating over ds from the observer to the point in question at r, the metric distance u and this coordinate distance r become related as $u = R\sin^{-1} r$, Rr, and $R\sinh^{-1} r$, for $k = 1$, 0, and -1, respectively.

2.6 Practical Geometry

If one asks a mathematician: what is geometry?, the reply may be unexpected, but instructive. So Veblen and Whitehead (1932) wrote: a part of mathematics is called geometry because a sufficient number of competent people think "on emotional and traditional grounds" that it is a good name.

This simply means that for mathematics itself it is not interesting how geometrical concepts relate to real space. However, for physics this is central. Working with abstract things, mathematical space cannot be a fully adequate picture of real physical space. *Mathematical* concepts coordinate, distance, and curvature are not equivalent to their *physical* counterparts. A "singularity" and the break-down of physical laws in it may only mean the inadequacy of the mathematical scheme leading to the singularity.

2.6.1 Geometry and Physics

Two opposite views on geometry and physics were put forward by Poincaré (1902) and Einstein (1921) when writing about the philosophy of science.

Poincaré vs. Einstein For Poincaré, geometry does not deal with real things: its notions belong to an ideal world. Only geometry together with physics, the geometry-physics unity, is open to empirical study. One may first choose geometry arbitrarily and then find the rest of the unity, physical laws, so that these will not contradict experiments. Or one may go other way round, to start from physical laws and find the geometry. Poincaré thought that it is preferable to leave the simplest Euclidean geometry intact and to change the physical laws. Modern physics gives another good reason not to forget flat geometry: the energy-momentum conservation laws directly follow from the maximal symmetry of Euclidean and Minkowski spaces (Noether's theorem).

Einstein's practical geometry assumes that geometry can be the subject of empirical study. The question whether the geometry of the universe is Euclidean or not has a clear meaning and may be answered by experience. Hence it is natural to formulate physical laws within Riemann's geometry. This is the way of general relativity.

Experiment is the final court in physics, and if the predictions for all experiments are identical, then both ways finally lead to the same geometry-physics unity. However, if predictions differ, crucial tests are required. General relativity and field gravity are modern examples (Chaps. 5 and 6).

Embedding Space In mathematics, a geometry of space may be built without assuming any embedding high-dimensional space (that is why one speaks about the internal geometry of curved spaces). In physics the concept of embedding space makes sense and deserves to be studied. In fact, the expression of the distance element requires that the internal distance u and the external radius R be measured in the same units. Thus the external standard of length in E^3 as if induces the unit in spherical space.[7] For instance, local lengths may be expressed in terms of the radius of curvature. Apparently having this in mind, in a 1824 letter to German scholar

[7]How numbers and physical quantities with units differ was extensively discussed by physicist Bridgman (1936) and mathematicians Menger (1959) and Whitney (1968).

Taurinus, Gauss wrote *I sometimes joke that it would be good, if Euclidean geometry were not true, because then we would have a priori absolute measure of length...*

In flat (Euclidean) space the radius of curvature is infinite and does not determine any absolute unit common with that of the embedding space. But another property is now related to the definition of the rigid length unit: Euclidean space is self-similar on all scales (congruency), and there is no preferred length which could be chosen as the unit. So, in $\mathcal{S}^2$ one may determine the absolute size of a triangle by measuring the sum of its angles, but in $\mathcal{E}^2$ all triangles have $a + b + c = \pi$—by looking at a triangle you cannot tell its size. One might surmise that a true scale-invariant fractal could exist only in Euclidean space, if one requires that the scale cannot be determined from the geometrical appearance of the fractal.

2.6.2 Measuring the Curvature

Non-Euclidean geometries—as mathematical models—do not contain internal contradictions. Hence, real space might be non-Euclidean (already Lobachevskij and Gauss made attempts to measure space curvature). Such an experimental approach probes the link between physics and geometry: is the curvature measurable, when one uses real units and procedures of length measurement? "If space curvature is real, it must make a difference in something we can measure" (Sandage 1992).

Rigid Sticks and Curvature The 2D beings of a spherical universe could walk around their world with a number of steps and establish its finite size. Locally, they could also measure the curvature R, e.g., using the angular excess formula. In analogy, such a measurement could be made also in our 3D world. However, the possibility to measure the curvature depends on a crucial assumption: There should be in curved space a length unit that can in principle be transferred from one place to another without changing it.

How to define a rigid rod in curved space is not obvious. In Euclidean space the ends of a free rigid rod draw parallel straight lines (geodesic curves of E^3), and there are no tensions between the internal parts of the rod, all moving along geodesics. So it is possible to imagine the rod made from independent, freely moving material points. In fact, one can also understand the unit length as the distance between two freely moving particles that were put into motion with equal velocities, e.g., perpendicularly to the line through them. In this way the length unit may be transferred into any point in space. Such a "soft" (free motion) standard gives the same results of measurement as an "absolutely hard" (rigid) stick. Indeed, the rigid stick is something that resists all forces trying to change the distance between its end points when it moves in space.

Things are different in curved space. There the Euclidean straight lines are replaced by geodesics, the shortest routes connecting two points. Then the two procedures (rigid stick, free motion) give different results. If one has the intuitively appealing rigid meter, its end points do not move along geodesic curves. Then the measurement of curvature is possible and its radius may be expressed in terms of the rigid meter. Or vice versa, lengths may be given with the fundamental radius of curvature as a unit.

2.6.3 The Deeper Value of Practical Geometry

The usual approach to curved space in physics has adopted the procedure, where the rigid meterstick ignores the geodesics of physical space. Thus distance measurement in astronomy has special significance beyond just knowing "how far away" a star or a galaxy is (needed for all that complicated astrophysics): it also probes our fundamental concepts of practical geometry and tests the validity of the rigid body hypothesis in our attempts to infer the curvature of space via non-local observations.

In this chapter we have briefly discussed some cosmic distance indicators and different kinds of distances which naturally arise in practice, notably angular size and luminosity distances. When one makes first steps in Euclidean space, indicators of different distance types are expected to give identical results. This is what happens, say, in the local galaxy universe within 100–200 Mpc where the Tully-Fisher relation both in the size and magnitude mode gives similar values for the Hubble constant.

In deep space, which means distances of the order of ~ 1000 Mpc and more, it is expected that different distance types start to diverge and only a good cosmological model will correctly link the distances to the underlying metric distance. We will see in later chapters how this happens with the Friedmann model. Above we have illustrated theoretical counterparts of practical distances in metric spaces using simple 2D spaces.

Chapter 3
Cosmic Distances and Selection Biases

Our ability to travel in space is badly limited. This is much felt in the process of extending distance measurements into deep space. Paul Hodge (1981) once began his review: "The determination of the extragalactic distance scale, like so many problems that occupy astronomers' attention, is essentially an impossible task". In fact, he was quite optimistic, and how else, the life is full of impossible things that nevertheless have been done.

Various types of Gregory's standard candles are still the main method to measure extragalactic distances, and also used in cosmological tests. One should be aware of fundamental difficulties accompanying this method. Here we give an overview of problems appearing in distance determination, when the astronomer works, as usually, with samples gathered from the sky (magnitude-limited), instead of samples obtained unrestricted from space.

3.1 Errors and Biases

All determinations of a physical quantity, including a cosmic distance, contain some error. In fact, one may speak about (1) random errors, (2) systematic errors, and (3) crude errors. A supposed distance indicator may be simply erroneous, leading to crude errors, an example being Hubble's brightest stars in galaxies, which were actually HII regions. Evolution of standard candles or their classes may cause systematic errors. And even if there were no inherent differences in the objects in different places and epochs, a sample of the *observed* objects may be much deformed by selection effects, leading to systematic errors in the inferred average distances.

Random peculiar velocities change the redshift from its ideal cosmological value, and generally these motions are not known for any individual galaxy. This adds always some error to the redshift distance.

Sometimes the method is sensitive to a factor whose effect must be modelled. Thus when using the time delay in gravitational lenses to derive the Hubble constant, one has to use a model for the mass distribution of the lensing galaxy. It also happens that a distance indicator may be made better when one notes the influence of an extra

Y. Baryshev, P. Teerikorpi, *Fundamental Questions of Practical Cosmology*, Astrophysics and Space Science Library 383,
DOI 10.1007/978-94-007-2379-5_3,

factor. For instance, the peak luminosity of the supernovae SN Ia is correlated with the decay rate of their light curves. Before such effects are known, they may give rise to errors that are not simply random, but systematic and distance-dependent.

Naturally, astronomer wants to see some measurable effect indicating the distance. Even if the observations are too inaccurate *something* is often seen and this is taken as a distance effect—usually leading to an underestimate.

An early example is the derivation of the Sun's distance by Aristarchus (310–230 B.C.). He knew that when the Moon appears exactly half full, then the Earth, the Moon and the Sun form a right-angled triangle with 90° at the Moon. He took the Moon–Sun angle to be 87 degrees and proved from the triangle that the ratio of the Sun's and the Moon's distances is between 18 and 20. In fact, the Moon–Sun angle in the half moon triangle is so close to 90 degrees (89.85°) that it was impossible to measure it and Aristarchus actually derived a *lower limit* to the Sun's distance. The use of standard candles can also be seen as a search for a measurable effect at a large distance. Something is seen, thought to be an "average standard candle", but which actually is an overluminous object.

3.1.1 The Concept of Bias

In mathematical statistics an estimator is some function of the sample, and in practice it is often the mean value of some parameters related to the members of the sample. An *unbiased estimator* $\hat{\mu}$ is such whose expected (mean) value is equal to the true value μ of the variable estimated: $E(\hat{\mu} \mid \mu) = \mu$. Then the bias $B(\mu)$ is the estimated minus the true value $B(\mu) = E(\hat{\mu} \mid \mu) - \mu$. In astronomy "bias" refers in a wider sense to situations resulting in a systematic error when one is working with observational data.

Different Ways of Speaking About Bias Sometimes the emphasis is how a sample is used for estimating something, sometimes one has in mind a selection effect deforming the sample. So, Smith (2003) speaks about (1) the truncation bias, (2) the modelling bias, and (3) the transformation bias. Of these the truncation bias "results from truncation of a sample according to a limit or limits on the observables when the latter are subject to random errors of measurement". The modelling bias arises from an incorrect model fitted to the data and the transformation bias results from a non-linearity, when one goes from a directly measured quantity with its associated error to some other quantity (say, from the distance modulus to the distance).

An example of the transformation bias appears with the log Hubble ratio. If the error in $\log H$ has a Gaussian distribution with the dispersion σ, (as may be expected if the error in the distance modulus is Gaussian), the calculated mean $\langle \log H \rangle$ gives the expectation of $\log H$. But the expectation of the usually desired quantity H is not $H_1 = 10^{\langle \log H \rangle}$, but $H_1 e^{(\ln 10\sigma)^2/2}$.

Lutz and Kelker (1973) discussed distances derived from trigonometric parallaxes. Assume that the parallaxes have been measured for a sample of stars and the measuring accuracy can be described as a Gaussian error law with dispersion σ_π. If one considers those stars for which the measured parallax is π_o, some of these have larger true parallaxes and some have smaller true parallaxes. Roughly speaking, there are more true parallaxes π available with $\pi < \pi_o$ than with $\pi > \pi_o$, so that the average true parallax for stars with observed parallax π_o is smaller than π_o. In other words, the average distance becomes underestimated due to this truncation bias.

Pfleiderer (1983) links the bias to the incompleteness of a sample, as "a data set has no bias or selection *per se*. It comes in only if it is compared to a hypothetical set filling some parameter domain completely, or if it is used as input for statistics or some other kind of conclusion". That is why one often speaks about a *selection* bias, which has affected the constitution of a sample—its properties and the wishes of the analyst do not always meet.

3.1.2 Observational Samples

The raw material of astronomy is often a magnitude-limited (ml) sample collected from an area in the sky, and containing all objects of the desired type brighter than a limiting magnitude m_l (in a certain wavelength band).

Magnitude- and Volume-Limited Samples The apparent magnitudes in the original ml-sample may have to be corrected for various effects and the resulting sample is generally no longer strictly ml-limited. This is relevant when it is good to have a sharp magnitude limit, e.g., for making volume-limited (vl) samples. Ideally, a vl-sample contains all objects within a given (corrected) absolute magnitude interval and within a distance range.

We define an effective limiting magnitude $m_{\mathrm{lim}}^{\mathrm{eff}}$ for a subsample with a correction A as $m_{\mathrm{lim}}^{\mathrm{eff}} = m_{\mathrm{lim}} - A$. Consider an extinction correction A. For unobscured objects with absolute magnitude M, a sample with the magnitude limit m_{lim} is complete up to the true distance modulus $m_{\mathrm{lim}} - M$. However, for objects needing the correction A, the sample is complete only up to $m_{\mathrm{lim}}^{\mathrm{eff}} - M$. So, be careful with corrected samples—each fixed correction A defines subsamples having their own effective magnitude limit.

Fundamental Equation of Stellar Statistics When looking at the sky, we at first see nothing but a distribution of apparent magnitudes $a(m)$, giving the number of stars within a narrow "magnitude window" $m \pm \frac{1}{2}dm$. With further information, this distribution may be determined separately for different kinds of stars, for instance according to the spectral type. How is this fundamental observation related to the distribution of stars in space? Obviously important is the number density $\rho(r)$ at distance r from us. Also, $a(m)$ depends on the luminosity function $\phi(M)$, which gives the probability of finding a star with absolute magnitude $M \pm \frac{1}{2}dM$, in a volume of space.

The number of objects with apparent magnitudes between m and $m + dm$ and lying in a spherical layer $(r, r + dr)$ within a solid angle ω is

$$N(m, r)\,dm\,dr = \omega\phi(M)\,\rho r^2 dm\,dr. \tag{3.1}$$

As $M(m,r) = m - 5\lg r/10$ pc, averaging over r yields the distribution

$$a(m) = \omega \int_0^{\infty} \phi(M)\,\rho r^2 dr. \tag{3.2}$$

This *fundamental equation of stellar statistics* for Euclidean transparent space was derived by Hugo von Seeliger (1849–1924) (von der Pahlen 1937).

A simple, but important case is when the radial distribution $\rho(r)$ is a power-law $\rho(r) = kr^{\alpha}$. When $\alpha = 0$, this represents the uniform spatial distribution, while various fractal distributions would be given with $-3 < \alpha < 0$ (the fractal dimension $D = \alpha + 3$). Substituting $\rho(r)$ into (3.2) gives

$$a(m) = \text{const} \times 10^{0.2(3+\alpha)m}. \tag{3.3}$$

The functional form of $a(m)$ does not depend on the luminosity function $\phi(M)$, at least in the ideal case where $\phi(M)$ contributes to the integral at each r (from 0 to ∞).

Number of Objects at Different Distances It may be useful to know how the objects in a magnitude limited sample are distributed at different distances. By integrating Eq. (3.1) the desired number for each r becomes

$$N(r) = \omega \int_{-\infty}^{m_1} \phi(M)\,\rho(r) r^2 dm. \tag{3.4}$$

The distribution $N(r)$ depends essentially on the luminosity function $\phi(M)$ of the objects considered. For the overall description of galaxies (all types together), Schechter (1976) proposed the function which is now often used:

$$\Phi(M)dM = \text{const} \times 10^{-0.4(\alpha+1)M} \mathrm{e}^{-10^{0.4(M^*-M)}} dM. \tag{3.5}$$

The standard values of the parameters are $\alpha = 1.25$, $M^* = -19.5 + 5\log h$.

The Eddington Bias At this point it is good to discuss a bias named after Eddington (1913, 1940). He posed the question: if the observed distribution function of quantities X with measured values x is $E(x)$, what is the true distribution of errorless quantities $T(x)$, when the errors of measurement ϵ have a Gaussian distribution with dispersion σ? He derived the following general relation between the two distributions:

$$T(x) = E(x) - \frac{1}{2}\sigma^2 d^2E(x)/dx^2 + \frac{1}{2}\left(\frac{1}{2}\sigma^2\right)^2 d^4E(x)/dx^4 - \cdots. \tag{3.6}$$

Such an inverse problem, inferring the true distribution from the observed one, has in its general form special mathematical difficulties. In practice, when we consider number counts of magnitudes, it is convenient to inspect the case $E(m) =$

$N_{obs}(m) = ke^{\beta m}$. Then the errorless count is $N(m) \approx N_{obs}(m)e^{-\frac{1}{2}\sigma^2\beta^2}$. Transforming to the usual presentation $N_{obs}(m) = k'10^{\alpha m}$ one obtains a link between the observed and the true counts:

$$\log N(m) = \log N_{obs}(m) - \frac{1}{2}\sigma^2\alpha^2/\log e. \qquad (3.7)$$

When the dispersion σ is constant, this formula shows that on a $\log N(m)$ vs. m diagram a linear counts curve with $\alpha > 0$ will shift upwards, preserving its slope. Then the slope α can be directly obtained from the counts. Vice versa, one may *predict* the observed counts from a theoretical law. This approach is especially useful in the realistic case when the accuracy of the measured magnitude depends on m. If σ increases with m, the slope of the number counts becomes steeper than the true one (Teerikorpi 2004).

One sometimes reads that the Eddington bias is important only close to the magnitude limit. Actually it operates at all magnitudes. Its cousin, the classical Malmquist bias, also works at all magnitudes. And a main result of Lutz and Kelker (1973) was that the bias in stellar parallaxes occurs at all parallaxes. Sometimes it is also said that the Eddington bias is due to the increase of the distribution towards faint magnitudes. Indeed, the bias vanishes when the measured distribution is constant (Eq. (3.6)). But it also vanishes for a linearly increasing (or decreasing) distribution, because then the second (and further) derivatives are zero (i.e. the convolution of a linear function with a symmetric Gaussian does not change the function).

The galaxies at a fixed measured magnitude originate either from brighter or fainter true magnitudes, due to the symmetric error distribution. Galaxies originally at $m \pm \frac{1}{2}dm$ are lost equally to the right and left. If the distribution is constant, it is clear that "incomers" from the right and the left compensate for those losses. But also if the distribution is linear (and increasing to the right), the smaller number of incomers from the left are exactly compensated by the larger number from the right—their sum is the same as the loss.

3.2 The Classical Malmquist Bias

Distant celestial bodies are observed, e.g., as traces recorded by a photographic plate or a CCD detector. Fortunately, luminous objects exist which may be seen from very far. This diversity in the cosmic zoo allows us to reach deep space, making cosmology possible. But it also involves problems, because we gather the objects from the sky, not from space.

Firstly, from large distances only very luminous objects are detectable, and we usually do not know how much the objects differ from the average one: there are no genuine Gregory's standard candles. Secondly, objects in the sky which are apparently similar (the same distance modulus!), have actually a complicated distribution of true distances. Distances larger than suggested by the distance modulus, are favoured because of the volume (r^2dr) effect. Thirdly, at large distances there is more space than within short distances around our position. Luminous galaxies are found throughout the cone, whereas dwarf galaxies concentrate into a small volume close to the vertex, as we simply cannot see those at large distance.

Very luminous objects are rare, so are not found in our vicinity. We perhaps see such objects that as a class might be useful distance indicators, but cannot calibrate them, which requires a distance ladder to reach a few of such objects. For instance, even if one could find standard candles among luminous quasars (Teerikorpi 2000), there is no known method to derive their distances independently of redshift (except, perhaps, assuming that they are radiating at the Eddington limit; Teerikorpi 2011).

When one uses "standard candles" (or "rods"), systematic errors related to the above problems creep into the distance estimates. The concepts of the different kinds of such biases are rather simple, though subtle.

3.2.1 A Unified Treatment of the Malmquist Bias

Following Butkevich et al. (2005), it is useful to consider two types of bias in a simple uniform manner. The first bias was discussed by Gunnar Malmquist (1920, 1922) and the second one has been studied more recently in connection with extragalactic distance estimates.

The Malmquist bias, $\Delta\langle M\rangle$, may be defined as the difference between an average absolute magnitude, $\langle M\rangle$, calculated in some way in a sample, and the true mean, M_0, for objects of the same class. Averaging over an entire sample gives the *integral* bias. Since, in absence of light extinction, an absolute magnitude is uniquely derived from an apparent magnitude and a distance, we can average over any of these variables keeping the other fixed. The result is a *differential* bias, either *magnitude-dependent*, when the mean is for the stars of a fixed apparent magnitude, or *distance-dependent*, when the averaging is done over the stars at the same *true* distance.

We derive here the Malmquist bias formulae in a few steps, considering both differential biases, and making the following assumptions:

1. There is no interstellar absorption.
2. The luminosity function $\phi(M)$ is independent of the distance r.
3. For any area in the sky, the spatial density of stars ρ depends only on distance.
4. At each m, the selection of a star into the sample does not depend on distance, but only on m (needed for the magnitude-dependent bias).
5. Completeness to a limiting magnitude $m_{\lim}$ (the distance dependent bias).

Basic Relations Averaging over m in (3.1) gives the fraction of stars that are brighter than $m_{\lim}$ and reside at the same distance r:

$$\psi\,(m_{\lim}, r) = \int_{-\infty}^{m_{\lim}} \phi\,(M)\,dm. \tag{3.8}$$

From the 5th assumption, at a distance r the absolute magnitude limit is $M_{\mathrm{lim}} = m_{\mathrm{lim}} - 5\lg\frac{r}{10\,\mathrm{pc}}$. Then we get the expression for ψ in terms of M_{lim}:

$$\psi\left(M_{\mathrm{lim}}\right) = \int_{-\infty}^{M_{\mathrm{lim}}} \phi\left(M\right) dM. \tag{3.9}$$

Now we can calculate two mean values of absolute magnitude. Averaging over distances, *the mean M for stars of a given apparent magnitude m* is

$$\langle M\rangle_m a\left(m\right) = \omega \int_0^{\infty} M\phi\left(M\right)\rho r^2 dr. \tag{3.10}$$

Integration over m (i.e. over M) gives *the mean for stars at a distance r*

$$\langle M\rangle_r \psi\left(M_{\mathrm{lim}}\right) = \int_{-\infty}^{M_{\mathrm{lim}}} M\phi\left(M\right) dM. \tag{3.11}$$

Gaussian Luminosity Function The above equations are valid for any luminosity function and density. If ϕ obeys a Gaussian law with the mean M_0 and the dispersion σ, then the following is true: $M\phi(M) = M_0\phi(M) - \sigma^2\phi'(M)$. Inserting this into (3.10) and taking (3.2) into account, we get

$$\left(\langle M\rangle_m - M_0\right) a\left(m\right) = -\sigma^2 \int_0^{\infty} \phi'\left(M\right)\rho r^2 dr \tag{3.12}$$

and substituting it into (3.11) gives

$$\left(\langle M\rangle_r - M_0\right) \psi\left(M_{\mathrm{lim}}\right) = -\sigma^2 \int_{-\infty}^{M_{\mathrm{lim}}} \phi'\left(M\right) dM. \tag{3.13}$$

The integrals are simply related to the derivatives of $a(m)$ and $\psi(M_{\mathrm{lim}})$:

$$\frac{da\left(m\right)}{dm} = \omega \int_0^{\infty} \phi'\left(M\right)\rho r^2 dr, \tag{3.14}$$

$$\int_{-\infty}^{M_{\mathrm{lim}}} \phi'\left(M\right) dM = \phi\left(M_{\mathrm{lim}}\right) = \frac{d\psi\left(M_{\mathrm{lim}}\right)}{dM_{\mathrm{lim}}}. \tag{3.15}$$

Substituting (3.14) and (3.15) into (3.12) and (3.13), respectively, we finally get the desired equations for the two Malmquist biases:

$$\Delta\langle M\rangle_m = -\sigma^2 \frac{d\ln a\left(m\right)}{dm} \quad \text{(magnitude dependent)}, \tag{3.16}$$

$$\Delta\langle M\rangle_r = -\sigma^2 \frac{d\ln \psi\left(M_{\mathrm{lim}}\right)}{dM_{\mathrm{lim}}} \quad \text{(distance dependent)}. \tag{3.17}$$

Equation (3.16) is the classical Malmquist bias for the stars of a given apparent magnitude (Eq. (17) in Malmquist 1922). The distance-dependent bias (3.17), which Malmquist did not consider, is equivalent to the formula derived by Teerikorpi (1975b) in a study of the Hubble law.

Spatial Distribution and Completeness An anisotropy in number density does not affect the final relations, because one may replace ρ by a mean $\overline{\rho}(r)$. In fact, the distance-dependent bias does not depend on the density ρ at all.

Generally the magnitude-dependent bias may be a complicated function of m. The term including the distribution of apparent magnitudes in (3.16) reduces to a simple form for the power-law space density $\rho(r) \propto r^{\alpha}$:

$$\langle M \rangle_m = M_0 - (3+\alpha)0.461\sigma^2. \tag{3.18}$$

With $\alpha = 0$, one obtains the celebrated Eddington-Mamlquist formula:

$$\langle M \rangle_m = M_0 - 1.382\sigma^2. \tag{3.19}$$

This relation, valid only for a uniform spatial distribution, was given already by Eddington (1914). The essence of the bias (selection from the sky or in space) was realized by Eddington in connection with a concrete problem (the reality of the division into dwarf and giant stars). This has been often the case with discoveries later generalized into wider applications.

Finally, we note that the bias at constant m (seen through the narrow window $m \pm \frac{1}{2}dm$) does not depend on the magnitude completeness at m, if the selection of the sample only depends on the magnitude and on nothing else. We emphasize that $a(m)$ is the distribution that would be observed in the sky for a complete sample, and the Malmquist bias is bound to occur, irrespective of whether we have or not such a complete sample at hand!

3.2.2 Two Kinds of Biases in Distance Determination

Eddington (1914) and Malmquist (1920, 1922) studied the difference between the derived and the true mean absolute magnitude of a stellar class. Later this bias was considered in the "inverse" problem of deriving distances for stars or galaxies assumed to be members of a class with the average magnitude M_0. The derived distance modulus for objects with apparent magnitude m $\langle \mu_{\mathrm{der}} \rangle_m$ and the true average modulus $\langle \mu_{\mathrm{true}} \rangle_m$

$$\langle \mu_{\mathrm{der}} \rangle_m = m - M_0, \qquad \langle \mu_{\mathrm{true}} \rangle_m = m - \langle M \rangle \tag{3.20}$$

differ by an amount given by the classical Malmquist bias $\Delta \langle M \rangle_m$

$$\langle \mu_{\mathrm{der}} \rangle_m = \langle \mu_{\mathrm{true}} \rangle_m + \Delta \langle M \rangle_m. \tag{3.21}$$

Behr (1951) noted, after comparing the width of the Local Group luminosity function to that of the field galaxies, that the standard candle method may lead to underestimated distances at large true distances. The idea was modelled by Scott (1957) for the selection of clusters of galaxies. It was sort of reinvented by Sandage and Tammann (1975) and Teerikorpi (1975a, 1975b), now in connection with field samples of luminosity classified spiral galaxies.

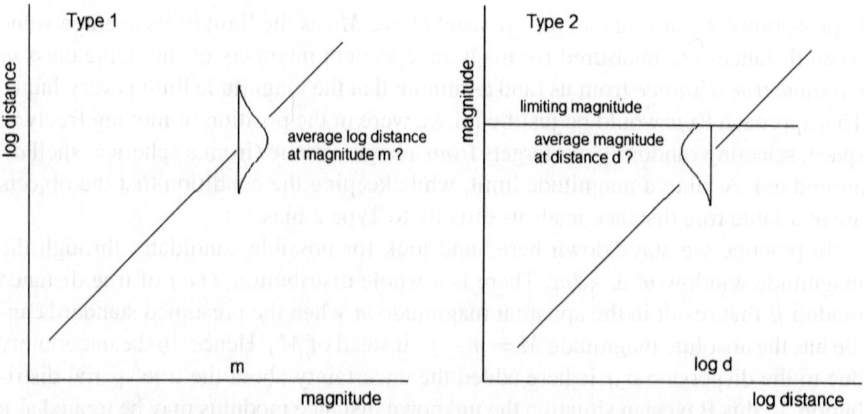

Fig. 3.1 The significance of the two bias types for a standard candle schematically shown in distance—magnitude diagrams. Note that if we identify the distance d with the redshift z, these diagrams represent the Hubble diagram, viewed in two different ways

Type 1 and Type 2 Biases We refer to the mentioned problems as two types of Malmquist biases (Teerikorpi 1997), the 1st type being intimately connected with the magnitude-dependent bias treated by Malmquist. One may define these two aspects in distance determination:

- Type 1 bias is the systematic average error in the distance modulus μ for a class of objects with derived $\mu = \mu_{\text{der}} = \text{const.}$
- Type 2 bias is the systematic error in the average derived $\langle \mu_{\text{der}} \rangle$ for the class of objects with true $\mu = \mu_{\text{true}} = \text{const.}$

With Type 1 bias one is interested in the distribution of true distance moduli, for a fixed derived modulus, while Type 2 bias is related to the derived distance moduli, at a given true distance. The objects of a standard candle class with the derived modulus μ all have the same apparent magnitude $m = \mu + M_0$. From this we see that the Type 1 bias has to do with the classical Malmquist bias. These two types have a link to the Hubble diagram of a standard candle. For instance, at a fixed redshift the average apparent magnitude deviates from the naively expected one by the amount given by the Type 2 bias. Thus Teerikorpi (1993) separated the study of the Hubble diagram into "distance against velocity" and "velocity against distance" (see Fig. 3.1).

Malmquist Biases and Probability Hendry and Simmons (1994, 1995) have formulated the distance estimation as a Bayesian inference problem. Indeed, one may connect the two kinds of bias to the two approaches to probability. This also touches the question arising when the sample contains only one member, i.e. the star or galaxy whose distance is being measured. What is the best value of the distance for *that* individual object?

Considering a standard candle, in the simplest case we have just its measured magnitude m. One might wish to ascribe to it the distance modulus $\mu = m - M_0$, with the error bars $\pm\sigma$. Here we are just following the usual "frequentist approach" to probability, regarding that $m - \mu$ would have M_0 as the limit of its average value when distances are measured for many independent members of this same class *at the same true distance* from us (and assuming that the magnitude limit is very faint). This approach to μ would be justified if we were in the position of moving freely in space, selecting randomly our targets from a large volume (from a spherical shell dr around us). Adding a magnitude limit, while keeping the condition that the objects are at a same true distance leads us directly to Type 2 bias.

In practice we stay "down here" and look for possible candidates through the magnitude window $m \pm \frac{1}{2}dm$. There is a whole distribution $f(\mu)$ of true distance moduli μ that result in the apparent magnitude m when the presumed standard candle has the absolute magnitude $M = m - \mu$, instead of M_0. Hence, to the uncertainty due to the dispersion σ_M is here added the uncertainty about the true spatial distribution. In this Bayesian situation the unknown distance modulus may be treated as a random variable with a more or less known distribution. This distribution is subject to modification upon acquisition of further information supplied by observational data.

The Scott Effect Elizabeth Scott (1957) concluded that at a given distance, a cluster with many members is more likely to be detected than a smaller cluster, or if a very distant cluster is seen, then it must be unusual. Especially, the brightest galaxies actually observed in the very distant clusters tend to be more luminous than the brightest galaxies in nearby clusters, and hence, for distant clusters, the brightest galaxy criterion leads to an underestimate of the distance. Scott used numerical simulations and an analytical model to show that the selection is bound to affect the Hubble z–m diagram for brightest cluster galaxies. She also pondered about how to decide between a real and selection induced deviation from linearity in the Hubble law: this may perhaps be solved by the accumulation of further data using telescopes reaching fainter objects. Such adding-a-fainter-sample test was applied, e.g., by Sandage (1988c).

3.2.3 The Behaviour of Biases for a Standard Candle

Let us consider a class of galaxies having a Gaussian luminosity function $G(M_p, \sigma)$. Such a class may be defined, e.g., via a fixed value of an observed parameter p, and it is assumed that the volume-limited value of M_p has been calibrated. A sample of these galaxies with apparent magnitudes measured up to some limit gives us a collection of derived distance moduli $\mu = m - M_p$.

Type 1 Bias The general Malmquist bias in distance moduli appears in the μ_{true} versus μ_{der} diagram. Because the derived modulus is simply $\mu_{\text{der}} = m - M_0$, the x-axis is practically the same as the m-axis, and an interval $d\mu_{\text{der}}$ is actually a narrow magnitude-window $m \pm \frac{1}{2}dm$. Looking through this window, the average systematic error in the derived distance modulus is given by the type 1 bias: $\Delta\langle\mu\rangle = \Delta\langle M\rangle_m$.

Table 3.1 The shift $\Delta\langle\mu\rangle$ for different values of $A' = M_{\mathrm{lim}} - M_0$ and σ_M

σ_M	0.3	0.4	0.5	0.6	0.7
$A' - \Delta\langle\mu\rangle$					
2.5	0.00	0.00	0.00	0.00	0.00
2.0	0.00	0.00	0.00	0.00	0.01
1.5	0.00	0.00	0.00	0.01	0.03
1.0	0.00	0.01	0.03	0.06	0.11
0.5	0.03	0.08	0.14	0.21	0.28
0.25	0.11	0.18	0.26	0.33	0.41
0.00	0.24	0.32	0.40	0.48	0.56
−0.25	0.41	0.49	0.57	0.65	0.73
−0.50	0.63	0.69	0.77	0.84	0.92
−1.00	1.09	1.13	1.18	1.25	1.32
−1.50	1.56	1.60	1.65	1.70	1.76

For a r^{α} density law, the bias $\Delta\langle\mu\rangle$ is constant, but generally the bias will depend on μ (i.e. on m), via the term containing $a(m)$ in the general Malmquist formula (3.16).

As the bias is often rather constant, its presence may be difficult to prove by studying how a parameter depends on the *derived* distance modulus. For example, in the past the observation that the Hubble ratio ($\log V/r_{\mathrm{der}} = \log V - 0.2\mu_{\mathrm{der}} + \mathrm{const}$) did not change with the derived modulus μ_{der} was sometimes taken, incorrectly, as indicating the absence of the bias.

Type 2 Bias This distance-dependent effect has been discussed in extragalactic astronomy, where the cosmological redshift allows one to assign *relative* distances to galaxies. When one observes progressively more distant standard candles with a dispersion σ_M, the distance modulus becomes more and more underestimated as the magnitude limit cuts away objects fainter than $M_{\mathrm{lim}} = m_{\mathrm{lim}} - \mu_{\mathrm{true}}$ from the Gaussian magnitude distribution (Fig. 3.1).

For a Gaussian luminosity function $G(M_0, \sigma_M)$ and a sharp magnitude limit, it is straightforward to calculate the amount of the distance-dependent bias $\langle\mu_{\mathrm{der}}\rangle - \mu_{\mathrm{true}}$ at each true distance modulus, due to the magnitude limit cutting galaxies from the fainter wing of the luminosity function (Teerikorpi 1975b; Sandage 1994; Butkevich et al. 2005; also Eq. (3.17) above). We use the parameter $A = (M_{\mathrm{lim}} - M_0)/2\sigma_M$ to express the position of the absolute limiting magnitude relative to the true average M_0. So at $A = 0$ the limit exactly halves the distribution. The bias becomes

$$\Delta\langle\mu\rangle = \sqrt{\frac{2}{\pi}}\sigma_M \frac{\mathrm{e}^{-A^2}}{1 + \mathrm{erf}(A)}, \tag{3.22}$$

where erf is the error function $\mathrm{erf}(A) = \frac{2}{\sqrt{\pi}}\int_0^A \mathrm{e}^{-t^2}dt$.

Table 3.1 gives calculated values of the bias for different values of $A' = (M_{\text{lim}} - M_0)$ and σ_M. It shows that standard candles with different means M_0, but the same dispersion, sampled up to the same limiting magnitude have a characteristic bias behaviour: for absolutely brighter classes the bias starts (or the same A' is reached) at a larger distance. This helps one to recognize the bias. We also see from the table that Type 2 Malmquist bias may well produce effects larger than perhaps expected from the classical formula $1.382\sigma^2$ for the average bias for a ml-sample (Bottinelli et al. 1988).

3.3 The Bias for a Distance Indicator $M = ap + b$

The period–luminosity relation of Cepheids, the rotational velocity–luminosity (Tully-Fisher) relation for spiral galaxies, and the decay-rate–maximum-luminosity relation for supernova explosions of type Ia, are important examples of distance indicators based on a relation $M = ap + b$. A general framework is relevant for understanding these indicators that continue to be the backbone of the cosmic distance ladder on different spatial scales.

3.3.1 The Direct and Inverse Relations

A standard candle (or rod) is often defined through a linear relation $M = ap + b$ where p is a parameter whose value can be measured without knowing the distance (e.g., the (log) period of a Cepheid). Behind the linear relation there is a bivariate distribution $\Phi(M, p)dMdP$ and the mentioned relation comes from the corresponding conditional probability distribution with p fixed. We write the *direct* regression form of this relation as

$$M = a \cdot p + b. \tag{3.23}$$

It is also useful to write down the *inverse* relation as:

$$p = a' \cdot M + b'. \tag{3.24}$$

The direct relation predicts the absolute photometric quantity from the measured parameter, as required for distance determination. Physically, the inverse relation tells how the often more fundamental quantity M (related to the total luminosity or mass) determines the value of the parameter p. How are these two forms connected? Assume that the parameter p has an intrinsic dispersion σ_p^i in galaxies with constant absolute magnitude M, and its distribution around the average value $p(M)$ of p at M is a Gaussian function $\Psi_M(p) \propto G(p(M), \sigma_p^i)$. We also need the luminosity function $\phi(M)$ of the galaxies in question: $\phi(M) = \text{const} \int_{-\infty}^{\infty} \psi_M(p)dp$. So the number of galaxies $N_p(M)$ in the element $M \pm \frac{1}{2}dM$, $p \pm \frac{1}{2}dp$ is

$$N_p(M)dMdp = \phi(M)\frac{\exp(-[p - p(M)]^2/2(\sigma_p^i)^2)dMdp}{\int_{-\infty}^{\infty}\exp(-[p - p(M)]^2/2(\sigma_p^i)^2)dMdp} \tag{3.25}$$

and the average value of M at a fixed p is obtained by integration over M.

3.3.2 *The Classical Bias*

The average values $\langle M\rangle_{p,\mathrm{d}}$ and $\langle M\rangle_{p,\mathrm{ml}}$ are easy to calculate when the luminosity function is exponential or Gaussian (Teerikorpi 1984). For the case $\phi(M)_\mathrm{d} \propto$ Gaussian $G(M_0, \sigma_M)$ the average absolute magnitudes become, in terms of the fundamental inverse relation parameters a' and b'

$$\langle M\rangle_{p,\mathrm{d}} = \left[\frac{a'}{(\sigma_p^i)^2}p - \frac{a'b'}{(\sigma_p^i)^2} + \frac{M_0}{(\sigma_M)^2}\right]\sigma_{M_p}^2, \tag{3.26}$$

$$\langle M\rangle_{p,\mathrm{ml}} = \langle M\rangle_{p,\mathrm{d}} - 1.382\sigma_{M_p}^2, \tag{3.27}$$

where σ_{M_p} is the dispersion of M at a fixed (errorless) value of p. It is related to the two other dispersions as $1/\sigma_{M_p}^2 = 1/\sigma_M^2 + 1/(\sigma_p^i/a')^2$. These results show explicitly that a distance indicator of the form $M = ap + c$ does not avoid the usual Malmquist bias.

In practice, the parameter p is measured with some error. Note that even with no intrinsic scatter ($\sigma_p^i = 0$) random errors in p (dispersion σ_p^o) give rise to the same bias as an intrinsic scatter of the same size. This reflects the decreased knowledge of M, when p is measured inaccurately, i.e. one does not actually find objects with a zero dispersion in M.

The General Formula When $M = ap + b$ is used to calculate distance moduli, to each p corresponds the modulus $\mu_p = m - M(p)$ and the bias in its general form is $\Delta\langle\mu\rangle = -\sigma_{M_p}^2 d\ln a_p(m)/dm = -\sigma_{M_p}^2 d\ln N_p(\mu_p)/d\mu_p$. Assuming that objects with different values of p are similarly distributed in space ($N_{p1}(\mu_{p2}) \propto N_{p2}(\mu_{p2})$) and the dispersion σ_{M_p} does not depend on p, the bias in the distance modulus μ_der can be conveniently expressed independent of p in a similar form as for a single standard candle class:

$$\Delta\langle\mu\rangle = -\sigma_{M_p}^2 d\ln N(\mu)/d\mu, \tag{3.28}$$

where the distribution $N(\mu)$ refers to a complete sample of distance moduli for the objects considered (not the same as the observed one; see below).

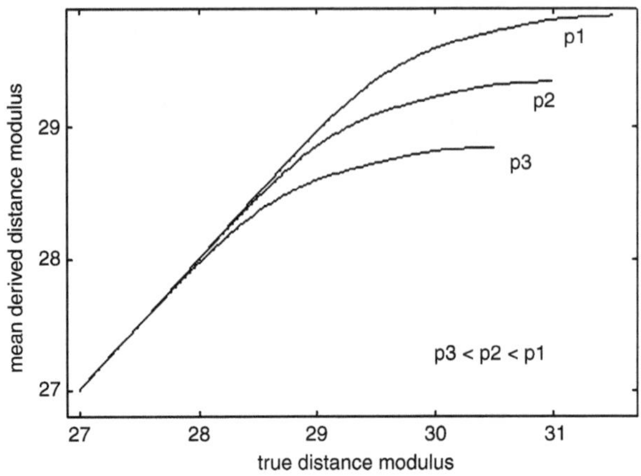

Fig. 3.2 The behaviour of the bias in the derived distance modulus versus the true modulus for three different values of p (p_1 corresponds to the most luminous class)

3.3.3 Type 2 Bias when $M = ap + b$

de Vaucouleurs (1983a) made a distinction between the "Malmquist effect" (the progressive truncation of the luminosity function at increasing distances) and the Malmquist bias in the distances derived from such a sample, arguing that the former may exist without the latter. Really, one might think that if there is a way of classifying galaxies into absolute magnitude bins using, say, a relation $M = ap + b$, the magnitude limit will certainly cut away fainter galaxies from the sample, but then the parameter p "glides" simultaneously, as if compensating for the distance dependent effect. However, the theory shows that such a compensation is not complete: average p shifts to larger values, but still, no matter what the value of p is, the corresponding distribution of true M is cut at a common $M_{\rm lim}$ which depends only on the distance. One cannot escape Type 2 bias, though at each true distance it is smaller by the factor $\sigma^2/(\sigma^2 + \sigma_M^2)$, as compared with the simple truncation effect of the luminosity function with the dispersion σ_M.

That a bias should exist also in the TF method was suspected by Sandage and Tammann (1984), when they derived the Hubble constant using a locally calibrated infrared TF relation, and the bias in the direct TF relation was introduced on a theoretical basis in Teerikorpi (1984).

Figure 3.2 shows schematically how the derived distance modulus behaves versus the true modulus for different values of the p-parameter when the apparent magnitude limit $m_{\rm lim}$ is the same. It is assumed that larger p corresponds to larger luminosity. For more luminous objects, the bias starts effectively at a larger true distance. More generally, it is clear that

- For given p and $m_{\rm lim}$, the bias increases with distance
- For given distance and $m_{\rm lim}$, the bias is smaller for larger p
- For given distance and p, the bias is smaller for fainter $m_{\rm lim}$

The Cluster Incompleteness Bias In this special case of Type 2 bias the galaxies are at the same distance, and the decreasing completeness of the sample toward fainter galaxies makes the distances to clusters of galaxies too short. In the past, there were assertions that there is no Malmquist bias in clusters (because there is no volume effect in clusters), which actually mixed up Type 1 and 2 biases. However, there were indications in the 1980s that the clusters, in comparison with unbiased field galaxies, give a too large Hubble constant. This led to the recognition of the cluster incompleteness bias (Teerikorpi 1987; Bottinelli et al. 1987; Kraan-Korteweg et al. 1988; Sandage and Tammann 2006b).

This bias produces several things for the TF relation (e.g., Sandage et al. 1995): (1) The derived slope of the TF relation is changed (less steep). (2) The apparent zero-point of the TF relation in the cluster is too bright. (3) The scatter becomes apparently low. The artificially decreased scatter is dangerous, because it may lead one to underestimate the selection bias. In this manner, the bias itself may produce an argument against its presence!

The Inverse Relation and the Zero Type 2 Bias Ideally, the parameter p is not restricted by any such observational limit as M_{lim}. Hence, at any distance, the distribution of observed p corresponding to a fixed M, and especially its average $\langle p \rangle_M$ is the same. Schechter (1980) thus realized that the inverse relation $p = a'M + b'$ has the nice property that it may be derived in an unbiased manner even from magnitude-limited samples.

In what manner could one use the inverse relation as a concrete distance indicator? Assume that there is a cluster of galaxies at a true distance modulus μ. Derive for each galaxy i which has p_i measured, its distance modulus using as a "predictor" of M the inverse relation: $\mu_i = m - (1/a')(p - b')$. Teerikorpi (1984) showed that the distance estimate $\langle \mu_i \rangle$ is unbiased, under the condition that there is no observational restriction to p. This result was confirmed by numerical simulations by Tully (1988).

Consider a special subsample, the galaxies having a constant absolute magnitude M. Deduce their magnitudes from p, using the inverse relation as $M(p) = (p - b')/a'$. Then the average bias for this subsample is

$$\langle \mathrm{bias} \rangle_M = \frac{\int_{-\infty}^{\infty} [M - M(p)] \psi_M(p) dp}{\int_{-\infty}^{\infty} \psi_M(p) dp} = M - \frac{1}{a'}(a'M + b') - \frac{b'}{a'} = 0. \quad (3.29)$$

As the bias for every M is zero, the *average* bias in the distance moduli μ deduced for the *whole* sample becomes zero. This result is independent of the (true) distance. However, there will be a p-dependent systematic error in the distance moduli deduced for individual objects, because generally the slope of the line $p = a'M + b'$ differs from the slope defined by $\langle M \rangle_{p,\mathrm{ml}}$. Hence it is important in this method that the parameter p can be measured for *all* the objects, *after* these have been selected from the sky, in order to obtain an unbiased average distance modulus.

Our ordinary idea about a distance indicator is close to the direct relation: measure p, take from the relation the expected $\langle M \rangle$, and calculate $\mu = m - \langle M \rangle$ for this one object. The use of the inverse relation may appear strange, because we tend to look at the predictor of M, $(1/a')(p - b')$ similarly as we look at the direct relation.

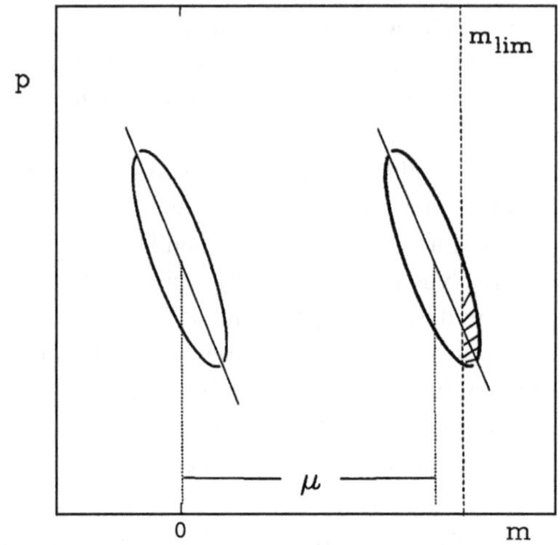

Fig. 3.3 An explanation of how the inverse Tull-Fisher relation (p vs. M) may under ideal conditions overcome the Type 2 Malmquist bias. The nearby calibrator sample is made to glide over the distant sample so that the regression lines overlap

Direct distance moduli are "individuals", while the inverse relation is collective: measure the average p for the sample and calculate from $\langle m \rangle$ and $\langle p \rangle$ the distance modulus.

The slope a and the zero-point b of the direct relation $M = ap + b$ should generally be derived from a local volume-limited sample with known distances. A practical example, for the TF relation, may be inspected in Theureau et al. (1997b). In contrast, in the ideal case the inverse relation does not require that its calibrators form a vl-sample. Further discussions on the inverse TF relation as a distance indicator may be found in Hendry and Simmons (1994, 1995), Triay et al. (1994), and Rauzy and Triay (1996).

The m–p diagram of Fig. 3.3, showing a "calibrator" cluster at 10 pc ($\mu = 0$) and a more distant cluster, reveals the secret of the inverse relation. The cluster to be measured is at the unknown distance modulus μ, and is cut by the magnitude limit m_1. Glide the calibrator cluster along the m-axis by the amount of μ. Then the inverse regression lines are superimposed. This means that the observed average of p at m is for the second cluster $\langle p \rangle_m$ which is the same as for the calibrator cluster at $M = m - \mu$. From this follows that $\langle \mu \rangle_m = m - (\langle p \rangle m - b')/a'$, and by averaging over all m that $\mu_{est} = \langle m \rangle - (\langle p \rangle - b')/a'$.

Problems The plus sides noted above for the inverse relation are balanced by a few problems that hamper its practical use as a distance indicator and make it difficult to solve simultaneously the problems of calibration and Type 2 bias.

Quite likely the p parameter is restricted by some upper and lower cut-offs (e.g., in the TF method narrow 21 cm line profiles are difficult to detect among the noise, while broad and low profiles tend to be missed, too). Expected features from such cut-offs were found in real data by Ekholm et al. (1999a). See also the Appendix of Sandage et al. (1995).

It was also pointed out (Teerikorpi 1990b; Teerikorpi et al. 1999) that since the measurement errors change the slope of the inverse TF relation for a distant sample, one should force that slope (the "relevant slope") through the calibrators, when calculating the distances. However, there is the nasty condition that the distribution of the parameter p for the calibrators should reflect the cosmic distribution (i.e. have the same average value, which is not the same as in the magnitude-limited distant sample). For example, for the diameter TF relation, the systematic error depends on the ratio of the calibrator and relevant (field) slopes a'_c/a', on the (Gaussian) dispersion of the distribution of the diameters σ_D and also on the space density law around the observer $\propto r^{-\alpha}$: $\langle \Delta \log H \rangle = (3-\alpha) \ln 10 \sigma_D^2 (a'_c/a' - 1)$. Since it depends on the width σ_D of the diameter function, the effect can be large even when the slopes a'_c and a' do not differ much (Ekholm et al. 1999a).

Subtle Biases in Direct and Inverse $ap+b$ Distances Landy and Szalay (1992) discussed the "inhomogeneous" Malmquist bias and its correction in the case of a general space density distribution. In Teerikorpi (1993) this correction was discussed with explicit reference to direct and inverse Tully–Fisher relations, which served to clarify certain points raised by Landy and Szalay. It was emphasized that Malmquist's formula with the term $d \log a(m)/dm$ was already a general one, and is best interpreted as applicable to a direct $M = ap + b$ relation, for a constant value of p (a "star class"). In that case, the distribution of m (or distance moduli) refers to *all* magnitudes which could be observed without any cut-off in the magnitudes.

Feast (1972) had already given a formula quite similar to (3.28), but now $N(\mu)$ was regarded as the distribution of the derived distance moduli in the *observed* sample. Feast's derivation contains the implicit assumption that the Type 2 bias is zero. One can conclude that Feast's and Landy & Szalay's variant of Malmquist's formula applies to the inverse distance moduli, having the Type 2 bias = 0 (Teerikorpi 1993; Feast 1994; Hendry and Simmons 1994).

Hence, Type 1 corrections for the direct distance moduli require data on the true spatial distribution of galaxies. Corrections for the inverse moduli depend on the distribution of apparent magnitudes (distance moduli) in the sample, hence on the selection function. Thus the Type 1 and 2 biases for the direct and inverse distances have curious complementary properties.

3.4 Some Other Finesses and Biases

To cite Sandage and Tammann (2006a): "Generally, the deeper a scientific problem is studied its solution becomes less simple. Although the first approximations made at the beginning can scout out a territory, as the data base expands, first approximations must often be replaced."

3.4.1 The Gould Effect: Original and Re-measured Samples

Gould (1993) pointed out a complication when a sample of galaxies, to be used for, say, infrared I-mag TF relation, is made from a sample originally based on selection criteria other than I-magnitude, e.g. apparent diameter. The Type 1 bias in the distance moduli from the I-mag relation does not now generally depend on the squared dispersion σ_I^2 of the I TF relation nor on σ_D^2 of the diameter relation, but on the covariance $\langle \epsilon_I \epsilon_D \rangle$ between the corresponding logarithmic distance errors ϵ. We refer to Gould (1993) for a mathematical treatment and give here some helpful heuristic explanations.

Explanations When one selects members of a standard candle class from the sky, the dispersion σ_1 in its absolute magnitude (or size) determines how the detected sample is distributed in space (for the fixed mean M_0 and a given space density law). When one uses these same objects in the detection band to derive their photometric distances $m - M_0$, the resulting average Malmquist bias arises from a convolution of the true spatial distribution and the distance error (also controlled by σ_1).

If one after the original selection re-measures the sample in another photometric band for which the standard candle dispersion is σ_2, then the spatial distribution of course remains the same, but now it is convolved with another distance error (not only σ_2 may differ form σ_1, but the errors may be only partially correlated). This is why the resulting mean systematic error in distance moduli now depends on the variance $\langle \epsilon_1 \epsilon_2 \rangle$.

An extreme case is when this covariance is zero, i.e. the deflections about the two TF relations are independent (or still more extremely, $\sigma_2 = 0$). Then there should be no Malmquist bias in the distance moduli from the I TF-relation. Let us explain. The original sample selected "from the sky" has a certain distance distribution. The second set of measurements produces symmetrical residuals around the standard candle mean, because of the independence on the original deviations from the mean. Hence, the average distance modulus inferred for the sample in question comes out unbiased.[1]

The Gould effect also means that if the original selection band is accompanied by a large dispersion in the distance indicator, the subsequent use of a smaller dispersion photometric band does not necessarily result in the perhaps naively expected significant decrease in the bias. This is again because it is the first band which determines the spatial distribution.

3.4.2 *Effects Caused by Dusty Medium*

Observations through a cloudy absorbing medium are affected by a selection effect: the "holes" in the medium offer a deeper sampling volume, hence such directions are preferred in a magnitude-limited sample. This results in an enhanced apparent clustering in the sky (Ambartzumian 1940); this effect will be also considered in

[1]In practice, it may be hard to find such pairs of observables which correlate with a common parameter (e.g., the TF parameter p), but have independent deflections (for instance, larger than average galaxies tend to be also more luminous than average).

Chap. 11. Also, the average extinction in a ml-sample is less than the all-sky average extinction (Holmberg 1974; Fesenko 1975).

Absorption Bias: Random Clouds In a simple example the dusty medium consists of identical clouds randomly distributed in space and each causing the extinction a mag. The mean absorption is $\langle \delta m \rangle = a\langle s \rangle$ where s is the number of clouds intersected by the line of sight. Let us take the distribution of the apparent magnitudes for the objects beyond the dust layer to be $\propto \mathrm{e}^{km}$. Then in a magnitude limited sample the number of objects observed through s dust clouds would be diminished by the factor of e^{sa}. Using the Poisson distribution one may calculate the mean number of clouds in front of the *observed* objects. For a uniform spatial distribution of the background objects ($\mathrm{e}^{km} = 10^{0.6m}$) the result is (Teerikorpi 1978)

$$\langle N \rangle = 10^{-0.6a}\langle s \rangle \approx \langle s \rangle (1 - 1.382a + \cdots). \tag{3.30}$$

Thus the mean extinction is $\langle N \rangle a$ and the second term in the Taylor series is essentially the one used by Holmberg (1974), $1.382\sigma^2$, where σ is the dispersion in extinction (the Poisson dispersion of s is $\sqrt{\langle s \rangle}$). Because we observe galaxies through the cloudy interstellar medium, systematic differences in its structure in different directions thus modify the mean extinctions in front of ml-samples and may lead to apparent anisotropies in photometric distances and the Hubble flow.

It should be noted that variable extinction does not change the relative distribution of luminosities (absolute magnitudes) in a ml-sample of field galaxies: a decreased extinction makes the sampling volumes deeper, but the relative change is similar for absolutely faint and bright galaxies. But if the galaxies are in a cluster, then the observable part of the luminosity function (the absolute magnitude limit) depends—in addition to the distance and the apparent magnitude limit—on the foreground extinction.

The Reddening Bias Another kind of effect, linked to the selection of stars, was originally pointed out by Holmberg (1974). It was shown to exist in real data by Teerikorpi (1990a) who studied its theory in more detail. This bias in stellar reddenings probably caused the almost zero reddenings measured for stars at high galactic latitudes, and was related to the debate on the average high latitude extinction.

To see the essence of the bias, suppose that a class of stars in a local vl-sample has the average intrinsic colour C_i. Then the average intrinsic colour of the same class of stars in a ml-sample will be bluer, because these stars are brighter, on the average, and because there is a colour vs. absolute magnitude relation among the stars. This bias, which makes the inferred reddenings progressively smaller, may be modelled as a function of distance. Thus Teerikorpi (1990a) concluded that hidden behind the bias there is significant non-zero reddening (caused by interstellar dust) at high galactic latitudes, which was later confirmed, e.g., by interstellar polarization measurements (Berdyugin and Teerikorpi 2002; Berdyugin et al. 2004) and from the IRAS 100 micron full-sky map sensitive to cool diffuse dust (Schlegel et al. 1998).

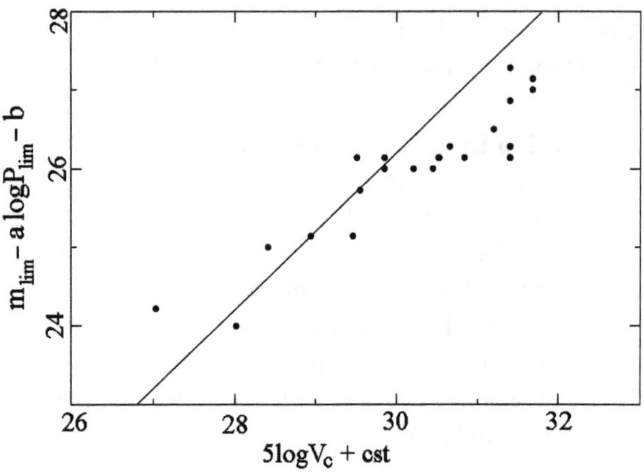

Fig. 3.4 One possible sign of the extragalactic Cepheid bias: the (normalized) apparent V magnitude limit of the Cepheid samples first grows with the distance (radial velocity V_c) and then tends to bend down there where the bias is suspected to appear in derived Cepheid distances (Paturel and Teerikorpi (2005); reproduced with permission © ESO)

3.4.3 *A Selection Bias in the Cepheid Method?*

At a given Cepheid period, the cosmic dispersion σ of the average absolute magnitude is so small (≈ 0.2 mag) that the systematic errors (taken to be proportional to σ^2) in the measured distance moduli have been usually considered as negligible. However, this may not be always the case.

Evidence for the Bias Paturel et al. (1997a) gave evidence for a slight bias in the Cepheid-based distances of a few galaxies and Lanoix et al. (1999) made calculations using the dispersion σ as a relevant factor for the incompleteness bias. They showed by excluding short period (low-luminosity) Cepheids, that in some cases the bias may be as high as 0.2–0.3 mag.

As a next step Teerikorpi and Paturel (2002) used the local Hubble law as an indicator of relative distances. It was then possible to test the behaviour of the Cepheid distances measured by the Hubble Space Telescope against relative kinematic distances. The tentative conclusion was that they were affected by an effect tending to produce too short distances, on the average (see also Fig. 3.4).

Contributing Factors Other factors than the small scatter σ in the *average* PL-relation are also relevant and can lead to a selection effect that is larger than suggested by the scatter in the PL-relation itself: the amplitude of variation, the longest observable period and dust extinction, together with the cut-off effect of the limiting absolute magnitudes of the Cepheid sample (Sandage 1988b) in V and I-bands, both needed for extinction corrections.

- *the amplitude of variation*: suppose that the Cepheid magnitude limit is $m_{\rm lim}$, the true distance modulus is μ, the average magnitude at the period P is $M(P)$ and the half-amplitude of variation is ΔM. Then one can measure only such Cepheids whose average absolute magnitude is, not brighter than $m_{\rm lim} - \mu$, but brighter than

$m_{\lim} - \mu - \Delta M$. In other words, at a given period P the effective magnitude limit for a Cepheid whose whole variation is observed, is brighter by ΔM than for a constant standard candle.

- *the longest observable period*: one may exclude shortest periods in order to decrease the selection bias at short periods where the Cepheids are absolutely fainter (Lanoix et al. 1999). But there is some maximum period, due to the finite observing time available.
- *the dust extinction*: there is some average extinction E obscuring the whole Cepheid population of the studied galaxy, arising in the Miky Way and in the galaxy itself. When the Cepheids are searched for, the extinction brightens the effective limiting magnitude by E. This does not refer just to the *observed* sample of Cepheids, affected by less extinction. It is the total extinction that influences the constitution of the observed sample.

Let us comment on the first point. One reasonably would like to see a "good" light curve and exlude those with the faintest parts missing. But this directly leads to the bias. For numerical simulations and other evidence for the relevance of the bias, see Paturel and Teerikorpi (2004, 2005, 2006).

One point deserving attention is that generally the number of well-observed Cepheids in a distant galaxy is of the order of a few tens at most, which must be a small fraction of the whole population. In order to assess better the bias, it would be important to know how the total number of Cepheids depends on the Hubble type and luminosity of the host galaxy.

3.4.4 Concluding Remarks

Large samples of classical photometric distance indicators have the advantage that one may investigate their completeness and composition, and recognize selection effects, and hence put the method on a safe basis.

Often in the forefront of scientific discoveries one rather likes to ignore selection effects, because in order to positively identify these, one should have enough of collected data, and this rarely happens in the avantgarde phase. Also, sometimes selection effects produce apparent phenomena which, if true, would certainly be more interesting than the mechanisms of selection and bias which naturally attract less attention.

Hence, it should be a part of the methodology to try to recognize all possible selection effects when one attempts to build an unbiased picture of the universe. When writing about the determination of the distance of the Sun, Gill (1877) warned about systematic errors in the methods and emphasized that "the observations themselves must afford the material for the determination and elimination of all errors to which it can be shown the method is liable". In the next chapter, the method of normalized distances, which has been applied to overcome Type 2 bias in the Tully-Fisher distances, also illustrates Gill's assertion.

One should emphasize that selection biases continue to take astronomers by surprise even in the local universe and they do not vanish anywhere, even though the astronomical data are accumulated beyond old magnitude limits. Problems are shifted towards larger distances (where the classical Malmquist bias turns into cosmological; Sect. 7.4.5) and fainter magnitudes, and new generations of astronomers have to learn how the biases reappear.

Chapter 4
Cosmological Redshift and the Distance Scale

Amidst the Great War raging in Europe, three remarkable articles appeared. Einstein (1917) presented the static, homogeneous, and finite cosmological model as a solution of his brand-new equations of general relativity. This required the introduction of the cosmological constant. de Sitter (1917) derived another solution: he made the matter density approach zero, and included the cosmological constant. This static world model predicted the "de Sitter effect" where light from remote objects is redshifted and the redshift has a non-linear dependence on the distance. In a novel manner, light became a tool to measure properties of the universe at large.

Slipher (1917) had managed to measure 25 nebular spectra for faint nebulae each needing exposures of tens of hours. The line shifts were mostly towards longer wavelengths, indicating speeds up to 1100 km s^{-1} when interpreted as Doppler effect. Slipher thus discovered cosmological redshifts.

4.1 The Distance–Redshift Law

Shapley and Shapley (1919) noted that fainter nebulae tend to have larger redshifts "indicating a relation of speed to distance or, possibly, to mass." After this hazy speculation, the next decade witnessed a race for the correct link between redshift and distance. Lundmark (1925) saw some relation, but his distance indicator, the angular size of a galaxy, was inaccurate (galaxies have a wide range of absolute sizes). Hubble (1929), using less dispersed indicators, could suggest a linear distance-redshift relation in the nearby galaxy universe. Within a few years Humason and Hubble showed that the relation exists still deeper in space.

Y. Baryshev, P. Teerikorpi, *Fundamental Questions of Practical Cosmology*, Astrophysics and Space Science Library 383,
DOI 10.1007/978-94-007-2379-5_4,

4.1.1 Empirical Properties of Cosmological Redshift

One directly measurable datum for a galaxy is the redshift z of its spectrum, defined as the relative shift of wavelengths λ or frequencies $\nu = c/\lambda$:

$$z = \frac{\lambda_{\rm obs} - \lambda_0}{\lambda_0} = \frac{\nu_0 - \nu_{\rm obs}}{\nu_{\rm obs}}, \tag{4.1}$$

where $\lambda_{\rm obs}$ is the observed and λ_0 is the rest wavelength of a spectral line. Hubble and Humason also used the quantity *apparent velocity*: $V_{\rm app} \equiv cz$.

Roberts (1972) showed that there is an excellent agreement between the redshifts of the optical lines and the million times longer 21 cm line of neutral hydrogen in galaxies. One has also compared different lines for the same distant cluster galaxies having radio, optical and X-ray lines. Now this basic property is established from radio to γ-rays, from 10^6 to 10^{20} Hz.

Photometric Redshifts can be derived for very faint galaxies without observable spectral lines, if the spectral energy distribution has been measured through several filters (see the web-site *hyperz*; Bolzonella et al. 2000).

Physical Interpretation Modern physics knows two experimentally verified ways to produce wavelength independent redshifts. The *Doppler effect* is caused by the mutual speed of a light source and the observer moving in space. The *gravitational redshift* appears when light is emitted by an atom sitting closer to a gravitating mass than the observer (the light from the Sun has $cz = 0.5$ km/s). Besides these mechanisms, there are theoretical ideas, in particular, *space expansion* making a photon's wavelength stretch. This widely accepted explanation of the cosmological redshift is a consequence of general relativity applied to the universe uniformly filled by matter.

The observed redshift $z_{\rm obs}$ may contain contributions from different physical causes so that (note: $1 + z = \lambda_{\rm obs}/\lambda_0$)

$$1 + z_{\rm obs} = (1 + z_{\rm cos})(1 + z_{\rm 3K})(1 + z_{\rm v})(1 + z_{\phi})(1 + z_{\rm other}), \tag{4.2}$$

where $z_{\rm cos}$ is the cosmological redshift, $z_{\rm 3K}$ is due to our velocity relative to the background radiation, $z_{\rm v}$ comes from the peculiar velocity of the observed galaxy, z_{ϕ} is the gravitational part caused by the local potential of the galaxy (e.g., in a cluster), and $z_{\rm other}$ is any unknown physical effect.

At large redshifts the cosmological redshift dominates over the Doppler shifts due to the motions of our Galaxy and the observed galaxy. The gravitational redshift is usually regarded as negligible in the light of galaxies.

Let us denote the line-of-sight component of our (the observer's) peculiar velocity as $v_{\rm 3K}$. Dropping other redshift sources, the result is

$$1 + z_{\rm obs} = (1 + z_{\rm cos})\left[\frac{c + v_{\rm 3K}}{c - v_{\rm 3K}}\right]^{1/2}. \tag{4.3}$$

For a small peculiar velocity, one may write $1 + z_{\rm obs} \approx (1 + z_{\rm cos})(1 + v_{\rm 3K}/c)$.

Corrections to the Redshift In relatively nearby space ($z < 0.01$), it is handy to express the redshift as a velocity $V = zc$. In galaxy catalogues one usually finds the velocity V as corrected to the Sun (heliocentric velocity) and the velocity V_{LG} as corrected to the barycentre of the Local Group (V_{LG}). The velocity V_{3K}, as corrected to the reference frame of the cosmic background radiation, is the velocity that an observer at rest "relative to the universe" (at our position) would measure for that galaxy.

The LEDA database makes a correction to the observed redshift, expressed as (heliocentric) velocity V, to obtain V_{3K}: $V_{3K} = V + 360\cos\theta$, where θ is the angle between the direction of the galaxy and our velocity relative to the background radiation and 360 (km s^{-1}) is its amplitude.[1] One might now think that $z_{cos} = V_{3K}/c$. As was noted by Harrison and Noonan (1979), if the galaxy is at rest, its cosmological redshift is actually

$$z_{cos} = \frac{V_{3K}/c}{1 - (V_{3K} - V)/c}. \tag{4.4}$$

Generally the error is tiny, but it is important to be aware of the difference between exact expressions and "fictitious" quantities in calculations involving observed and cosmological redshifts mixed with peculiar velocities.

For the very local Hubble flow, the centre of expansion is usually taken to be the barycentre of the Local Group, often placed on the line between M31 and our Galaxy at 2/3 of the distance to M31 (assuming the mass ratio of the galaxies is 2). It is simple to correct the distance to this barycentre, but what about the radial velocity? For instance, Yahil et al. (1977), the LEDA database choice, make the correction to the radial velocity V referred to the Sun as $\Delta v = +308\cos\lambda$ km/s. Here λ is the angle between the direction of the galaxy and the direction of the solar motion apex ($l = 105°$, $b = -7°$).

In fact, the observer at the centroid would measure a slightly different velocity than $V_{Yah} = V + \Delta v$, depending on the location of the galaxy and the nature of the nearby velocity field (Ekholm et al. 2001). If there is a Hubble flow relative to the centroid (or if the deviation from it is spherically symmetric), one may show from the first principles of the Hubble velocity field that the centroid would see the velocity $V_{Yah}/\cos\theta$, where θ is the angle between the centroid and the Galaxy as seen from the measured galaxy. This angle (and the correcting factor $1/\cos\theta$) is large only for nearby galaxies located about perpendicular to the M31–Galaxy line.

To derive this result we first note that the "corrected" observer at the Sun is at rest relative to the LG centroid, as supposed in the Yahil et al. correction. Relative to the Hubble flow, the observer at the Sun has a peculiar velocity Hr_c towards the centroid (in order to keep the observer fixed). Then the corrected velocity of the galaxy at the distance r_g from us is its Hubble velocity Hr_g minus the projection of our peculiar velocity along the line-of-sight to the galaxy and the projection of the peculiar velocity of the galaxy itself (v_{pec} assumed radial relative to the centroid): $V_{Yah} = Hr_g - Hr_c\cos\alpha - v_{pec}\cos\theta$. Here α

[1] The 1950.0 equatorial coordinates of the direction are $\alpha(3K) = 11.25$ h and $\delta(3K) = -5.6°$, so $\cos\theta = \sin\delta(3K)\sin\delta_{gal} + \cos\delta(3K)\cos\delta_{gal}\cos(\alpha(3K) - \alpha_{gal})$.

is the angle between the galaxy and the centroid as seen from our position. Deviding by $\cos\theta$ and noting that $(r_g - r_c \cos\alpha)/\cos\theta$ is the galaxy-to-centroid distance $r_{c,g}$, we see that $V_{Yah}/\cos\theta$ gives the true expression for the Hubble law plus the peculiar velocity as observed from the centroid: $V_{Yah}/\cos\theta = Hr_{c,g} - v_{pec}$.

The K Correction The K-effect is a technical phenomenon caused by the redshift. It appears when one measures apparent magnitudes through filters defining a fixed wavelength band. Photometric studies of distant objects generally use a fixed finite bandwidth defined by a response function S.

Two things happen when a spectrum is redshifted: (1) the spectrum is stretched: $\Delta\lambda \rightarrow (1+z)\Delta\lambda$, and (2) it is shifted towards longer wavelengths: $F(\lambda_{obs})$ originates from the rest wavelength $\lambda_{obs}/1+z$. Putting these two things together the observed magnitude is obtained from the formula:

$$m_{obs} = -2.5 \log \int_0^\infty S(\lambda) F\left(\frac{\lambda}{1+z}\right) d\lambda/(1+z) + \text{const.} \tag{4.5}$$

The K-correction is defined so that the corrected magnitude (the expression for m_{obs} with $z = 0$) is the observed magnitude minus the K-correction:

$$m_{cor} = m_{obs} - K(z). \tag{4.6}$$

To remember the sign in the definition imagine that the "K-effect" is added to the "true" magnitude m_{cor} to give the observed one: $m_{obs} = m_{cor} + K(z)$.

A quasar's continuum spectrum may often be approximated as a power-law $F(\lambda) \propto \lambda^{-\alpha}$. Then the K-correction is simply $K = 2.5(1-\alpha)\log(1+z)$. When the spectral index $\alpha = 1$, the correction is zero—the stretching of the spectrum and its shift exactly compensate each other.[2] Generally the K-correction for quasars (also affected by strong emission lines) is relatively small in comparison for galaxies whose corrections may reach several magnitudes depending on the Hubble type and the observing band (Fig. 4.1).

The K-corrections for galaxies of different types are needed for various purposes such as calculating the luminosities and analysing the magnitude-redshift relation. For evaluating the K-correction at an observed wavelength λ for an object at redshift z the spectral energy distribution of the radiation $F(\lambda)$ should be known at the rest-frame wavelength $\lambda/(1+z)$. Modern calculations of K-factors use a model for the SED of a galaxy (Poggianti 1997).

What is the Distance in the Hubble Law? Observations suggest a linear distance-redshift relation, something that Hubble saw for his handful of nebulae in 1929. In fact, what one observes for a standard candle is

$$m = 5 \log z + M + \text{constant}. \tag{4.7}$$

[2] If the spectrum is expressed using frequency units, $f(\nu) \propto \nu^{-\alpha_\nu}$ then the exponents are related to each other as $\alpha_\nu = 2 - \alpha$ and the K-correction is $K = 2.5(\alpha_\nu - 1)\log(1+z)$.

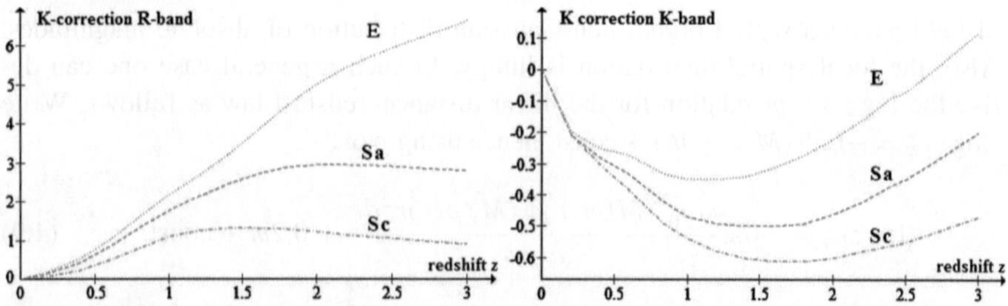

Fig. 4.1 K-corrections for galaxies of type E, Sa, and Sc for two wavelength bands R (effective wavelength 6400 Å) and K (22 000 Å) (the data from Poggianti 1997)

The constant $(5 \log c/H + 25)$ contains the Hubble constant H. In terms of the luminosity distance r_{lum} the empirical Hubble law becomes

$$cz = H r_{\mathrm{lum}}. \tag{4.8}$$

One may write analogous expressions for a standard rod and angular size distance (then $cz = H r_{\mathrm{ang}}$). Modern cosmology suggests that behind the observed Hubble law is a cosmological linear velocity–distance relation. Here velocity V_{exp} refers to space expansion causing the redshift by the Lemaître effect (Chap. 7), and distance is the momentary metric distance r_{m}. Do not confuse the law $V_{\mathrm{exp}} = H_0 r_{\mathrm{m}}$ with the common form of (4.8) as $V = Hr$ where $V = cz$ refers to an inaccurate Doppler interpretation of the redshift.

4.1.2 The Linearity of the Redshift Law

In the expanding Friedmann model, the linearity reflects the homogeneity of cosmic matter. Thus in our inhomogeneous world there can well be local deviations from linearity, and such indeed are seen. Quite another thing is that in the 1960s to 90s there were claims for non-linearity in a more universal sense. These controversies have calmed down, and now the arguments usefully illustrate selection effects on the derived distance-redshift law.

Hawkins's Result for Field Galaxies Hawkins (1962) argued that the Hubble ($\log z$ vs. m) diagram for field galaxies supported the quadratic law $z = kr^2$. This was suggested to be expected from the gravitational redshift in a static uniform universe.

The quadratic law predicts for a standard candle $\log z \propto 0.4m$, instead of the linear Hubble law $\log z \propto 0.2m$ (assuming a constant Type 1 bias which would appear, e.g., for a uniform galaxy distribution). With a good standard candle, a sufficient range of the magnitude m and no selection limit to the redshift, one should easily make observationally a difference between these two cases. However, the sample used by Hawkins was not made of good standard candles, but was a collection of field galaxies with a broad, non-Gaussian distribution of absolute magnitudes. Also, the local spatial distribution is lumpy. In such a general case one can derive the $\log z$ vs. m relation for the linear distance–redshift law as follows: Write $\langle \log r \rangle_m = -0.2(\langle M \rangle_m - m) + \text{const}$, hence using $r \propto z$

$$\langle \log z \rangle_m = -0.2 \frac{\omega \int_0^\infty M(m,r)\phi(M)\,\rho(r) r^2 dr}{a(m)} + 0.2m + \text{const}. \tag{4.9}$$

The distribution of apparent magnitudes $a(m)$ is $\omega \int_0^\infty \phi(M) \rho r^2 dr$. It is clear from our discussion of the Malmquist bias (Chap. 3) that the first factor is a constant if the luminosity function is Gaussian and if the radial number density is constant (or $\propto r^{-\alpha}$). Then the relation is linear with the classical slope 0.2. Contrary to Soneira (1979), this factor does depend on the luminosity function (as also pointed out by Nicoll and Segal 1982). That it depends, is at the heart of the treatment of the classical Malmquist bias.

Without such a discussion of the expected $\langle \log z \rangle_m$ vs. m relation, one cannot use a general wide-luminosity Hubble diagram to derive the correct slope of the redshift law (for an application see Koranyi and Strauss 1997).

Quadratic Law from Galaxy Groups de Vaucouleurs (1972) derived distances to galaxy groups from the angular sizes and magnitudes of the five brightest members. The resulting V/R vs. V diagram showed an increasing trend, roughly as expected from a quadratic law $V \propto R^2$. Such a law seemed to support the hierarchical cosmology which he studied at the time.

Luminosity classified field galaxies also showed an increasing V/R (Teerikorpi 1975a, 1975b), which was as expected from the Type 2 selection effect due to the magnitude limit m_1. If R is the distance inferred from a supposed standard candle (M_0), a lower envelope in the V/R vs. V diagram is defined by $V/R = V/r_0$ (where $M_0 = m_l - 25 - \log r_0$). Hence, the Hubble ratio increases with velocity (distance) and the slope depends on r_0 (i.e. on M_0).

For the groups identified from a galaxy catalog, its magnitude limit will produce a pressure towards brighter n:th brightest magnitudes depending on the distance r: $M_n < M(r) = m_{\text{lim}} - 25 - 5 \log r$. This pressure also means brighter members M_i, $i \leq n$, on average. Due to this selection, related to the Scott effect and Type 2 bias, the first, second etc. brightest members in the *selected* groups are not standard candles—their average luminosities change together with the limit $M(r)$. The observed behaviour of the Hubble ratios V/R_i derived from the 1st, 2nd etc. brightest member galaxies can be understood using a statistical model similar to that of Peebles (1968) who studied the luminosities of the brightest cluster galaxies, though in this case the analysis includes the magnitude limit (Teerikorpi 1975b, 1981d).

Segal's Law Segal (1972) proposed his chronometric redshift theory, producing in static Einstein space the redshift that increases as the distance squared, in agreement with Hawkins and de Vaucouleurs. Segal also wanted to show that the $z \propto r^2$ law explains the redshift and magnitude data better than the linear law. His use of galaxy samples with a wide luminosity range led to lengthy treatments requiring the construction of the luminosity function and consideration of the magnitude limit (e.g., Segal and Nicoll 1996). Due to the involved analyses, it is not so transparent why Segal continued to obtain results favouring the quadratic law (see Koranyi and Strauss 1997). We note that Segal (1976) derived for local galaxies the redshift—number relation $\log N(z) = \frac{3}{2} \log z + \text{const.}$, as expected from the quadratic law if the local spatial distribution is homogeneous (however, it is far from that).

The results claiming a quadratic redshift law had in common that they were based on distance indicators quite vulnerable to selection effects, e.g. having a broad luminosity function. The modern evidence for linearity comes from a variety of indicators with a narrow luminosity function.

Brightest Cluster Galaxies In his 1929 discovery paper Hubble allowed for possible non-linearity at larger distances. However, the work on brightest cluster galaxies soon showed that the relation continues as linear as was possible to measure by that method. Sandage (1995b) tells the story of the campaign set up by Hubble together with Humason to measure redshifts for clusters. In their 1931 paper the largest redshift was 19700 km s^{-1} and in 1936 the record was 42000 km s^{-1} for the Boötis cluster. With the new Palomar 200-inch telescope, Humason measured for the Hydra cluster $z = 0.2$ (60000 km s^{-1}). New observations, together with better defined apparent magnitudes again showed the linear Hubble law (Humason et al. 1956).

Extension of the Hubble diagram to $z = 0.46$ (by Minkowski in 1960 for the cluster containing the radio source 3C295) showed that the linearity continues to this limit as well. In fact, at such redshifts one already may expect some deviation from linearity; in the Friedmann model the linearity is basically between the underlying expansion velocity and metric distance, neither of which is proportional to the redshift and the luminosity distance, except at short distances. In other words, the test for linearity goes over to the search for effects predicted by Friedmann models (as will be discussed in Chap. 7).

The proper way to measure the size and magnitude of the brightest cluster galaxies, suitable for cosmological tests (clusters being "hard points" in the Hubble flow; Giovanelli et al. 1997), has been much discussed. Sandage (1972) noted that the size measurements of the brightest elliptical cluster galaxies from the Palomar Sky Survey plates approximated an isophotal diameter. The redshift–angular size diagram showed a small scatter around a slope close to -1, instead of -2 expected from the quadratic law. Actually, one would like to measure angular sizes corresponding to the same linear size. The Petrosian (1976) size measure has such a goal.

If all galaxies had the same axially symmetric shape of the intrinsic surface brightness profile $I(r)$, one realizes that there should be a way to compare (angular size) distances by inspecting the angular behaviour of the surface brightness (say, considering where $I(\theta)$ has dropped into half of its central value). A practical way of doing this is to use the growth curve $l(\theta) = 2\pi \int_0^\theta I(\theta)\theta^2 d\theta$ in the Petrosian

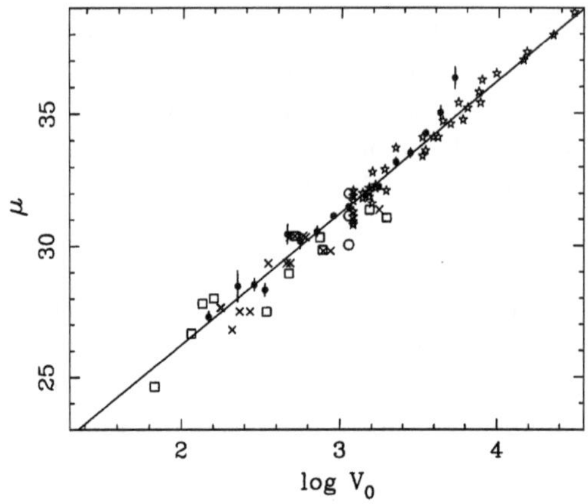

Fig. 4.2 Distance modulus μ vs. (log) radial velocity diagram: SNIa (*stars*), the B-mag TF distance from unbiased plateau (*dots*: averages of several galaxies), Cepheids (*crosses*: in individual galaxies, *squares*: groups with a Cepheid galaxy). The distance modulus μ is related to the (luminosity) distance d_{lum} (in Mpc) as $\mu = 5\log d_{\mathrm{lum}} + 25$ (Courtesy of M. Hanski)

function η as defined by Kron (1995):

$$\eta(\theta) = \frac{1}{2}\frac{d\ln l(\theta)}{d\ln\theta}. \tag{4.10}$$

The function η has the simple meaning that it is the surface brightness at θ divided by the average surface brightness up to θ. For an elliptical galaxy it is generally a monotonically decreasing function and for identical galaxies its fixed value corresponds to a specific distance from the centre, even if there is a cosmological dimming or even an evolution leaving the profile shape intact. This formalism is useful in attempts to detect the strong surface brightness dimming predicted by space expansion (Chap. 8).

Djorgovski and Spinrad (1981) constructed the angular size–redshift relation for the brightest cluster galaxies using such a Petrosian size measure. It agreed well with the linear Hubble law (angle $\propto 1/z$) in the range $0.01 < z < 0.2$ where the Friedmann models predict little deviation. Keeping at relatively small redshifts $z \le 0.2$, to avoid the problem with evolution, all the ways to define the size and magnitude have resulted in a linear redshift relation, with no hint at the quadratic one. This is so also for star-like sources like supernovae.

A Composite Diagram Figure 4.2 shows a Hubble diagram up to $z \approx 0.1$. It contains, from different sources, individual Cepheid-measured galaxies, groups containing a Cepheid-galaxy, unbiased TF data, and supernovae Ia data. The linear Hubble line with the expected slope of 5 is shown.

Only when the Hubble law could be studied at redshifts approaching unity, using the very luminous Ia class supernovae, one could detect a deviation "upwards", redshift increasing quicker than the luminosity distance. It was this effect that was interpreted as an acceleration of the Friedmann universe and led to the adoption of Einstein's cosmological constant or dark energy as a part of the standard world model (Chaps. 7 and 8).

4.2 The Value of the Hubble Constant

The linearity of the Hubble law means that the Hubble constant H_0 exists. Within the Friedmann model H_0 fixes the distance scale and its inverse is the Hubble time T_H characterizing the age of the expanding universe. In other cosmological frameworks the Hubble constant may have other physical significance. The unit km s^{-1}/Mpc well conveys the meaning in expanding space models: the increase of the recession velocity per each megaparsec. In astrophysical formulae, the Hubble constant often appears in the dimensionless form $h = h_{100} = H_0/100$ km s^{-1}/Mpc.

4.2.1 The Start at 625 km s^{-1}/Mpc

The first estimation of the Hubble constant (and the age of the expanding universe) was made by Georges Lemaître (1927, 1931), even before the Hubble law was discovered! In 1927 he predicted that in an expanding universe the redshift should at small distances grow directly in proportion to distance. From apparent magnitudes for a few tens of nearby galaxies, he estimated an average distance of 0.95 Mpc. Combining this number with the average radial velocity (from the redshifts) 600 km s^{-1}, he obtained 625 km s^{-1}/Mpc for the ratio between velocity and distance.[3] The corresponding Hubble time for Lemaître's universe is about 1.6 milliard years.

When was the Malmquist bias first applied in study of the Hubble constant? In fact, Hubble's (1936b) old value of $H_0 = 526$, canonical for years, already included an attempt to correct for the Malmquist bias (he applied the $1.382\sigma^2$ formula for the brightest stars of galaxies).

The Fall Down of H_0 Hubble (1936b) thought that the error in his 526 km s^{-1}/Mpc was at most 15 percent. The age argument gave this estimate some support. Namely, the corresponding Hubble time $1/H_0$ is about 1.8 milliard years, conveniently close to the age of the oldest Earth rocks as at the time determined from radioactive dating. De Vaucouleurs (1983b) noted with some hilarity that "this agreement was often presented to students of that era as a brilliant confirmation of both theory and the observations"!

It took only two decades for the measured value of H_0 to drop from 526 to 75. The main reasons were linked to the distance scale: (1) Baade's work on M31 showing Cepheid luminosities must be increased. (2) The brightest stars/HII regions confusion was cleared up and (3) Hubble's magnitude scale was corrected by Sandage (1958).

As to the Type 2 bias, it seems that Behr (1951) was the first to point out, after comparison of the width of the Local Group luminosity function to that of the field

[3] For more information on the chequered history of the attempts to measure H_0 see Sandage (1995b) and Tammann (2006).

galaxies, that the standard candle method may lead to systematically short distances at large true distances. He concluded that $H_0 = 260$, i.e. a half of Hubble's value. This already gave a rough idea of how important a selection bias may be, but the result went largely unnoticed.

4.2.2 Towards the Unbiased Value of H_0

The 1970s and 80s were marked by the efforts of Sandage and Tammann, on one hand, and by de Vaucouleurs and collaborators, on the other hand, to measure the Hubble constant. De Vaucouleurs's work led to the "short" distance scale, with $H_0 \approx 100$. Sandage and Tammann preferred the "long" distance scale ($H_0 \approx 55$), which led to a debate on the correct value of H_0.

Apart from some other differences in the distance ladders, the major single reason for the broad gap was the fact that Sandage and Tammann paid more attention to the selection bias affecting magnitude-limited samples. Especially, both teams had as the last step large samples of field galaxies. Their distances were determined using distance indicators calibrated at shorter distances. Such data are very vulnerable to the selection bias.

Where Type 2 Bias Appears If you want to see Type 2 bias in action, just take any large magnitude-limited galaxy sample, calculate for each galaxy $H = V/R$ using a photometric distance indicator (say, the TF relation), and plot H against V. You will see that H stays first roughly constant and then starts to increase (Fig. 4.3). So clear is this phenomenon and so much expected from simple reasoning that one cannot but repeat the words by Tammann et al. (1980): "If an author finds H_0 to increase with distance he proves in the first place only one thing, i.e., he has neglected the Malmquist effect!" One should add that here "distance" means the true distance or at least a true relative distance (e.g., redshift). If "distance" is the inferred distance, H does not necessarily change with the distance, though it may have a wrong average value—we have moved from Type 2 to Type 1 bias. This serves as a warning that a simple comparison of the photometric distances from two methods may hide a common bias.

In their "Steps toward the Hubble constant", Sandage and Tammann (1975) studied the H vs. $\log V$ diagram for luminosity classified spirals. The diagram showed an increase of H, caused by the magnitude limit. At small true distances there is an unbiased region where the Hubble constant could be derived. In fact, each luminosity class has its own limiting distance (Teerikorpi 1976; a modern discussion was given by Sandage 1999).

For a Gaussian luminosity function $G(M_0, \sigma)$ and a sharp magnitude limit m_l, it is easy to calculate the amount of the bias $\langle\mu\rangle - \mu_{\rm true}$ at each *true* distance (Teerikorpi 1975b; Sandage 1994), or how much $\langle\log H\rangle$ will increase with the kinematic distance; this happens differently for different values of $m_{\rm lim}$ and p exactly as expected from the selection bias (Fig. 4.3). This behaviour also makes it possible to use so-called normalized distances to study the value of the Hubble constant.

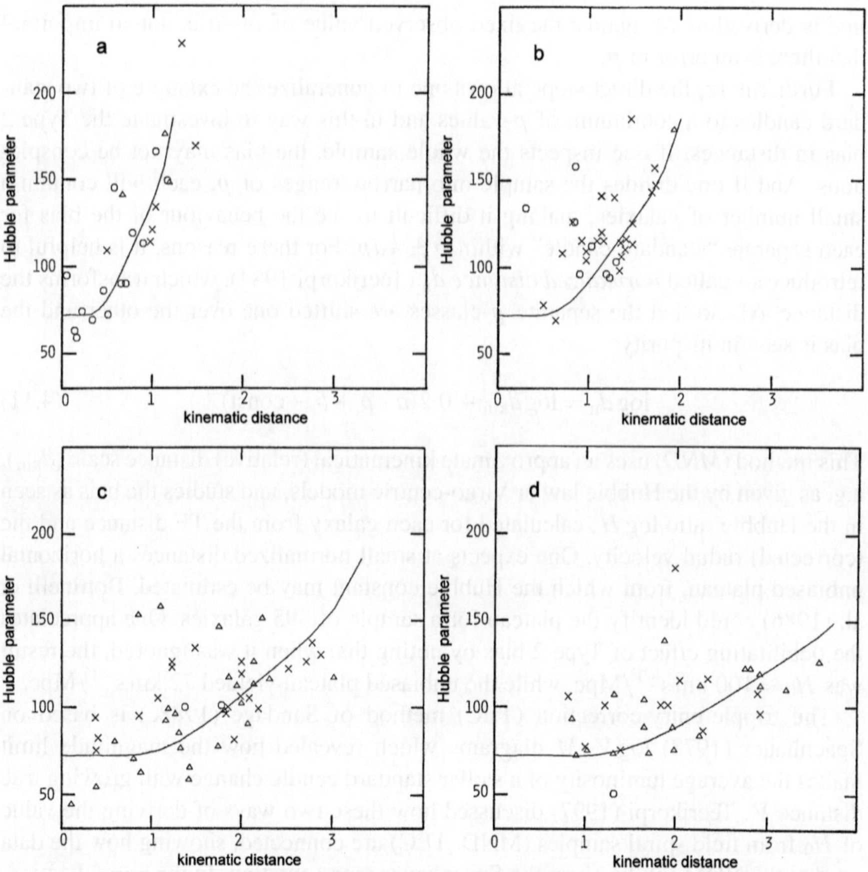

Fig. 4.3 The Hubble parameter from the Tully-Fisher B-magnitude relation as a function of kinematic distance (in units of the Virgo cluster distance) for a magnitude limited sample. The maximum rotational velocity increases from (**a**) to (**d**). Based on the data and analysis in Bottinelli et al. (1988)

Normalized Distances Take two standard candle classes, with luminosity functions $G(M_1, \sigma)$ and $G(M_2, \sigma)$ so that $\Delta M = M_1 - M_2$. If both are sampled from the sky up to a sharp limit m_l, it is easy to see that along the true distance modulus axis μ, the Type 2 bias suffered by these two candles is depicted by curves of the same form, but separated by constant $\Delta\mu = -\Delta M$. The curve of the brighter candle achieves only at larger distances the bias suffered by the fainter one. In this way, inspection of two or more standard candle classes gives a way to recognize the bias. Van den Bergh's luminosity classes clearly showed this effect (Teerikorpi 1975b).

When the TF relation $M = ap + b$ between the magnitude and maximum rotational velocity of spiral galaxies entered the scene, there was some uncertainty about which slope to use in distance determinations to individual galaxies—direct, inverse, or something between? Bottinelli et al. (1986) and also Lynden-Bell et al. (1988) argued that the direct slope is in a sense the natural one so that the regression line is derived as M against the fixed observed value of p—it is not so important that there is an error in p.

Furthermore, the direct slope allows one to generalize the example of two standard candles to a continuum of p-values and in this way to investigate the Type 2 bias in distances. If one inspects the whole sample, the bias may not be conspicuous. And if one divides the sample into narrow ranges of p, each will contain a small number of galaxies, making it difficult to see the behaviour of the bias for each separate "standard candle" within $p \pm \frac{1}{2}dp$. For these reasons, it is helpful to introduce so-called *normalized distance* d_n (Teerikorpi 1984), which transforms the distance axis so that the separate p-classes are shifted one over the other and the bias is seen in its purity:

$$\log d_n = \log d_{\mathrm{kin}} + 0.2(a \cdot p + b - \mathrm{const}). \tag{4.11}$$

This method (*MND*) uses an approximate kinematical (relative) distance scale (d_{kin}), e.g. as given by the Hubble law or Virgo-centric models, and studies the bias as seen in the Hubble ratio $\log H$, calculated for each galaxy from the TF distance and the (corrected) radial velocity. One expects at small normalized distances a horizontal unbiased plateau, from which the Hubble constant may be estimated. Bottinelli et al. (1986) could identify the plateau for a sample of 395 galaxies. One appreciates the debilitating effect of Type 2 bias by noting that when it was ignored, the result was $H_0 \approx 100\ \mathrm{km\,s^{-1}/Mpc}$, while the unbiased plateau yielded $72\ \mathrm{km\,s^{-1}/Mpc}$.

The tripple-entry-correction (TEC) method of Sandage (1988c) is based on Spaenhauer (1978) $\log V$–M diagrams which revealed how the magnitude limit makes the average luminosity of a stellar standard candle change with growing true distance V. Teerikorpi (1997) discussed how these two ways of deriving the value of H_0 from field spiral samples (MND, TEC) are connected, showing how the data used in the NDM can be given the Spaenhauer representation. In the normalization, the unbiased parts of the separate Spaenhauer patterns amalgamate together to form a common unbiased plateau.

In a nutshell, the TEC leaves the p-classes in separation, so their individual behaviour can be inspected (cf. Fig. 4.3), while the MND unites them into one ensemble, and the common bias behaviour is seen. Both methods are empirical in essence.

The Unbiased Plateau The normalized distance formula should also include terms describing how the effective magnitude limit changes due to inclination and galactic extinction (Bottinelli et al. 1995). One must also include the Hubble type dependence in the TF relation. If the magnitude limits differ from a fixed one m_0, then one must add a term $0.2\Delta m_{\mathrm{lim}} = 0.2(m_0 - m_{\mathrm{lim}})$ to the definition of d_n (note the sign: a brighter (lower) m_{lim} increases the normalized distance, i.e. makes the bias start at a shorter true distance). The sum of the corrections A_{incl}, A_{type}, and A_{gal} is used as one factor in the expression for the normalized distance:

$$\log d_n = \log d_{\mathrm{kin}} + 0.2(ap + b - \mathrm{const} + \Delta m_{\mathrm{lim}} + A_{\mathrm{incl}} + A_{\mathrm{type}} + A_{\mathrm{gal}}). \tag{4.12}$$

This form of d_{norm} was used by Theureau et al. (1997a) in their study of H_0.

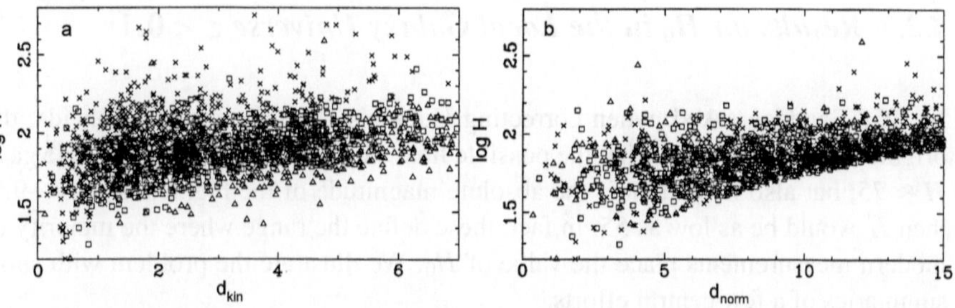

Fig. 4.4 The Hubble parameter from the Tully-Fisher diameter relation as a function of the kinematic distance (*left*) and the normalized distance (*right*) for the KLUN sample. After normalization, the typical pattern appears, with the "plateau" and the biased part with the sharp lower envelope that reflects the effect of the limiting angular size (from Theureau et al. 1997a; reproduced with permission © ESO)

The number of galaxies remains regrettably small in the unbiased region. In Bottinelli et al. (1986) the total number of galaxies was 395, while the size of the adopted plateau was = 41. In Theureau et al. (1997a), the size of the KLUN sample was 4164, and the adopted plateau contained 478 galaxies (see Fig. 4.4). Generally the empirical plateau contains about 10% of the total sample—in Theureau et al. (1997a) the cumulative error of $\langle \log H \rangle$ was seen to reach about 2% when the fraction of the sample is 0.1.

The Cluster Incompleteness Bias Because of the large number of galaxies in a small volume of space, one hopes to derive the distance of a cluster with a higher accuracy than the distance to an individual galaxy (so that $\sigma_\mu \to \sigma_\mu/\sqrt{n}$). However, when one applies the direct TF relation, there is always a systematic error on some level, due to the incompleteness bias: for small $p = \log V_{\max}$ the observed galaxies are brighter than the average (Teerikorpi 1987). For an individual cluster this is seen as a change in the residual $m - (ap + b)$ as a function of p. For a collection of clusters it is recommended to apply a normalization on $\log V_M$ (Bottinelli et al. 1987) and inspect the combined data from the clusters in one diagram.

In principle, the use of TF relations with small scatter may reduce the incompleteness bias. However, be aware of the Gould effect (Chap. 3) which can increase the bias from that expected from the small dispersion.

The inverse TF-relation seems to be a solution to the cluster incompleteness bias. In practice, there are problems. It is essential that there should be no selection according to the HI-linewidth. Another thing is that the calibrator slope for the inverse relation, derived from bright nearby galaxies, is not necessarily the correct slope for distant galaxies, if their magnitudes are less accurate than for the calibrators. If one ignores this problem, the inverse relation will give too small distances (Teerikorpi 1990b), as concretely shown for the Virgo cluster by Fouqué et al. (1990).

The correct slope for the inverse relation is especially important, because the aim is to reach large distances, i.e. to use extreme values of m and p. A small error in the slope causes large errors at large distances.

4.2.3 Results on H_0 in the Local Galaxy Universe $z < 0.1$

In 1958 Sandage asked, when correcting Hubble's 1936 distances (Cepheids, the brightest-star criterion), "what the possible final value of H is likely to be". He gave $H \approx 75$, but also noted that if the absolute magnitude of the bightest star is -9.5, then H would be as low as 55. In fact, these define the range where the majority of modern measurements place the value of H_0. We illustrate the problem with short summaries of a few central efforts.

The KLUN Sample and the TF Method Bottinelli et al. (1986) found $H_0 = 72\ \mathrm{km\,s^{-1}/Mpc}$ using the method of normalized distances, i.e. extracting a sample cleaned of galaxies suffering from the Type 2 bias. This value was based on the de Vaucouleurs calibrator distances. The calibrator distances adopted by Sandage and Tammann at the time led to $H_0 = 63$.

A developed version of the MND was applied by Theureau et al. (1997b) to the KLUN (Kinematics of the Local Universe) sample constructed on the basis of the Lyon-Meudon extragalactic data base and containing 5171 galaxies with diameters D_{25} larger than 1.6 arcmin. Its 400 unbiased galaxies (ten times more than Bottinelli et al. 1986 used) reached up to 2000–3000 km/s. Using the calibration from the HST observations of extragalactic Cepheids available at the time gave: $H_0 = 53 \pm 5\ \mathrm{km\,s^{-1}/Mpc}$ (the B-magnitude relation), $H_0 = 57 \pm 5\ \mathrm{km\,s^{-1}/Mpc}$ (the diameter relation).

Ekholm et al. (1999b) extended the measurement deeper in space with the inverse relation—a task harder than might have been expected from its "unbiased" nature. A simple application of the inverse relation, calibrated with 15 galaxies with Cepheid distances, yielded $H_0 \approx 80\ \mathrm{km\,s^{-1}/Mpc}$ for the diameter relation and $H_0 \approx 70$ for the magnitude relation. However, it was concluded that the calibrator selection bias affects the derived value of H_0 even when the *relevant* slope is used (as briefly discussed in Chap. 3). After the required correction for the calibrator bias, which also depends on the radial space distribution of galaxies, the sample reaching 10000 km/s (i.e. three times deeper than the sample in the MND approach) yielded:

$H_0 = 59 \pm 5\ \mathrm{km\,s^{-1}/Mpc}$ (the inverse diameter relation)
$H_0 = 58 \pm 6\ \mathrm{km\,s^{-1}/Mpc}$ (the inverse B-magnitude relation)

The HST Key Project In this program, Cepheids were measured in many local spiral galaxies in order to calibrate several long-distance indicators. The distances were based on the PL relation of the Large Magellanic Cloud Cepheids with the zero-point corresponding to $(m - M)^0_{\mathrm{LMC}} = 18.50$.

Freedman et al. (2001) (also Freedman and Madore 2010) summarized the results from the TF (I-band) and the Fundamental Plane (FP) methods, the surface brightness fluctuations (SBF) method and the SNII and SNe Ia distance indicators. These averaged to (systematic errors included):

$H_0 = 72 \pm 8\ \mathrm{km\,s^{-1}/Mpc}$.

The TF galaxy sample extends to about 9000 km/s and the type Ia supernovae extend to about 30000 km/s, where the peculiar velocities should not be a problem at all.

The HST Project for the Calibration of SNe Ia With Sandage as Principal Investigator this HST project measured Cepheids in galaxies where a SN Ia event had been well observed. Thus Cepheid distances for 10 host galaxies were determined and these were used for calibrating the maximum luminosity in the explosions of Ia supernovae (Sandage et al. 2006).

The Hubble diagram for the SNIa host galaxies constructed by the HST team contained 62 SNe Ia in the range $3000 < V_0 < 20000$ km/s, defining a Hubble line with the exact expected slope of 0.2 (for B, V, and I bands) and a small scatter of $\sigma = 0.15$ mag. Together with the locally calibrated luminosities the Hubble line delivered the value of the Hubble constant:
$H_0 = 62.3 \pm 5.3\ \mathrm{km\,s^{-1}/Mpc}$.

As to the possible (non)universality of the Cepheid PL relation, Sandage and his associates (e.g., Tammann et al. 2003) have argued that there are differences also in the slopes, so that low metallicity Cepheids have flatter slopes, and the derived distance would depend on what relation is used. After a correction of the host galaxy distances for the assumed metallicity dependence (as measured by [O/H]), the Sne Ia were calibrated to be somewhat brighter than in the HST Key Project work. "Failure to take the slope differences in the P-L relation into account as a function of metallicity using Cepheids as distance indicators results in incorrect Cepheid distances. Part of the 15% difference between our long distance scale [...] and that of the HST Key Project short scale is due to the effect of using an inappropriate P-L relation." (Sandage et al. 2009). For an analysis of this question by the HST Key Project, see Freedman and Madore (2010).

The last three results (the inverse TF relation, the HST Key Project, the SNIa calibration project) which reach distances corresponding to $V \approx 10000$–$20000\ \mathrm{km\,s^{-1}}$, are similar within the error ranges, though there may still be systematic errors, e.g., due to the suspected Cepheid distance bias (Chap. 3; Paturel and Teerikorpi 2005). In fact, the calibrators which Ekholm et al. (1999b) used in the inverse TF work, had their average adopted distance modulus about 0.2 larger than those used by Freedman et al. (2001) which would shift their value from 58 to about 64.

Figure 4.5 shows how the best value of the measured Hubble constant has decreased during the several decades of observational cosmology and has now finally levelled off around 60 to 70.

4.3 On Physical Methods in the Galaxy Universe

Along with the major distance indicators for large samples, there are special methods which may provide distance estimates for a number of individual objects. Especially interesting are physical methods which bypass the local calibration process

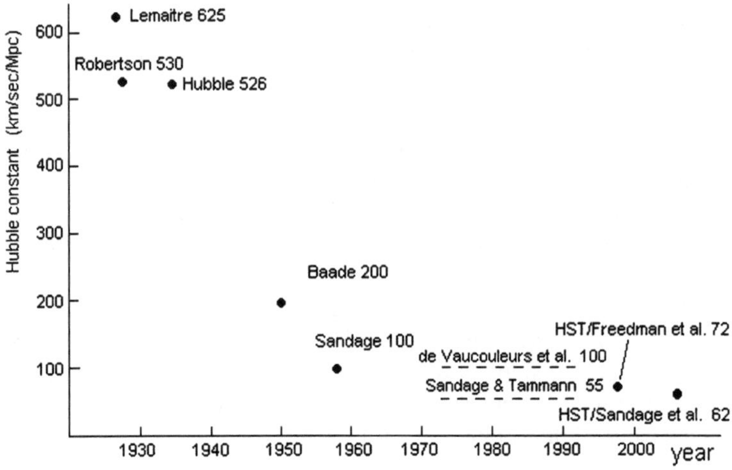

Fig. 4.5 The evolution of the measured value of the Hubble constant in the past decades mainly tells about the difficulty of measuring distances of galaxies. The last two points come from the Hubble Space Telescope observations

and may reach large distances. On the other hand, systematic errors due to selection effects or simplified physical models may be difficult to control because of the relatively small number of objects. There is also a general bias due to the fact that one "knows what to expect" and thus some important effect may go unnoticed.

4.3.1 The Sunyaev-Zeldovich Effect

The physics of this method involves the interaction of the cosmic microwave background radiation and the hot (10^8 deg) intergalactic gas in rich clusters of galaxies. Sunyaev and Zeldovich (1972) described the effect expected when the CBR is observed in the direction of such a hot gas. The inverse Compton scattering will give photons below the peak of the thermal distribution (i.e. at wavelengths longer than about 2 mm; the Rayleigh-Jeans part of the spectrum) extra energy, turning them into higher-frequency photons. Hence, at the cluster the sky will be a little cooler than around it (the relative decrement is about 10^{-4}). The measurement of this SZ effect can be used to infer the angular size distance to the cluster (Silk and White 1978).

The Basic Idea Assume that the cluster and the gas cloud is spherical, with radius R, and the gas is of a uniform density, with the electron number density n_{e} and temperature T_{e}. The drop in the temperature $\frac{\Delta T}{T}$ depends on the size of the cloud (determining the average line-of-sight column density), the electron number density, and also the temperature:

$$\frac{\Delta T}{T} = -aT_e n_e R. \tag{4.13}$$

Measuring the temperature deficit would give us the radius of the cluster R if we knew the electron temperature and density of the hot gas. Information about these quantities is given by the X-ray spectrum $I(\nu)$ of the bremsstrahlung emission from the gas. It can be written as

$$I(\nu) = bn_e^2 T_e^{-1/2} e^{-h\nu/kT_e} R. \tag{4.14}$$

The unknown temperature T_e is reflected in the slope of the spectrum ($\log I(\nu) \propto -T_e^{-1}+$ const., leaving the unknowns n_e and R to be solved from two equations, as the factors a and b are known. From the size R and the observed angle follows the angular size distance. This simplified description sets aside many technical and physical details.

Silk and White (1978) viewed it realistic that one might measure distances to nearby cluster to 40% accuracy. Only recently has such precision become in reach. There are tens of clusters measured in this way in the redshift range of 0.1 to 0.9, and depending on the sample and analysis the Hubble constant H_0 has been derived to be from about 60 to 75 (Carlstrom et al. 2002; Bonamente et al. 2006). Systematic errors due to density and temperature inhomogeneities in the gas, departures from isothermality, and asphericity may explain part of the scatter (Kawahara et al. 2008).

4.3.2 The Time Delay in Gravitational Lensing

In his book *Morphological astronomy* Zwicky (1957) correctly predicted several future applications of "the discovery of images of nebulae which are formed through the gravitational fields of foreground nebulae", such as testing the general theory of relativity and measuring dark masses of galaxies. Furthermore, gravitational lensing could be used as big natural telescopes to "throw very welcome light on a number of cosmological problems".

As one important application, Refsdal (1964) showed how the time difference Δt between the changes in the lensed images of a distant supernova could lead to the scaling (the Hubble constant) of the used Friedmann model. Here the redshifts of the lens and the background object (z_L, z_S) are needed. The first gravitational lens was discovered in 1979 (the quasar 0957 + 561). In the 1990s astronomers began to apply Refsdahl's idea to this and other lensed variable quasars.

If the angles between the lens and the images are θ_1 and θ_2, one can calculate the difference between the two light path lengths from the light source to the observer: $L_1 - L_2 = [(\theta_1^2 - \theta_2^2)/2][D_S D_L/(D_S - D_L)]$. Here D_S and D_L are the angular size distances to the background source and the lensing object, respectively. Writing at small redshifts $D = zc/H_0$, the desired time difference becomes

$$\Delta t = \Delta L/c = [z_S z_L/(z_S - z_L)] \times H_0^{-1} \times (\theta_1^2 - \theta_2^2)/2. \tag{4.15}$$

The derived value of H_0 is inversely proportional to the time difference and the velocity of light is the only physical constant in this formula. In practice the time difference (typically between 10 and 100 days) must be analysed using the Friedmann model and instead of a point-mass model one has to use a mass distribution for the lens.[4] The results have varied in the range from 50 to 80 (Jackson 2007).

After a slow start, both the SZ-effect method and the gravitational lensing method have began to deliver useful complementary information on the value of the Hubble constant at large distances in the not-so-well studied redshift range 0.2–0.8.

4.3.3 The Distance Scale and the Eddington Luminosity

In the era of the Great Debate, Öpik (1922) made a remarkable determination of the distance to the Andromeda galaxy, using its rotation velocity, angular size, and optical flux (as reported already in 1918 in Moscow). From the data available, Öpik first calculated the distance of M31 to be 785 kpc, a value that he proposed instead to be 450 kpc in the 1922 paper based on a different value of the mass-to-luminosity ratio of M31.

Öpik's method belongs to a wider class of dynamical distance evaluation methods. Consider the relation between the gravitating mass of a system and an observable velocity quantity for a test particle:

$$M = arV^2 = a\theta d V^2. \tag{4.16}$$

This formula contains a size r within the object (seen at angle θ at distance d) and the relevant velocity V (or dispersion σ). For an orbit around the mass M, the constant of proportionality a is simply G^{-1}. Then, omitting projection factors, the observations of θ and V give the quantity

$$M/d = a\theta V^2. \tag{4.17}$$

If one can express M/d in another form containing observable quantities and known constants plus the unknown d, then one can bypass the M/d degeneracy and derive the distance d (Teerikorpi 2011). For instance, if the mass M can be given as $\propto d^n$ where $n = 0, 2$ or 3, then one can solve for d, but if $M \propto d$, then any prior dependence on d disappears.

As a case of $n = 2$, if one expresses the mass M using the luminosity L and the mass-to-luminosity ratio γ, so that $M = \gamma L = \gamma f d^2$, then the distance can be solved as

$$d = \frac{a}{\gamma} \times \frac{\theta_0 V^2}{f}. \tag{4.18}$$

[4]Often the trend is towards more interest in the astrophysics of a phenomenon than in its possible use as a distance indicator. For instance, in the gravitational lense method one may study H_0, assuming a mass model for the lense, or perhaps better, one may study the mass model, assuming H_0 (regarded to be more firmly derived with several other methods) (Jackson 2007).

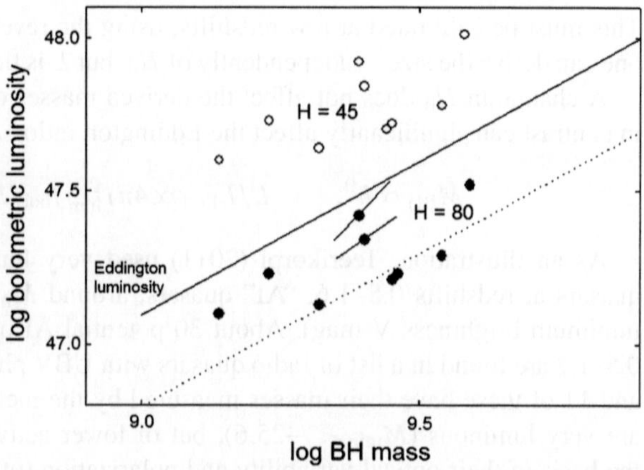

Fig. 4.6 Luminosity vs. compact nucleus mass (in logarithmic scales and solar units) for very luminous yet quiescent quasars ($0.6 < z < 1.5$) for two values of H_0. The upper line is $L_{\rm Edd}$, and the lower one is $0.5L_{\rm Edd}$. Varying H_0 causes only a vertical shift (Eq. (4.19)). Two quasars show the (small) effect of $\Delta\Omega_\Lambda = \pm 0.15$ ($\Omega = 1$). (Based on data in Table 1 in Teerikorpi 2011; reproduced with permission © ESO)

This is Öpik's method, which requires an estimate for the mass-to-luminosity ratio γ. Using modern data, where the horizontal part of the rotation curve starts at a distance of about 2.5 deg from the centre of M31, one has to adopt the mass-to-luminosity ratio $M/L \approx 17$ (in the solar unit) in order to derive the currently accepted distance of 0.77 Mpc. Such a high value of M/L simply means the presence of a lot of dark matter in the halo of the Andromeda galaxy. Why then could Öpik obtain such an accurate value for the distance without knowing about dark matter? Apart from Öpik having had good luck, he also had data that covered the innermost nebula (within 2.5 arcminutes and not 2.5 degrees from the centre!), where the dark matter is not important.

It would be difficult to apply Öpik's distance indicator to distant galaxies. Teerikorpi (2011) considered an analogous method, not for galaxies but for hypothetical Eddington radiators for which M/L has a known value from physics. Such objects might exist among active galactic nuclei, where the energy generation may be either linked to or limited by the Eddington luminosity, even though we are presently unable to use either the spectrum or some other property that can be measured without knowing the distance, to identify AGNs (such as quasars) shining close to $L_{\rm Edd}$.

In its simplest form, the Eddington luminosity is the maximum value of luminosity that can be powered by spherical accretion and its value depends only on the mass of the radiating object and on physical constants: $L_{\rm Edd} = 1.26 \times 10^{38}(M/M_\odot)$ erg s^{-1}. The corresponding M/L ratio is small, $\gamma_{\rm Edd} = 3 \times 10^{-5}$. These "Eddington radiators" would be among the most luminous stable sources within a class of a fixed mass.

Recent years have made it possible to infer masses of the compact nuclei in distant galaxies and quasars. In such determinations one may have to relate the size to measurements other than the angular size (while the needed velocity value is given by the line width $\Delta\lambda$ of some emission line (Hβ, CIV, or MgII)). One can infer the size of the dynamically relevant broad line region from the optical luminosity (Koratkar and Gaskell 1991), using in practice a luminosity–size (L–r) relation.

This must be calibrated at low redshifts; using the reverberation mapping method, one can derive the size r independently of H_0, but L is tied to H_0 (Kaspi et al. 2000).

A change in H_0 does not affect the derived masses of high-redshift objects, but in contrast can significantly affect the Eddington ratios $L/L_{\rm Edd}$:

$$M_{\rm BH} \propto h^0, \qquad L/L_{\rm Edd} \propto 4\pi r_{\rm lum}^2 f_{\rm bol}/M_{\rm BH} \propto h^{-2}. \tag{4.19}$$

As an illustration, Teerikorpi (2011) used very luminous yet quiescent radio quasars at redshifts 0.5–1.6, "AI" quasars, around $M_{\rm min} \approx -26.0 + 5\log h_{100}$ (a minimum brightness V mag). About 30 potential AI objects in the redshift range 0.5–1.7 are found in a list of radio quasars with UBV photometry (Teerikorpi 2000) and 11 of these have their masses measured by the method described above. They are very luminous ($M_{\rm min} < -25.6$), but of lower activity than fainter quasars on the basis of their optical variability and polarization (other properties in Teerikorpi 2001, 2003).

It turned out that if one assumes that this sample of possible Eddington radiator candidates indeed has their luminosity around $L_{\rm Edd}$, one may infer which value of H_0 leads this efficiency (Fig. 4.6). Using standard values for the needed parameters led to $H_0 \approx 65$ km s^{-1}/Mpc. However, the systematic errors arising from the various steps of calculation are not yet under control and we also do not know if these objects are really Eddington radiators; if they are fainter (which is typical for the active galactic nuclei in general), then the derived value of the Hubble constant is just a lower limit.

This example also shows that it is often interesting to consider how a cosmic phenomenon is related to the distance, even when its practical value as a distance indicator may not be high.

4.3.4 Precision Cosmology from the Background Radiation

The angular power spectrum of the thermal background radiation can be interpreted in terms of the physics of the photon, baryon and dark matter mixture around the recombination epoch when electrons and protons combined to form neutral hydrogen. Before the recombination photons and baryons were tightly coupled and oscillations were set up under the effect of gravity and radiation pressure. After the recombination, photons streamed freely and from their angular distribution now on the celestial sphere we can extract information about the conditions some 400 000 years after the big bang (according to the standard cosmology), corresponding to the redshift ≈ 1500 (as reviewed by Hu and Dodelson 2002).

Especially, the first peak in the angular power spectrum, observed at a scale of about 1 deg, is a projection of the sound horizon at the recombination epoch. The angle of this size is predicted by a flat spatial geometry ($\Omega = 1$) and the prediction does not depend much on Ω_Λ (or $\Omega_{\rm m}$). The smaller angular-scale peaks contain additional information about the Friedmann model. Especially, the heights of the peaks depend inversely on the quantity $\Omega_{\rm m} h^2$. This means that one cannot derive the value of the Hubble constant independently; it is coupled with the matter density parameter (degeneracy), so that different combinations of the matter densities and the Hubble constant can correspond to similar angular power spectra. For instance, from the WMAP observations Spergel et al. (2003) inferred that $H_0 = 72 \pm 5$ km s^{-1}/Mpc when $\Omega_\mathrm{m} = 0.27$ and $\Omega_\Lambda = 0.73$.

An accurate measurement of the Hubble constant in the large-scale galaxy universe would provide an important outside constraint for the multi-parametric models used to interpret the anisotropies of the background radiation. The methods which currently can be extended, say, to $z > 0.5$ still have considerable uncertainty in the derived value of H_0.

Chapter 5
Gravitational Physics for Cosmic Scales

World models have been always related to views on the reason why an apple falls. At the heart of a cosmological model there must be a theory of that universal force that acts between all masses and rules the dynamics of the whole universe. In current cosmology, it is general relativity. This is not a quantum theory and developments in theoretical physics suggest that also other possibilities exist to construct a gravity theory to be used in world models—a good reason to pay attention to trends in gravity physics.

Even our Solar System has become an arena of new gravity effects challenging explanations. In his lucid report Anderson (2009) asks "Is there something we don't know about gravity?". Anomalies in the observed local motions include an unexplained change in our Moon's orbit, the increase in the length of the Astronomical Unit, the deceleration of the spacecraft Pioneer 10 and 11, and the spacecraft speedups after Earth flybys. These oddities in our cosmic backyard made Anderson conclude "Maybe astronomy is already experiencing a transformation to … a new theory of gravity." This reminds one of the tiny anomaly in Mercury's orbit, whose explanation required the birth of a new gravity theory a century ago.

5.1 The Nature of the Gravitational Interaction

The new physics of the 20th century radically changed the basis of cosmological models. In the same decade when the galaxy universe and the Hubble law were discovered, the young Russian physicists George Gamow, Dimitrij Ivanenko and Lev Landau introduced the "*Ghc*"-classification of physical theories and predicted the appearance of a future relativistic quantum gravity theory (Gamow et al. 1928). Then Matvej Bronstein (1934), whose short life ended as a victim of Stalin's terror, noted that cosmological models should include both relativistic and quantum physics.

Old world models were based on classical Newtonian gravity ("*G*"-theory). The current cornerstone, general relativity, is a "*Gc*"-theory. Modern physics has the vision of quantum relativistic gravity. However, in spite of efforts, up to now there is still no such "*Ghc*"-theory, on which future cosmology is expected to be based. In

Y. Baryshev, P. Teerikorpi, *Fundamental Questions of Practical Cosmology*, Astrophysics and Space Science Library 383,
DOI 10.1007/978-94-007-2379-5_5, © Springer Science+Business Media B.V. 2012

this chapter we pay brief visits to Newton's theory, Einstein's general relativity, and Poincaré-Feynman's field gravity, a possible low-energy limit of the *Ghc* theory.

5.1.1 Newton's Gravity Theory

In 1987 scientists celebrated three triumphal centuries of Newton's gravity theory. Indeed, even now most works in celestial mechanics use the classical theory that allows one to calculate with a high accuracy the motions in our Solar System. Dynamical studies of stars and galaxies, as well as the large cosmological N-body simulations, are also based on Newtonian gravitation.

Initial Principles In modern terms, Newton made a non-relativistic non-quantum theory of gravity. It is based on Euclidean 3D space and 1D time. Their homogeneity and isotropy involve more than simplicity or beauty; those symmetries lie behind the conservation laws in classical physics.

An inertial system of reference is defined by means of the free motion of a reference body (plus the units of measurement). The centre of mass of a body moves with a constant velocity if there are no external forces or *their sum is zero*. The principle of least action is valid for a particle's motion.

A mass distribution with density $\rho(\vec{r})$ defines the gravitational potential $\varphi_N(\vec{r})$. A change at one point instantaneously alters the potential at all other points. This "action at a distance" is typical of classical theories with no "field" as a mediator of the interaction and with no upper velocity limit.

Main Equations In a nutshell, Newtonian gravity theory is contained in two equations. Poisson's equation for the gravitational potential is

$$\Delta\varphi_N = 4\pi G\rho. \tag{5.1}$$

The equation of motion of a particle under a Newtonian gravity force $\vec{F}$ is

$$\vec{F} = m_{\mathrm{I}}\left(\frac{d\vec{v}}{dt}\right) = -m_{\mathrm{G}}\vec{\nabla}\varphi_{\mathrm{N}}. \tag{5.2}$$

The ratio of the gravitational and inertial masses $m_{\mathrm{G}}/m_{\mathrm{I}}$ is not restricted by Newton's theory itself. The *weak equivalence principle* states $m_{\mathrm{G}} = m_{\mathrm{I}}$.

Solutions Poisson's equation for a spherically symmetric static body with mass M, radius R and constant density ρ_0 has the following solutions outside and inside the body (r is the distance from the centre of the ball):

$$\varphi_{\mathrm{N}}(r) = -\frac{GM}{r} \quad (r > R), \qquad \varphi_{\mathrm{N}}(r) = -2\pi G\rho_0\left(R^2 - \frac{r^2}{3}\right) \quad (r < R). \tag{5.3}$$

The Newtonian gravity force acting on a test particle with mass m at the point defined by the radius-vector $\vec{r}$ from the centre of the sphere with mass M is

$$\vec{F}_{\mathrm{N}}(\vec{r}) = -m\vec{\nabla}\varphi_{\mathrm{N}} = -\frac{GmM}{r^2}\frac{\vec{r}}{r}, \tag{5.4}$$

and its orbit around the sphere is an ordinary conic section.

In an important solution of Newton's equation of motion a test particle is passing by a much more massive central mass M at the impact distance b. If its speed is v, it will suffer a small deflection

$$\theta_{\mathrm{N}} = \frac{2GM}{v^2 b}. \tag{5.5}$$

Interestingly, light ($v = c$) passing by the Sun would be deflected an angle $\theta_{\mathrm{N}} = 2GM_{\odot}/R_{\odot}c^2 = 0.875''$. The observed value is two times larger.

5.1.2 Modern Physics of Fundamental Interactions

Modern physics deals with four presently known fundamental interactions: the electromagnetic, the weak, the strong and the gravitational.

Electromagnetic, Weak and Strong Forces The first three interactions are all described within a single Lagrangian formalism of the relativistic quantum field theory. This theory contains on a fundamental level:

- flat Minkowski space-time & inertial reference frames
- the energy-momentum (EM) tensor of the field
- the uncertainty principle & the principle of superposition

These forces are viewed as an exchange of force mediator particles, bosons of various types (photons, W and Z bosons, gluons) carrying energy-momentum between fermions. The Standard Model of elementary particles includes quantum electrodynamics, Weinberg-Salam-Glashow electroweak model, and quantum chromodynamic. The particles are quanta of fields, either fermion (half-integer spin) or bosons (integer spin) and the physical interactions are imagined as the exchange of force carriers (e.g., Bogolubov and Shirkov 1976).

Gravitational Interaction This fourth force is much weaker (Table 5.1) and it is described in a dramatically different way, as a curvature of space-time itself. General relativity, a relativistic non-quantum theory, contains:

- curved Riemannian space-time & non-inertial reference frames
- no energy-momentum tensor of the gravity field
- the equivalence principle

Table 5.1 The fundamental forces of Nature

Force	Relative strength	Range	Phenomena
Strong	2000	10^{-13} cm	atomic nuclei
Weak	10^{-8}	10^{-15} cm	neutron decay
Electromagnetic	1	infinite	atoms, molecules, solids
Gravitational	10^{-43}	infinite	stars, galaxies, universe

These lists make it clear why general relativity and quantum field theory represent very different physical interactions and why there are great difficulties in the quantization of the geometric gravity. The absence of the energy-momentum tensor for the gravity field is striking. Noether's theorem links conservation of energy-momentum to Minkowski space. Hence, no conservation laws exist for the EM of the matter plus gravity field in general relativity, precisely because there is no global Minkowski space. For the same reason, general relativity is not quantizable in an ordinary sense (Yilmaz 1997); the energy of the field quanta can not be defined properly.

5.1.3 Geometrical and Field Approaches to Gravitation

The geometrical way to gravity physics was developed by Einstein (1915, 1916) in his general relativity and he quickly applied his theory to cosmology.

Einstein's Geometrical Approach Here geometry becomes a dynamical physical entity that may be deformed, stretched and even spread (gravitational waves). So gravity is related to the curvature of space and has a singular position among interactions. In conditions of weak gravity, general relativity has passed all available tests in the Solar System and in binary pulsars. More accurate weak field tests are still needed as well as new tests of strong-gravity effects (Will 2005; Bertolami et al. 2006a).

How to Unify General Relativity with Quantum Physics Conceptual tensions between quantum mechanics and general relativity continue to complicate the most pressing question in today's theoretical physics: how to merge quantum theory and gravitation (e.g., Yilmaz 1997; Amelino-Camelia 2000; Chiao 2003). The standard scheme of quantization applied to general relativity gives a theory that is not "renormalizable" (i.e. leads to inevitable infinities in physical quantities). Also other attempts, based on the string/M theory, canonical/loop quantum gravity, and noncommutative geometry, encounter big problems and after decades of work we still have no quantum geometrical gravity theory.

Poincaré–Feynman Field Approach Meanwhile, a rather different approach exists. As early as 1905 Poincaré in "On the dynamics of the electron" advanced a relativistic theory for all physical interactions, in flat 4D space-time (now called Minkowski space). He pointed out that analogously to electrodynamics, gravitation should propagate with the speed of light, and there should exist mediators of the interaction—gravitational waves, or *l'onde gravifique* as he called them (Poincaré 1905, 1906).

Poincaré thus could be viewed as the father of what is now called the relativistic quantum field of gravitational interaction. This field theory approach to gravity (gravidynamics) was examined by a number of physicists (e.g., Birkhoff 1944; Moshinsky 1950; Thirring 1961; Kalman 1961).[1]

The strategy and principles of the field gravity theory were discussed by Feynman, who wrote that "geometrical interpretation is not really necessary or essential for physics" (Feynman et al. 1995). He pointed to the central role of the energy of the gravity field for a reasonable theory of gravitational force. For him gravitons were particles carrying the energy-momentum of the field: "the situation is exactly analogous to electrodynamics—and in the quantum interpretation, every radiated graviton carries away an amount of energy $\hbar\omega$". A consistent field gravity theory, where the inertial frames, Minkowski space, and energy of the gravity field have the central role, has been partly developed by Sokolov and Baryshev (Sect. 5.3).

Some physicists have concluded that general relativity and field gravity are essentially identical, while others argue it is impossible to derive general relativity from a field without assuming something similar to geometrization. Appendix B briefly reviews this important and debated item.

5.2 Einstein's General Relativity

General relativity is a relativistic non-quantum theory ("Gc" in Gamow-Ivanenko-Landau grouping), and a prototype of the geometrical approach to gravitation. It is a mathematically exact non-linear theory without inner limitations to its physical applications. Its predictions include extraordinary objects—singularities within black holes and of the whole universe.

5.2.1 Initial Principles and Basic Equations

Geometrization According to general relativity all gravity phenomena can be described by the metric of a Riemannian space. It has no "prior geometry", like

[1]A quantum description of the field was attempted by Bronstein (1936), Fierz and Pauli (1939), Ivanenko and Sokolov (1947), Feynman (1963, 1971), Weinberg (1965), Zakharov (1965), and Ogievetsky and Polubarinov (1965). Others tried to combine geometry and field; this has led to three different theories with different predictions (Sect. 5.2.3).

Minkowski space in other fundamental interaction theories. Gravity is not a material field in space, but the property of the curved space itself. The role of the gravitational potential is played by the metric tensor g^{ik}. It determines the 4-interval of the corresponding Riemannian space:[2]

$$ds^2 = g_{ik} dx^i dx^k. \tag{5.6}$$

A test particle moves on a *geodesic line* of the Riemannian space—this is one form of the *equivalence principle* important for the birth of general relativity.

The Principle of Least Action Einstein's field equations are obtained from the principle of least action by the variation of the metric tensor g_{ik} in the action S of the system matter + gravity field. The action has the form

$$S = S_{(\mathrm{m})} + S_{(\mathrm{g})} = \frac{1}{c} \int \left(\Lambda_{(\mathrm{m})} + \Lambda_{(\mathrm{g})}\right) \sqrt{-g} d\Omega, \tag{5.7}$$

where S_{m} and S_{g} are the actions for the matter and gravitational field, $\Lambda_{(\mathrm{m})}$ is the Lagrangian for the matter, and the Lagrangian for the field is

$$\Lambda_{(\mathrm{g})} = -\frac{c^4}{16\pi G} \Re, \tag{5.8}$$

where $\Re$ is the scalar curvature of the Riemannian space.

Einstein's Field Equations Variation δg_{ik} in (5.7) gives $\delta(S_{(\mathrm{m})} + S_{(\mathrm{g})}) = 0$ for the field equations:

$$\Re^{ik} - \frac{1}{2} g^{ik} \Re = \frac{8\pi G}{c^4} T^{ik}_{(m)}, \tag{5.9}$$

where $\Re^{ik}$ is the Ricci tensor. $T^{ik}_{(\mathrm{m})}$ is the energy-momentum tensor (EMT) of the matter. It includes all kinds of material substances, such as particles, fields, radiation, dark energy, and the vacuum $T^{ik}_{(\mathrm{vac})} = g^{ik}\Lambda$ (Λ is Einstein's cosmological constant). $T^{ik}_{(m)}$ does not contain the EMT of the gravity field itself, as gravitation is not a material field in general relativity.

The Equation of Motion of Test Particles A mathematical consequence of the field equations (5.9) is that as the covariant derivative of the left side equals zero (due to Bianchi identity), so for the right side we also have the continuity equation

$$T^{ik}_{(\mathrm{m});k} = 0. \tag{5.10}$$

---

[2] We use main definitions and notations similar to Landau and Lifshitz (1971), so 4D tensor indices are denoted by Latin letters $i, k, l, \ldots$ which take on the values 0, 1, 2, 3, and the metric has signature $(+, -, -, -)$.

This continuity equation also gives the equations of motion for the considered matter. It implies the geodesic equation of motion for a test particle:

$$\frac{du^i}{ds} = -\Gamma^i_{kl} u^k u^l. \tag{5.11}$$

$u^i = dx^i/ds$ is the 4-velocity of the particle and Γ^i_{kl} is the Christoffel symbol.

5.2.2 *The Weak Field Approximation*

All relativistic gravity effects that have been directly tested by observations, happen in the weak field, where the potential $|\varphi| \ll c^2$. This is why the weak field approximation plays an important role in gravity physics.

The Metric Tensor In the case of a weak gravity field the metric tensor is usually expressed in the form

$$g_{ik} = \eta_{ik} + h_{ik}, \qquad g^{ik} = \eta^{ik} - h^{ik}, \qquad g^i_k = \delta^i_k. \tag{5.12}$$

The quantities $|h_{ik}| \ll 1$, as well as the Minkowski space metric η_{ik} are not tensors of the Riemannian space. The different signs for covariant (in g_{ik}) and contravariant (in g^{ik}) components are caused by the exact identity (valid for the metric tensor of a Riemannian space)

$$g_{ik} \cdot g^{ik} = 4. \tag{5.13}$$

Equation (5.12) implies that the tensor g^{ik} is represented by the sum of two non-tensor quantities, as η^{ik} and h^{ik} are not tensors of the curved Riemannian space. The situation is fundamentally different in the consistent field approach where the sum of these two tensors is a tensor of Minkowski space.

Field Equations In linear approximation, Einstein's equations (5.9) are

$$\left(\Delta - \frac{1}{c^2}\frac{\partial^2}{\partial t^2}\right) h^{ik} = \frac{16\pi G}{c^4}\left[T^{ik}_{(m)} - \frac{1}{2}\eta^{ik} T_{(m)}\right]. \tag{5.14}$$

In the important case of a static spherically symmetric weak gravity field the solution of these equations gives the metric tensor in isotropic coordinates:

$$g_{ik} = \eta_{ik} + \frac{2\varphi_N}{c^2}\mathrm{diag}(1,1,1,1), \qquad g^{ik} = \eta^{ik} - \frac{2\varphi_N}{c^2}\mathrm{diag}(1,1,1,1), \qquad g^i_k = \delta^i_k, \tag{5.15}$$

where $\varphi_{\mathrm{N}} = -GM/r$ is the Newtonian potential. This shows transparently the metric tensor in the role of the gravitational potential.

The Equation of Motion in the Weak Field The post-Newtonian approximation of the weak field includes all terms of order v^2/c^2 or φ_N/c^2. PN-geodesic equations are frequently used in relativistic celestial mechanics. The 3-acceleration of a test particle in the static spherically symmetric weak gravity field is given by the equation

$$\left(\frac{d\vec{v}}{dt}\right)_{\rm GR} = -\left\{1 + (1+\alpha)\frac{v^2}{c^2} + (4-2\alpha)\frac{\varphi_{\rm N}}{c^2} - 3\alpha\left(\frac{\vec{r}}{r}\cdot\frac{\vec{v}}{c}\right)^2\right\}\vec{\nabla}\varphi_{\rm N} + (4-2\alpha)\frac{\vec{v}}{c}\left(\frac{\vec{v}}{c}\cdot\vec{\nabla}\varphi_{\rm N}\right), \tag{5.16}$$

where $\vec{v} = d\vec{r}/dt$, $\varphi_{\rm N} = -GM/r$, and $\vec{\nabla}\varphi_{\rm N} = GM\vec{r}/r^3$ (Brumberg 1991). The important parameter α is determined by the choice of the coordinate system: $\alpha = 2$ for the Painlevé coordinates, $\alpha = 1$ for the Schwarzschild coordinates, and $\alpha = 0$ for harmonic and isotropic coordinates. Hence the orbit of a particle will depend on this choice. However, one can suggest that an *observable* physical quantity should not depend on the coordinate parameter α and a procedure of measuring the quantity must be formulated, so that there will be no coordinate dependence in the final results.[3]

Modified Lagrangian It is well known that the observations of distant Ia supernovae are usually interpreted as indicating the acceleration of the universal expansion due to the repulsive effect of the cosmological constant or dark energy. It has also motivated the study of the consequencies of modified Lagrangians of general relativity (Carroll et al. 2005) so that one might understand the global acceleration without an unknown substance, dark energy. Viewed in this way, the high redshift Hubble diagram becomes a test of classical general relativity. This approach has been studied, including the effects of the modified Lagrangian on the Solar System, the stellar structure and the cosmological models, by several authors (see, e.g., Multamäki and Vilja 2008). Clifton et al. (2011) have written an extensive review of different modified gravity theories and their cosmological consequencies.

5.2.3 *The Problem of Energy-Momentum of a Gravity Field*

The problem of the energy of the gravity field in general relativity has a long history, it was, in fact, born together with Einstein's equations.

Hilbert (1917) noted that "in general relativity there are no equations of energy … corresponding to those in orthogonally invariant theories [i.e. theories in Minkowski space]". Then Emmy Noether (1918), his pupil, proved that the symmetry of Minkowski space is the cause of the conservation of the EM of a physical

---

[3]The problem of the meaning of coordinate systems in general relativity has been debated for a long time and there is no commonly accepted solution (see, for instance, Misner et al. 1973, p. 1097; Mitra 2002, 2006).

field. Many results of modern quantum field theories are based on this profound theorem. The Minkowski "prior geometry" conveniently guarantees the tensor nature of the energy-momentum and its conservation for the fields.

Einstein and Grossmann (1913) had come close to Noether's result: "remarkably the conservation laws allow one to give a physical definition of the straight line, though in our theory there is no object or process modelling the straight line, like a light beam in ordinary relativity theory". In other words, they said that the conservation laws imply Minkowski geometry. They also emphasized that the gravity field must have an EM tensor as other physical fields. However, in the final general relativity Einstein dropped this requirement in order to have a generally covariant theory with no prior Minkowski geometry.

Schrödinger (1918) showed that the object t^{ik} offered by Einstein for describing the energy-momentum of the gravity field in general relativity may be made to vanish for the Schwarzschild solution transformed to Cartesian coordinates. Bauer (1918) pointed out that Einstein's EM object, when calculated for a flat space-time but in a curvilinear coordinate system, leads to a nonzero result. So, t^{ik} can be zero when it should not be, and nonzero when it should! Einstein (1918) replied that Nordström had already informed him about this problem, and noted that in his theory t^{ik} is not a tensor, and "there may very well be gravitational fields without stress and energy density".

The "pseudo-tensor" character of the gravity field in general relativity has been discussed from time to time, causing surprises for each new generation of physicists. Rejecting Minkowski space leads to deep difficulties with the definition and conservation of the energy-momentum for the field.

Pseudo-tensor of the Gravity Field Einstein's equations (5.9) imply that the covariant derivative of the EM tensor of matter vanishes (the continuity equation):

$$T^{ik}_{(\mathrm{m});k} = 0. \tag{5.17}$$

One is tempted to see in this expression a usual conservation law, but let us cite the remarkable statement by Landau and Lifshitz (1971, Sect. 101): "however, this equation does not generally express any conservation law whatever. This is related to the fact that in a gravitational field the four-momentum of the matter alone must not be conserved, but rather the four-momentum of matter plus gravitational field; the latter is not included in the expression for $T^{ik}_{(\mathrm{m})}$". To define a conserved total four-momentum for a gravitational field plus the matter within it, Landau and Lifshitz suggested the expression

$$\frac{\partial}{\partial x^k}(-g)(T^{ik}_{(\mathrm{m})} + t^{ik}_{(\mathrm{g})}) = 0. \tag{5.18}$$

Here $t^{ik}_{(\mathrm{g})}$ is called the energy-momentum pseudo-tensor. It is important that the quantities $t^{ik}_{(\mathrm{g})}$ do not constitute a tensor, i.e. they depend on the choice of the system of coordinates. There are many variants of the expressions suggested for the pseudo-tensor, among them Einstein's (non-symmetric) and Landau and Lifshitz's (symmetric) pseudo-tensors.[4] Modern discussions of the energy-momentum pseudotensor problem in general relativity and its modifications may be found in Xulu (2003) and Multamäki et al. (2008).

Non-localizability Could one give a physical reason for the non-tensor nature of the energy of a gravity field? E.g., it could be the non-localizability of the gravity field in the geometrical approach (Misner et al. 1973, p. 467): "It is not localizable. The equivalence principle forbids." They also noted other properties of the pseudo-tensor: "There is no unique formula for it, ..., 'local gravitational energy-momentum' has no weight. It does not curve space. It does not serve as a source term ... It does not produce any relative geodesic deviation of two nearby world lines ... It is not observable." So the actual cause of the absence of the gravity energy, i.e. the pseudotensor nature of the EMT of the gravity field in general relativity, is geometrization. In contrast, other fields may be localizable, i.e. detectable with a local transformation of the field energy to the energy of a particle.

If there is no local field energy density, there is no energy in a finite volume, either. In Friedmann cosmologies this leads to the perplexing continuous creation/vanishment of energy within comoving volumes (Chap. 12).

Solving the Energy Problem Using "Effective" Spaces? Some efforts combine general relativity with a field by accepting some Lorentz-covariant properties of an "effective" Minkowski space in "effective" Riemannian space. Babak and Grishchuk (2002) claimed the field approach and general relativity are identical: "GR may be formulated as a strict non-linear field theory in flat space-time. This is a different formulation of the theory, not a different theory." They present the metric tensor $g^{ik}(x^l)$ of a curved space-time as a sum of two non-tensors $g^{ik} = (\eta^{ik} + h^{ik})\sqrt{\gamma/g}$ with the condition $g^{ik}g_{il} = g^k_l = \delta^k_l$ (the Minkowski metric η^{ik} is not a tensor of curved space). Their Lagrangian theory contains an energy-momentum tensor (close to LL-pseudotensor), black holes, quadrupole radiation and expanding space cosmology. However, Straumann (2000) emphasized that in general relativity there is a non-trivial topology, unlike Minkowski space.

Logunov and Mestvirishvili (1989) developed another field theory, called the relativistic theory of gravitation. Besides the metric tensor g^{ik} of the effective Riemann space, they introduced a "causality principle" as an additional restriction on g^{ik}. Because of this there is no black hole solution in this theory. The scalar part of gravitational tensor potentials exists only in a static field and can not be radiated. The cosmological solution is the Friedmann expanding space with critical matter density.

Yilmaz (1992) constructed a field theory where the right-hand side of the field equation contains the EMT of the gravity field. The metric of the effective Riemann

[4]It has been argued that this way of importing the energy into the gravity field is physically inconsistent (Logunov and Folomeshkin 1977; Logunov and Mestvirishvili 1989). Moreover Yilmaz (1992) has shown that due to the Freud identity for any pseudo-tensor $\partial_i(\sqrt{-g}t^i_k) = 0$, creating a difficulty with the definition of the gravitational acceleration.

space has an exponential form and excludes the event horizon and singularity. The existence of the EMT of the gravity field allows one to consider consistently N-body solutions, unlike in other theories.

Lacking but Required Properties of the Gravity Field EMT In the quantum relativistic field theories of other interactions the EMTs of boson fields have the following features:

- symmetry, $T^{ik} = T^{ki}$;
- positive energy density, $T^{00} > 0$;
- zero trace for massless fields, $T = 0$.

The cited attempts to introduce the energy-momentum tensor of the gravity field within geometrical and effective "field" approaches do not possess the last two features. These are necessary for a consistent field theory for both static and free fields, similarly as in the case of the electromagnetic field.

For the case of a spherically symmetric weak static gravity field, one can easily calculate the value of the energy density of the gravity field for different pseudo-tensors. For instance, in harmonic coordinates the Landau-Lifshiz symmetric pseudotensor gives $t^{00}_{\mathrm{LL}} = -\frac{7}{8\pi G}(\vec{\nabla}\varphi_N)^2$. Also the "final" energy-momentum tensor, found by Grischuk et al. (1984), has a negative energy density of the weak static field: $t^{00}_{\mathrm{GPP}} = -\frac{11}{8\pi G}(\vec{\nabla}\varphi_{\mathrm{N}})^2$. Hence, both the LL-pseudo-tensor and the GPP-tensor give a negative energy density for the static gravitational field. Also their traces do not vanish for static fields.

One may conclude that all theories which introduce a metric $g_{ik} \approx \eta_{ik} + h_{ik}$ of an "effective" Riemannian space lose some essential properties of the field approach (e.g., the scalar part of the gravity field) and acquire some features of the geometrical approach (such as the negative energy of the field, the event horizon etc.). All these derivations of "geometry" from "gravitons" explicitly or implicitly contain propositions reducing the field approach to geometry (Padmanabhan 2008). Hence, the principal question is still there—how to make a consistent quantum gravidynamics, which is based on relativistic quantum principles and has a geometrical interpretation only as an approximation of reality, like geometrical optics is in quantum electrodynamics.

5.3 Poincaré-Feynman's Field Approach to Gravity Theory

Poincaré (1905, 1906) suggested that all physical forces, including gravitation, could be considered within the same physical principles (especially the Lorentz invariance). In the Poincaré-Feynman field approach, the gravity force between Newton's apple and the Earth is caused by the exchange of gravitons. Gravitons (real and virtual) are mediators of the gravitational interaction and represent quanta of a relativistic tensor field ψ^{ik} in Minkowski space η^{ik}. This path offers a natural solution to the energy problem. Minkowski space implies the invariance under the Poincaré group transformation and hence the usually defined energy-momentum tensor of the gravity field, as follows from Noether's theorem.

The construction of field gravity is not yet complete and important questions, like the quantization of the gravity field, remain open. The strategy is not to write down the final exact non-linear equations, but to control the physical sense of all theoretical quantities used in the description of the gravitational interaction. The consistent Lagrangian field gravity theory was started in the works by Thirring (1961) and Kalman (1961), and continued by Sokolov, Baryshev and others.[5]

Up to now, only the weak field approximation at the post-Newtonian level has been studied in detail, but this is enough to show that the field approach is feasible and that it is not experimentally equivalent to general relativity. Below we give a compact summary following Baryshev (2003).

5.3.1 Initial Principles

In *Lectures on Gravitation* (Feynman et al. 1995) gravitation is described as a relativistic tensor field in Minkowski space. Feynman discussed a standard quantum field description of gravity "just as the next physical interaction". Minkowski space-time, inertial reference frames and the Lagrangian formalism appear in this description, as well as the energy-momentum tensor and positive energy density of the field, the zero trace of the EMT for massless fields, the energy quanta of the field, and the uncertainty and superposition principles. These form a natural starting point for understanding the physics of gravity phenomena similarly to other fundamental forces.

Consistent Iterations The gravity field has a positive energy which, in turn, becomes a source of an additional gravity field. This non-linearity is taken into account by an iteration procedure. It is usual in physics to start with a linear approximation and then add non-linearity by means of iterations. In this way, the field gravity theory is constructed step by step using an iteration process so that at each step all physical properties of the EMT of the gravity field are under control. Each step of iteration is described by linear gauge-invariant field equations with fixed sources in the right side. An important outcome of this procedure is that the superposition principle can be reconciled with the non-linearity of the gravity field.

The Principle of Least Action Within the Lagrangian formalism of the relativistic field theory, one derives the equations of motion for the gravity field and for the matter, using the principle of least action. It states that for the true motion the variation of the action $\delta S = 0$.

[5]These include: Sokolov and Baryshev (1980), Baryshev and Sokolov (1983, 1984), Sokolov (1992a, 1992b, 1992c, 1992d), Baryshev (1982, 1995, 1996, 2003), Baryshev and Kovalevski (1990), Baryshev and Raikov (1995), Baryshev and Paturel (2001), Paturel and Baryshev (2003a, 2003b), and other refs. in this book.

The action integral for the whole system of a gravitational field and particles (matter) within it, consists of three parts:

$$S = S_{(g)} + S_{(int)} + S_{(m)} = \frac{1}{c} \int \left(\Lambda_{(g)} + \Lambda_{(int)} + \Lambda_{(m)} \right) d\Omega. \tag{5.19}$$

The notations (g), (int), (m) refer to the actions for the free gravity field, the interaction, and the free particles (matter). The physical dimension of each part of the action is [S] = [energy density]×[volume]×[time], meaning that the definition of energy is assumed within the principle of least action.

In general relativity the action integral (5.7) has only two parts S_g and S_m. There is no interaction part, because of geometrization.

Lagrangian for a Free Gravitational Field Within the Poincaré-Feynman approach the gravity field is presented by symmetric 2nd rank tensor potentials ψ^{ik} in Minkowski space with metric η^{ik}. The Lagrangian for a free gravitational field we take in the form:

$$\Lambda_{(g)} = -\frac{1}{16\pi G} \left[\left(2\psi^{;n}_{nm} \psi^{lm}_{,l} - \psi_{lm,n} \psi^{lm,n} \right) - \left(2\psi^{;l}_{ln} \psi^{,n} - \psi_{,l} \psi^{,l} \right) \right]. \tag{5.20}$$

This differs from Thirring's (1961) choice by a divergent term, which does not change the field equations, but has the advantage that the canonical energy momentum tensor is symmetric. Here $\psi^{ik}_{,l} = \partial \psi^{ik} / \partial x^l$ is the ordinary partial derivative of the symmetric second rank tensor potential.

Lagrangian for Matter The Lagrangian for matter depends on the physical problem in question (particles, fields, fluid or gas). Gravity is also a kind of matter and at each iteration step it is considered as a source fixed by the preceding step. For relativistic point particles the Lagrangian is

$$\Lambda_{(p)} = -\eta_{ik} T^{ik}_{(p)}, \tag{5.21}$$

where $T^{ik}_{(p)}$ is the energy-momentun tensor of the particles

$$T^{ik}_{(p)} = \sum_a m_a c^2 \delta(\mathbf{r} - \mathbf{r}_a) \left\{ 1 - \frac{v_a^2}{c^2} \right\}^{1/2} u^i_a u^k_a. \tag{5.22}$$

Here $m,\ v,\ u^i$ are the mass, 3-velocity, and 4-velocity of a particle.

For a relativistic macroscopic body the EMT is

$$T^{ik}_{(m)} = (\varepsilon + p)\, u^i u^k - p \eta^{ik}. \tag{5.23}$$

The energy density ε and pressure p refer to a comoving volume element.

The Principle of Universality and Interaction Lagrangian In the field approach the *principle of universality* states that the gravitational field ψ^{ik} interacts with all kinds of matter via their energy-momentum tensor T^{ik}, so the Lagrangian for the interaction has the form:

$$\Lambda_{(\text{int})} = -\frac{1}{c^2}\psi_{ik}T^{ik}. \tag{5.24}$$

The principle of universality of gravitational interaction (UGI), (5.24), was introduced by Moshinsky (1950). It replaces the equivalence principle of general relativity[6] and tells that the free fall acceleration of a body does not depend on its rest mass—gravity "sees" only the energy momentum tensor of any matter. The UGI and least action principles imply those consequences of the equivalence principle which do not create paradoxes.

The equivalence principle cannot be a basis of field gravity, because it accepts the equivalence between the inertial motion and the accelerated motion under gravity. For example, this principle creates a puzzle when considering an electric charge resting in the Earth's gravity field on a laboratory table. Due to the equivalence of the laboratory frame (together with the table) and an accelerated frame with $a = g$, the charge on the table is equivalent to an accelerated charge and should thus radiate energy according to $P = (2/3)(e^2/c^3)a^2$ ergs/s. In the field theory this charge does not radiate, as it is at rest in an inertial system.

5.3.2 Basic Equations of the Field Gravity

Field Equations Using the variation principle to obtain the field equations from the action (5.19) one must assume that the sources T^{ik} of the field are fixed (or the motion of the matter given) and vary only the potentials ψ^{ik} (serving as the coordinates of the system). On the other hand, to find the equations of motion of the matter in the field, one should assume the field to be given and vary the trajectory of the particle (matter). So keeping the EMT of matter in (5.24) fixed and varying $\delta\psi^{ik}$ in (5.19) we get

$$-\psi^{ik,l}{}_{l} + \psi^{il,k}{}_{l} + \psi^{kl,i}{}_{l} - \psi^{,ik} - \eta^{ik}\psi^{lm}{}_{,lm} + \eta^{ik}\psi^{,l}{}_{l} = \frac{8\pi G}{c^2}T^{ik}. \tag{5.25}$$

The field equations (5.25) are identical to the linear approximation of Einstein's field equations and that is why there are many similarities between general relativity and field gravity in the weak field regime.

However, the difference is that ψ^{ik} and η^{ik} (and their sum, too) are true tensors in Minkowski space. But, as we mentioned earlier, in general relativity the quantities h^{ik} and η^{ik} are not tensors of a general Riemannian space.

---

[6]This point is also different from all "effective geometry" theories where the universality of gravity is understood as geodesic motion in Riemannian space.

Remarkable Features of the Field Equations First, the divergence of the left side of the field equations (5.25) is zero, implying the conservation law

$$T^{ik}{}_{,k} = 0, \tag{5.26}$$

in the approximation corresponding to the step in the iteration procedure. In the zero-th approximation it does not include the EMT of the gravity field, but the first approximation contains the gravity field of the zero-th approximation and expresses the conservation laws and the equations of motion at the post-Newtonian level.

Second, Eqs. (5.25) are gauge invariant, i.e. they do not change under the following transformations of the potentials:

$$\psi^{ik} \Rightarrow \psi^{ik} + \lambda^{i,k} + \lambda^{k,i}. \tag{5.27}$$

An important difference between this gauge transformation and the general covariant transformation of coordinates in general relativity is that (5.27) is performed in a fixed inertial reference frame.

Third, the gauge freedom (5.27) allows one to put four additional conditions on the potentials, in particular the Hilbert-Lorentz gauge:

$$\psi^{ik}{}_{,k} = \frac{1}{2}\psi^{,i}. \tag{5.28}$$

With the gauge (5.28) the field equations take the form of wave equations:

$$\left(\Delta - \frac{1}{c^2}\frac{\partial^2}{\partial t^2}\right)\psi^{ik} = \frac{8\pi G}{c^2}\left[T^{ik} - \frac{1}{2}\eta^{ik}T\right]. \tag{5.29}$$

The trace of this equation gives the field equation for the scalar part $\psi = \eta^{ik}\psi_{ik}$ of the gravitational potentials:

$$\left(\Delta - \frac{1}{c^2}\frac{\partial^2}{\partial t^2}\right)\psi(\vec{r}, t) = -\frac{8\pi G}{c^2}T(\vec{r}, t). \tag{5.30}$$

Note the opposite signs in the right-hand sides of Eqs. (5.29) and (5.30). We shall see that this corresponds to the important fact that the pure tensor part of the field represents attraction, while the scalar part gives repulsion.

The Tensor-Scalar Structure of the Field Gravity The multi-component structure of the tensor potential has very important consequences in the quantum field theory. It is well known that the symmetric 2nd rank tensor field ψ^{ik} can be decomposed under the Poincaré group into a direct sum of subspaces. This represents one particle with spin 2, one particle with spin 1, and two particles with spin 0 (Barnes 1965):

$$\{\psi^{ik}\} = \{2\} \oplus \{1\} \oplus \{0\} \oplus \{0'\}. \tag{5.31}$$

The tensor ψ^{ik} contains $n = 10$ independent components. The relation between the number of components n and the value of the spin s ($n = 2s + 1$) is fulfilled for the four particles as $10 = 5 + 3 + 1 + 1$ in Eq. (5.31).

As the field equations (5.25) are gauge invariant under the transformation $\psi^{ik} \to \psi^{ik} + \lambda^{i,k} + \lambda^{k,i}$ one may use 4 additional functions λ^i to delete the 4 components corresponding to spin 1 and the first spin 0, leaving only the spin 2 and the second spin 0 parts of the tensor potential. Hence, after the Hilbert-Lorentz gauge (5.28), the field equations (5.29) will describe the mixture of two fields with spin 2 and spin 0, generated by two corresponding parts of the source of the gravity field:

$$\{\psi^{ik}\} = \{2\} \oplus \{0\} \Longleftrightarrow \{T^{ik}\} = \{2\} \oplus \{0\}. \tag{5.32}$$

Now we can present the initial tensor potential and the EM tensor of the source as the sum of pure tensor spin 2 and pure scalar spin 0 parts:

$$\psi^{ik} = \psi^{ik}_{\{2\}} + \psi^{ik}_{\{0\}} = \left(\psi^{ik} - \frac{1}{4}\eta^{ik}\psi\right) + \frac{1}{4}\eta^{ik}\psi, \tag{5.33}$$

$$T^{ik} = T^{ik}_{\{2\}} + T^{ik}_{\{0\}} = \left(T^{ik} - \frac{1}{4}\eta^{ik}T\right) + \frac{1}{4}\eta^{ik}T. \tag{5.34}$$

For the traces of each part we have the expressions: $\eta_{ik}\psi^{ik}_{\{2\}} = 0$, $\eta_{ik}\psi^{ik}_{\{0\}} = \psi$, $\eta_{ik}T^{ik}_{\{2\}} = 0$, $\eta_{ik}T^{ik}_{\{0\}} = T$. Hence, Eq. (5.29) can be written in the form

$$\Box\psi^{ik}_{\{2\}} = \frac{8\pi G}{c^2}T^{ik}_{\{2\}} \quad \text{or} \quad \Box\phi^{ik} = \frac{8\pi G}{c^2}\left[T^{ik} - \frac{1}{4}\eta^{ik}T\right] \tag{5.35}$$

and

$$\Box\psi^{ik}_{\{0\}} = -\frac{8\pi G}{c^2}T^{ik}_{\{0\}} \quad \text{or} \quad \Box\psi\frac{1}{4}\eta^{ik} = -\frac{8\pi G}{c^2}T\frac{1}{4}\eta^{ik}, \tag{5.36}$$

where $\psi^{ik}_{\{2\}} = \phi^{ik}$ and $\eta_{ik}\phi^{ik} = 0$.

This means that field gravity theory is actually a tensor-scalar theory, where the scalar part of the field is simply the trace of the tensor potentials $\psi = \eta_{ik}\psi^{ik}$ generated by the trace of the energy-momentum tensor of the matter $T = \eta_{ik}T^{ik}$. According to the wave equations for spin 2 and spin 0 fields, both kinds of gravitons are massless particles.

Zakharov (1965) showed that the interacting gravitational field ψ^{ik} in Eq. (5.25) is described by spin 2 and spin 0 gravitons. From quantum field considerations (the condition of transversality of the gravitational vertex) he concluded that only spin 2 gravitons may be emitted, which corresponds to quadrupole gravitational waves. However, according to the wave equation (5.36), the trace $T(\vec{r}, t)$ of the EMT of matter will generate gravitational radiation, e.g. via spherical pulsations of a gravitating system. The radiated scalar wave is monopole and has a longitudinal character in the sense that a test particle in the wave moves along the direction of the wave propagation.

Equations of Motion for Test Particles Variation of the action integral (5.19) with respect to particle coordinates gives the equation of motion in a fixed gravitational field (Kalman 1961; Baryshev 1986):

$$A^i_k \frac{d(mcu^k)}{ds} = -mcB^i_{kl}u^k u^l, \tag{5.37}$$

where $mcu^k = p^k$ is the 4-momentum of the particle, and

$$A^i_k = \left(1 - \frac{1}{c^2}\psi_{ln}u^l u^n\right)\eta^i_k - \frac{2}{c^2}\psi_{kn}u^n u^i + \frac{2}{c^2}\psi^i_k, \tag{5.38}$$

$$B^i_{kl} = \frac{2}{c^2}\psi^i_{k,l} - \frac{1}{c^2}\psi_{kl}{}^{,i} - \frac{1}{c^2}\psi_{kl,n}u^n u^i. \tag{5.39}$$

The rest mass m of the particle cancels out. This shows how the principles of least action and universality imply the principle of equivalence in the form: the rest mass m of a body is equal to its inertial and gravitating masses ($m = m_{\mathrm{I}}$, $m = m_{\mathrm{G}}$). The rest mass includes all contributions from all interactions. So a test of the equivalence principle, using masses of different chemical materials, in fact checks the universality of the rest mass.

Repulsive Force of the Scalar Part Inserting to the equation of motion (5.37) the scalar part of the gravitational potentials $\psi^{lm}_{\{0\}} = (1/4)\psi\eta^{lm}$, we get the equations of motion of a particle in the scalar field $\psi = \psi_{lm}\eta^{lm}$ as

$$\left(1 + \frac{1}{4}\frac{\psi}{c^2}\right)\frac{dp^i}{ds} = \frac{m}{4c}\left(\psi^{,i} - \psi_{,l}u^l u^i\right). \tag{5.40}$$

In the case of a weak field ($\psi/c^2 \ll 1$) this equation gives for spatial components ($i = \alpha$) the expression for the gravity force

$$\frac{d\vec{p}}{dt} = -\frac{m}{4}\vec{\nabla}\psi. \tag{5.41}$$

As the trace of the weak static field (5.47) is equal to $\psi = -2\varphi_{\mathrm{N}}$, we get

$$\frac{d\vec{p}}{dt} = +\frac{1}{2}m\vec{\nabla}\varphi_N. \tag{5.42}$$

This means that the scalar spin 0 part of the tensor field leads to repulsion; only together with the tensor spin 2 part the result is Newton's force.

The Energy-Momentum Tensor of the Gravity Field The standard Lagrangian formalism and the Lagrangian of the gravity field (5.20) give the following expression for the canonical energy-momentum tensor:

$$T^{ik}_{(g)} = \frac{1}{8\pi G}\left\{\left(\psi^{lm,i}\psi_{lm}{}^{,k} - \frac{1}{2}\eta^{ik}\psi_{lm,n}\psi^{lm,n}\right) - \frac{1}{2}\left(\psi^{,i}\psi^{,k} - \frac{1}{2}\eta^{ik}\psi_{,l}\psi^{,l}\right)\right\} \tag{5.43}$$

Two important remarks should be made about this expression.

First, the EMT has an ordinary tensor character. However, the Lagrangian formalism cannot give a unique expression for an EMT of any field (e.g., Bogolubov and Shirkov 1976) because a term with zero divergence can always be added. For the final determination of the EMT of the field additional physical requirements must be used, like the positive energy density, the symmetry, and the zero-trace in the case of a massless field.

Second, the negative sign of the scalar part (the 2nd term in brackets) does not mean the spin 0 field has negative energy. It reflects the repulsive force produced by the scalar when the field is interacting with the sources.

For the free field the energy is positive for the pure tensor (spin 2) and scalar (spin 0) components. Indeed, the free-field total Lagrangian (5.20) can be divided into two independent parts that correspond to two particles with spin 2 (ϕ^{ik}) and spin 0 (ψ). Their free field Lagrangians are

$$\Lambda_{\{2\}} = \frac{1}{16\pi G}\phi_{lm,n}\phi^{lm,n}, \quad \text{and} \quad \Lambda_{\{0\}} = \frac{1}{64\pi G}\psi_{,n}\psi^{,n}. \tag{5.44}$$

Both signs are plus, due to the positive energy density condition for integer spin free particles. Corresponding EMTs for the tensor and scalar free fields

$$T^{ik}_{\{2\}} = \frac{1}{8\pi G}\phi_{lm}{}^{,i}\phi^{lm,k}, \quad \text{and} \quad T^{ik}_{\{0\}} = \frac{1}{32\pi G}\psi^{,i}\psi^{,k} \tag{5.45}$$

are symmetric, have a positive energy density and a zero trace for the case of plane monochromatic waves.

The Role of the Scalar Part of the Field The scalar ψ is an intrinsic part of the gravitational tensor potential ψ^{ik} and is not related to extra scalar fields such as introduced in the Jordan-Brans-Dicke theories. So the observational constraints existing for this extra scalar field do not restrict the scalar part ψ of the tensor field ψ^{ik}. Moreover, without the scalar ψ it is impossible to explain the classical relativistic gravity effects.

The most intriguing consequence of the field gravity theory is that the scalar part (spin 0) corresponds to a repulsive force, while the pure tensor part (spin 2) corresponds to attraction (Eq. (5.57) in the next section). This explains the "wrong" sign for the scalar part in the Lagrangian for the gravity field (5.20).

5.3.3 Post-Newtonian Approximations

The field equations (5.29) and the equations of motion (5.37) lead to various observable consequences of the field gravity. Some simple cases demonstrate how to calculate weak-field predictions within field gravity.

Static Spherically Symmetric Weak Field For a spherically symmetric static weak field of a body with mass density ρ_0 and total mass M, the zero approximation of the total EMT equals that of the matter

$$T^{ik}_{(m)} = \mathrm{diag}(\rho_0 c^2, 0, 0, 0) \tag{5.46}$$

and the solution of the field equations (5.29) is the Birkhoff's potential

$$\psi^{ik} = \varphi_{\mathrm{N}} \mathrm{diag}(1, 1, 1, 1), \tag{5.47}$$

where $\varphi_{\mathrm{N}} = -GM/r$ is the Newtonian potential outside the gravitating body. We note again that ψ^{ik} is a true tensor quantity in Minkowski space.

The gravitational field (5.47) can be expressed as the sum of the pure tensor and scalar parts $\psi^{ik} = \psi^{ik}_{\{2\}} + \psi^{ik}_{\{0\}}$ so that

$$\psi^{ik} = \frac{3}{2}\varphi_{\mathrm{N}} \mathrm{diag}\left(1, \frac{1}{3}, \frac{1}{3}, \frac{1}{3}\right) - \frac{1}{2}\varphi_{\mathrm{N}} \mathrm{diag}(1, -1, -1, -1). \tag{5.48}$$

This corresponds to attraction by spin 2 and repulsion by spin 0 potentials.

In the first (post-Newtonian) approximation the total EMT of the system is equal to the sum of the EMT for the matter, interaction and gravity field (Kalman 1961; Thirring 1961; Baryshev 1988):

$$T^{ik}_{(\Sigma)} = T^{ik}_{(\mathrm{p/m})} + T^{ik}_{(\mathrm{int})} + T^{ik}_{(\mathrm{g})}. \tag{5.49}$$

From the solution (5.47) and the expressions for the interaction EMT

$$T^{ik}_{(\mathrm{int})} = \rho_0 \varphi_{\mathrm{N}} \mathrm{diag}(1, 1/3, 1/3, 1/3) \tag{5.50}$$

and the EMT of the gravity field

$$T^{ik}_{(\mathrm{g})} = +\frac{1}{8\pi G} (\nabla\varphi_{\mathrm{N}})^2 \mathrm{diag}(1, 1/3, 1/3, 1/3) \tag{5.51}$$

we find the total energy density for the system gas + gravity in the form

$$T^{00}_{(\Sigma)} = T^{00}_{(\mathrm{p/m})} + T^{00}_{(\mathrm{int})} + T^{00}_{(\mathrm{g})} = \left(\rho_0 c^2 + e\right) + \rho_0 \varphi_{\mathrm{N}} + \frac{1}{8\pi G} (\nabla\varphi_{\mathrm{N}})^2. \tag{5.52}$$

Here $(\rho_0 c^2 + e)$ gives the rest mass and kinetic (or thermal) energy densities, $\rho_0 \varphi_{\mathrm{N}}$ is the negative interaction energy density, and $\nabla\varphi^2_{\mathrm{N}}/8\pi G$ is the positive and localizable energy density of the gravitational field.

Physical Sense of the Potential Energy The total energy of the system is in the PN approximation will be

$$E_{(\Sigma)} = \int T^{00}_{(\Sigma)} dV = E_0 + E_{\mathrm{k}} + E_{\mathrm{p}}, \tag{5.53}$$

where $E_0 = \int(\rho_0 c^2)dV$ is the rest-mass energy, $E_k = \int(e)dV$ is the kinetic energy, and E_p is the classical potential energy that equals the sum of the interaction and gravitational field energies:

$$E_p = E_{(int)} + E_{(g)} = \int\left(\rho_0\varphi_N + \frac{1}{8\pi G}(\nabla\varphi_N)^2\right)dV = \frac{1}{2}\int \rho_0\varphi_N dV. \qquad (5.54)$$

The PN Correction due to the Energy of Gravity Field A gravitating body is surrounded by a material gravity field ψ^{ik} whose mass-energy density is given by the 00-component of the EMT of this field in Eq. (5.51). In the PN level this leads to a nonlinear correction for the gravitational potential.

Considering the energy density of the gravitational field (the last term in Eq. (5.52)) as the source in the field equation of the second order, we get a non-linear addition to Birkhoff's ψ^{00} component

$$\psi^{00} = \varphi_N + \frac{1}{2}\frac{(\varphi_N)^2}{c^2}. \qquad (5.55)$$

Corrections to other components do not influence the motion of particles in this approximation.

The PN Equations of Motion Substituting Birkhoff's potential (5.47) into the equation of motion (5.37) and taking into account the nonlinear PN correction (5.55) one gets the 3-acceleration for a test particle:

$$\left(\frac{d\vec{v}}{dt}\right)_{FG} = -\left(1 + \frac{v^2}{c^2} + 4\frac{\varphi_N}{c^2}\right)\vec{\nabla}\varphi_N + 4\frac{\vec{v}}{c}\left(\frac{\vec{v}}{c}\cdot\vec{\nabla}\varphi_N\right). \qquad (5.56)$$

This equation coincides with the PN equation of motion in general relativity only when in Eq. (5.16) the isotropic or harmonic coordinates are used (i.e. $\alpha = 0$). It is important that within the field gravity theory the equations of motion do not depend on the choice of the coordinate system.

The Newtonian Limit Substituting (5.48), which gives the gravitational potential, into (5.37) and neglecting all terms of the order v^2/c^2 we get the Newtonian force as the sum of two parts: the attractive force due to the spin 2 part and the repulsive force due to the spin 0 part:

$$F_N = F_{\{2\}} + F_{\{0\}} = -\frac{3}{2}m_0\nabla\varphi_N + \frac{1}{2}m_0\nabla\varphi_N = -m_0\nabla\varphi_N. \qquad (5.57)$$

This calculation shows that even on the Newtonian level the physics of the field gravity theory dramatically differs from general relativity.

Chapter 6
Predictions of Gravity Theories

Celestial bodies and the entire universe itself offer many ways to test theories of gravitation. General relativity, the well-known basis of current cosmology, can explain a wide spectrum of phenomena from the deflection of light by the Sun to the Hubble law of redshifts within the Friedmann model. At the same time it is a non-quantum theory and still requires testing in strong gravity. As we saw, a quite different approach, the relativistic field theory, is also interesting as it aims to describe the gravitational interaction in the same way as other fundamental forces are treated in physics.

In weak-gravity conditions, general relativity and field gravity usually deliver similar predictions, but these diverge in some tests and when one goes towards strong gravity. Thus different cosmic situations may allow one to test two cardinally different concepts of gravitation, either strengthening the present foundations or opening routes to new ideas in cosmology.

6.1 Gravitation at Different Scales and in Diverse Conditions

The Newtonian gravity force is given as $F_{\mathrm{N}} = ma = GmM/r^2$. This familiar expression seems universal, but actually the various quantities have wide ranges, much of which territory has not been experimentally explored and may hide secrets of gravitation. Such quantities are the distance R between two masses, the time of interaction t, the masses m and M, their velocities V, the acceleration a, and the strength of the force F (Unzicker 2007).

Modified Newtonian Dynamics Milgrom (1983) suggested that for accelerations below a critical value of about 10^{-8} cm/s^2 there is a deflection in the classical Newtonian law in this region where ordinary non-relativistic physics would otherwise work. This was motivated by the possibility to explain the horizontal rotation curves of spiral galaxies without dark matter.

Y. Baryshev, P. Teerikorpi, *Fundamental Questions of Practical Cosmology*, Astrophysics and Space Science Library 383,
DOI 10.1007/978-94-007-2379-5_6,

Milgrom writes the acceleration (a) vs. force law in a modified form:

$$f(a/a_0)a = \frac{GM}{r^2}. \tag{6.1}$$

In this expression the function f approaches zero for small values of the acceleration $a < a_0$ and approaches unity for higher accelerations. The quantity a_0 is considered as a constant of nature and its value could be of the order of 10^{-10} m/s^2. The accelerations at the rims of spiral galaxies are less than such values. Then at large enough distances from the centre of a galaxy the acceleration becomes so small that one may approximate $f(x) \approx x$. Putting this and $a = V(r)^2/r$ in Milgrom's law (6.1) one obtains

$$V(r)^4/r^2 \approx a_0 GM/r^2. \tag{6.2}$$

We see that (1) at large distances from the mass M the rotation $V(r)$ keeps constant: $V^4 = Ga_0M$, and (2) for a constant mass-to-luminosity ratio M/L, the result is $L \propto V^4$, not unlike the Tully-Fisher relation (Sect. 2.3.1).

Bekenstein (2007) has developed a relativistic version of MOND. A critical test will be an actual measurement of the gravitational force for accelerations $< 10^{-8}$ cm/s^2, which can be done in near future satellite experiments.

Gravitation on Different Distance and Mass Scales After MOND, studies of the validity of the inverse square law on different scales of distance, mass, and acceleration have appeared. This interest has also been motivated by the string theory, where certain ideas predict deviations from the inverse square law at small, but still rather close to "everyday" distances (mm-scales or less). On short scales, gravity could fall off as $1/r^{2+n}$ where n is the number of large extra dimensions extending over macroscopic scales, within those variants of the string theory. No deflection from the Newtonian gravity has been found at short distances down to μm scales (Adelberger et al. 2003; Nesvizhevsky and Protasov 2004; Kapner et al. 2007).

Baryshev and Raikov (1995) suggested from quantum considerations that for small masses the Newtonian gravity law may fail. One may compare the gravitational interaction energy $E_{\text{int}} = GmM/r$ between two particles (masses m and M) at a mutual distance r, with the uncertainty principle $\Delta E \Delta t > \hbar$. Here the accuracy in measuring the energy is $\Delta E \approx E_{\text{int}}$ during the interaction time $\Delta t \approx r/v$, which implies a condition on the product of masses:

$$mM > \frac{v}{c}\frac{\hbar c}{G} = \frac{v}{c} m_{\text{Pl}}^2. \tag{6.3}$$

Hence the geodesic motion will be violated if the product of masses is less than $v/c\times$ the square of the Planck mass m_{Pl}, and it is expected that for *small masses* Newton's gravity law is not valid. If one of the particles is a photon, then light will not be deflected as predicted if the wavelength of the photon is longer than the gravitational radius of the deflecting mass $R_{\text{g}} = 2GM/c^2$. Such a photon will not move along a geodesic line. Radioastronomical observations could test in this way the nature of gravitation (Baryshev et al. 1996a, 1996b).

From Weak to Strong Fields In the Solar System the Newtonian gravitational potential is small compared to the speed of light squared $\phi/c^2 = GM/Rc^2 \ll 1$. This is called the weak field situation. In binary pulsars and neutron stars and in active galactic nuclei one meets gravitational potentials $\phi \approx c^2$. Such environments permit us to test gravity theories in strong field conditions where classical relativistic effects are more spectacular and other predicted phenomena become observationally achievable.

6.2 Einstein's General Relativity: Predictions

When Einstein (1915, 1916) presented the final mathematical formulation of general relativity, his theory made three predictions that became classical: the gravitational shift of spectral lines, the gravitational bending of light, and the advance of Mercury's perihelion.

As to the bending of light by the Sun, three values competed for the deflection angle at the solar radius: Newtonian gravity predicted $0.875''$, the scalar gravity theory of the Finnish physicist Gunnar Nordström anticipated $0.0''$, and Einstein's general relativity predicted $1.75''$. The bending was first detected in 1919 from eclipse observations made by two British expeditions, and the deflection favored Einstein's theory. This scientific result was reported by newspapers as a major story.

Since then many tests in the Solar System and observations of binary pulsars have confirmed the weak field predictions of general relativity. Effects of strong gravity are an important future subject. Even now black hole candidates and the loss of energy by gravitational radiation are studied.

6.2.1 Major Predictions for Observations

Einstein's field equations and the equations of motion give many predictions for both strong and weak fields. The weak field effects have been tested with an accuracy of about 0.1% (Will 1993, 2005).

The Deflection of Light A photon moving at impact distance b from a point mass M is deflected, in the weak field region, by the angle:

$$\theta_{\mathrm{GR}} = \frac{4GM}{c^2 b}. \tag{6.4}$$

In comparison, a particle passing the same mass with velocity v will experience a deflection given by

$$\theta_{\mathrm{GR}} = \left(1 + \frac{v^2}{c^2}\right)\theta_{\mathrm{N}}, \tag{6.5}$$

where $\theta_N = 2GM/v^2 b$ is the Newtonian value for the deflection angle. Note that for the velocity $v = c$ we get Eq. (6.4) with $\theta_{GR} = 1.75''$ for the Sun. The reader may verify that in order to derive this formula from the equation of motion (5.16) one should use isotropic or harmonic coordinates (i.e. $\alpha = 0$).

Gravitational Frequency-Shift A spectral line with frequency ν_{em} radiated by an atom at the distance r from the surface of a massive body with radius R and mass M, will be observed at infinity to have a frequency ν_{obs}. In the weak field approximation ($R \gg R_g$) this redshift is

$$z_{GR} = \left(\frac{\nu_{em} - \nu_{obs}}{\nu_{obs}}\right)_{GR} = \frac{GM}{c^2 r}. \tag{6.6}$$

For the Sun the value of $GM_\odot/R_\odot c^2$ is 1.9×10^{-10}.

An accurate interpretation of the gravitational redshift effect within general relativity considers a clock in a gravity field. The clock is running faster when it is farther from the gravitating body. The general relation is $dt = d\tau/\sqrt{g_{00}}$, so Einstein's gravitational redshift for the Schwarzschild metric is

$$1 + z_{GR} = \left(1 - \frac{2GM}{c^2 r}\right)^{-1/2}. \tag{6.7}$$

Will (1993) and Okun et al. (2000) emphasized that the energy of the photon does not change during its radial motion in the gravity field, i.e. it does not lose or gain energy. This differs from the usual mechanical behaviour of a particle.

The Time Delay of Light Signals The time delay phenomenon, or the Shapiro effect, occurs when an emitter at a distance r_1 sends a light signal to a mirror at a distance r_2 from a gravitating mass. If R is the distance between the emitter and the mirror, then the additional travel time is

$$(\Delta t)_{GR} = \frac{4GM}{c^3} \ln\left(\frac{r_1 + r_2 + R}{r_1 + r_2 - R}\right). \tag{6.8}$$

For the Sun, $4GM_\odot/c^3$ amounts to 20 microseconds. This effect occurs due to the curvature of space near the massive body, where it takes for the light signal a longer time to traverse a given distance as the time flows slowly.

The Perihelion Shift of a Planet The rate of the secular pericentre shift on an elliptical orbit of a planet (around a star having mass M) is

$$(\dot{\omega})_{GR} = \frac{6\pi GM}{c^2 a(1 - e^2)P}, \tag{6.9}$$

where a is the semi-major axis, e is the eccentricity, and P is the period of the orbit. For Mercury this gives $43''$/century, while for the binary pulsar PSR1913 + 16 the effect is much larger, about 4°/year.

The Lense-Thirring Effect An elliptical orbit of a non-rotating test particle, moving in the field of a central massive *rotating* body, will revolve as a whole about the direction of the axis of the central body with the rate

$$\Omega_{\mathrm{LT}} = \frac{2GJ}{c^2 a^3 (1-e^2)^{3/2}} (\vec{j} - 3\vec{l}(\vec{l}\cdot\vec{j})), \tag{6.10}$$

where $\vec{j} = \vec{J}/J, \vec{l} = \vec{L}/L, \vec{L}$ is the orbital angular momentum of the particle, and $\vec{J}$ is the angular momentum of the central body.

This Lense-Thirring precession is also called the "dragging of inertial frames". For an Earth-orbiting satellite it is about 0.1″/year, meaning that the orbit will make a whole rotation in about 13 million years.

The Geodetic Precession of a Gyroscope The rate of precession of a gyroscope orbiting a rotating massive body is:

$$\Omega_{\mathrm{WS}} = \frac{3GM}{2c^2 R_o^2} \vec{n} \times \vec{V}_o + \frac{GJ}{c^2 R_o^3} (3\vec{n}(\vec{n}\cdot\vec{j}) - \vec{j}). \tag{6.11}$$

Here $\vec{R}_o$ is the radius vector of the centre of inertia of the gyroscope, $\vec{n} = \vec{R}_o/R_o$, $\vec{V}_o$ is the orbital velocity, $\vec{J}$ and M are the angular momentum and the mass of the central body, and $\vec{j} = \vec{J}/J$.

The effect is a sum of two independent parts, one due to the gravitational potential of the central body, effectively non-Newtonian (the Weyl-effect), and the second due to its rotation (the Schiff-effect). For a gyroscope orbiting the Earth over the poles this precession amounts to about 7″/year.

The Quadrupole Gravitational Radiation The weak field approximation of Einstein's equations allows one to infer that a system of moving bodies will radiate energy in the form of gravitational waves. The total radiation in all directions gives the quadrupole luminosity (ergs/sec)

$$L_{\mathrm{GR}} = \frac{G}{45c^5} \dddot{D}_{\alpha\beta}^{\,2}. \tag{6.12}$$

$D_{\alpha\beta}$ is the quadrupole moment of the system. Gravitational waves in general relativity are transversal and correspond to a particle with spin 2. For instance, a binary system (with the component mass M_i and the semi-major axis a) will lose orbital energy via quadrupole gravitational radiation with power[1]

$$L_{\{2\}} \approx 2 \times 10^{32} \left(\frac{M_1}{M_\odot}\right)^2 \left(\frac{M_2}{M_\odot}\right)^2 \left(\frac{M_1+M_2}{2M_\odot}\right) \left(\frac{a}{R_\odot}\right)^{-5} \text{erg/s}. \tag{6.13}$$

[1] It is important to note that to calculate the loss of energy (6.12) one should use an expression for the energy-momentum "pseudotensor" of the gravitational field, not defined uniquely in general relativity. This has originated a long discussion about the reality of gravitational waves and their ability to carry energy in general relativity.

6.2.2 Strong Gravity, Black Holes, Quantum Gravity

An exact solution of Einstein's equations (5.9) for any centrally symmetric mass distribution is called the *Schwarzschild metric*. It has the following form for the 4-interval in the Schwarzschild system of coordinates (t, r, θ, ϕ):

$$ds^2 = \left(1 - \frac{r_g}{r}\right)c^2dt^2 - \frac{dr^2}{1 - \frac{r_g}{r}} - r^2(\sin^2\theta d\phi^2 + d\theta^2). \tag{6.14}$$

In other systems of coordinates the interval has different form. Note that the metric in Eq. (6.14) depends only on the total mass M of the gravitating body. The quantity r_g is called the Schwarzschild radius for the mass M:

$$r_g = \frac{2GM}{c^2} = 3\frac{M}{M_\odot}\text{ km}. \tag{6.15}$$

At $r = r_g$ the 00-component of the metric equals zero and the 11-component is infinite. Hence the gravity becomes so strong that nothing, not even light, can escape a body whose whole mass M is inside r_g (a black hole).

The Black Hole An external observer within a static system of coordinates will see matter collapsing eternally on the black hole. But in a free-falling coordinate system, co-falling matter will within a finite (and rather short) proper time cross the gravitational radius. Then the matter inevitably falls into the centre of the field $(r = 0)$, the true singularity of the metric.

Another important result is the equation of hydrostatic equilibrium

$$\frac{dp}{dr} = -\frac{G(\rho + p/c^2)(M + 4\pi pr^3/c^2)}{r^2(1 - r_g/r)}. \tag{6.16}$$

In this *Tolman-Oppenheimer-Volkoff equation* the factor $1/(1 - r_g/r)$ leads to an infinite pressure gradient for $r \to r_g$. This has a deep consequence: there is an upper limit for the mass of static compact relativistic stars, around 2–3 $M_\odot$. In the standard general relativity, compact objects with larger masses may exist only as black holes.

General Relativity Without Black Holes? Theoretical studies of the gravitational collapse have led some authors to conclude that there are solutions describing relativistic compact objects with sizes close to the Schwarzschild radius but with no event horizon.

If a substance has an unusual equation of state $p = p(\rho)$, like that of the physical vacuum and dark energy, it is possible to obtain non-singular static general relativistic solutions for arbitrary large masses, which are stable, and have no singularity, no event horizon and no information paradox (Dymnikova 2002; Mazur and Mottola 2004; Chapline 2005).

An analysis of the physical meaning of the coordinate transformation in general relativity led Mitra (2000, 2006) to argue that a black hole should have zero mass and that instead of black holes, Einstein's equations have as a solution an "eternally collapsing object" (ECO). Its size is close to R_g and it emits energy all the time so that an event horizon never originates.

Robertson and Leiter (2005) adhered to the strong principle of equivalence so that "special relativity must hold locally for all time-like observers in all of space-time". They found solutions of Einstein's equations satisfying the requirement for time-like world line completeness. The resulting "magnetospheric eternally collapsing objects" (MECO) possess an intrinsic magnetic moment and do not have any event horizon and curvature singularity.

These works show that additional conditions on the equation of state or coordinate transformations or the metric tensor of Riemannian space can change the physical contents of geometrical gravity theory.

Testing Quantum Geometry Quantum geometry predicts violation of the equivalence principle, possible violation of the Lorentz invariance, and time-varying fundamental physical constants at such a level that their detection may be realistic in near future (Amelino-Camelia et al. 2005). However, up to now increasingly strong limits have been derived on variations of fundamental constants (Chand et al. 2004). Also first observations of sharp images of a very distant supernova did not confirm the predicted quantum structure of space-time at Planck scales (Ragazzoni et al. 2003). However, time delays between gamma-photons with different energies from active galactic nuclei could be interpreted as a quantum gravity effect, assuming that there is no intrinsic difference in the emission time at the source (reviewed by Wagner 2009 and Amelino-Camelia and Smolin 2009).

6.3 Poincaré-Feynman Field Approach to Gravitation: Predictions

In general relativity, Minkowski space is a tangent space at each point of curved space (the local Lorentz invariance). The field approach uses the global Minkowski space to describe all four fundamental interactions as fields in space. Though the classical relativistic gravity effects in the weak field are identical in both theories, some specific effects of the field gravity may distinguish it from general relativity. Such are scalar gravitational waves, the translational motion of rotating bodies, and the surface and the magnetic field of the relativistic compact bodies in "black hole candidates"

6.3.1 Newtonian Force

In the field approach, the force between Newton's apple and the Earth is caused by the exchange of gravitons. Gravitons (real and virtual) mediate the gravitational interaction and represent quanta of a relativistic tensor field ψ^{ik} in Minkowski space η^{ik}. Thus the gravity force has an ordinary quantum nature. Experiments by Nesvizhevsky et al. (2002) point to this direction. Using freely falling ultra-cold neutrons it was shown that the gravity acts like the usual electric force producing quantum energy levels in the micro-particles moving in the gravity field (Westphal et al. 2007).

Attraction and Repulsion As we discussed earlier, within the field gravity the usual Newtonian force can be thought of as the sum of an attractive and a repulsive force ($\vec{F}_N = (\vec{F}_{\{2\}} + \vec{F}_{\{0\}}) = \frac{3}{2}\vec{F}_N - \frac{1}{2}\vec{F}_N$), corresponding to the pure tensor part (spin 2) and the scalar part (spin 0) of the tensor potential. This new understanding of the Newtonian potential opens new ways for experiments on the nature of gravitational interaction, e.g., to measure the scalar "antigravity" even in weak-field laboratory conditions.

6.3.2 Post-Newtonian Predictions for Observations

Only the weak field approximation at the post-Newtonian level has been studied in detail for the field gravity, but this already gives predictions, which differ from general relativity. In Chap. 5 we gave the PN equations of motion (5.57) for a particle in a static spherically symmetric weak field.

The Pericentre Shift and Positive Gravity Energy The rate of the pericentre shift of the orbit of a test particle with semi-major axis a, eccentricity e and period P, can be directly calculated from Eq. (5.57):

$$(\dot{\omega})_{\text{FG}} = \frac{6\pi GM}{c^2 a(1-e^2)P}. \tag{6.17}$$

This formula is the same as in general relativity, but the interpretation is different. For instance, the nonlinear contribution (the 2nd term in Eq. (5.56) due to $T^{00}_{(\text{g})}$ provides 16.7% of the total value (6.17). Therefore in the field theory the pericentre shift is directly affected by the positive energy density of the gravity field, making this physical quantity experimentally measurable.

Light in the Gravity Field Within the field gravity theory the deflection of light and the time delay of light signals are consequences of the gravity-electromagnetic field interaction, described by the Lagrangian $L_{\text{int}} = \psi_{ik} T^{ik}_{(\text{elm})}$. This gives the effective refraction index in the PN approximation:

$$n(r) = 1 + \frac{2GM}{c^2 r}. \tag{6.18}$$

Hence the velocity of a light signal will have the value

$$c_g(r) = \frac{c}{n} = c\left(1 - \frac{2GM}{c^2 r}\right), \tag{6.19}$$

and the direction of light changes and the time delay appears, both fully consistent with observations.

Atom in Gravity Field The gravitational redshift of spectral lines is caused by the shift of atomic levels. It is universal, as gravitation changes the total energy and all energy levels of an atomic system. In the PN approximation $E_{\rm obs} = E_0(1 + \varphi_{\rm N}/c^2)$ and hence $h\nu_{ik}^{\rm obs} = \Delta E_{ik}^0(1 + \varphi_{\rm N}/c^2)$. The same result as from this energy argument was derived by Moshinsky (1950) who calculated the interaction of the gravity field with the spinor and electromagnetic fields of a hydrogen atom. More generally, the gravitational redshift is

$$1 + z_g = \left(1 + \frac{2\Phi}{c^2}\right)^{-1/2}. \tag{6.20}$$

This formula gives the correct PN result $z \approx |\varphi|/c^2$.

6.4 Astrophysical Tests of Gravity Theories

Remarkably, all post-Newtonian classical relativistic gravitational effects—from bending of light to precessions (Table 6.1)—have the same values in general relativity and field gravity, which are very different theories physically.

6.4.1 Rotating Bodies and Binary Systems

Though one can not clarify the nature of gravity just by measuring classical relativistic effects in the Solar System and binary pulsars, there are even in the weak field regime untested effects, which may offer crucial experiments.

The Equivalence Principle Modern tests have shown that the inertial and gravitational masses are equal to within about 10^{-13}, and new tests have also been suggested (Haugan and Lämmerzahl 2001; Bertolami et al. 2006b).

In the field gravity theory the basic concept is the rest mass m of a particle or a body consisting of interacting particles. The major problem is how to give proper relativistic definitions for inertial and gravitating masses without referring to the non-relativistic Newtonian equation of motion.

According to the PN equation of motion (5.57) the 3-acceleration of a test particle (1) does not depend on the rest mass m of the test body, and (2) does depend on its velocity v and the value of the gravitational potential $\varphi_{\rm N}$ at the location of the

Table 6.1 Comparison of gravity theories

Physics and predictions	Newtonian gravity	General Relativity	Field Gravity
Geometry of space	Euclidean	Riemannian	flat Minkowskian
Basic reference systems	inertial	any	inertial
Gravitational potentials	ϕ_N	metric tensor	symmetric tensor, trace
Positive field energy density	No	No	Yes
Gravity force	Yes	Curvature	Attraction + repulsion
Gravitational frequency shift	Yes	Yes	Yes
Bending of light θ	$\frac{1}{2}\theta$	Yes	Yes
Relativistic perihelion shift	No	Yes	Yes
Time delay (Shapiro effect)	No	Yes	Yes
Lense-Thirring effect	No	Yes	Yes
Precession of gyroscope	No	Yes	Yes
Relativistic compact objects	(Yes)	Yes	Yes
Event horizon	Yes	Yes	No
Singularity	Yes	Yes	No
Quadrupole radiation	No	Yes	Yes
Scalar gravitational waves	No	No	Yes
Free fall of a rotating body	Yes	Yes	No

particle. This means that there are different ways in relativistic regime to define the inertial and gravitational masses, which gives new possibilities to test their equality.

A new test of the equivalence principle could utilize the motion of a rotating body. According to general relativity, such a body will have the same translational motion as the non-rotating one (if tidal effects can be neglected). Within field gravity one should integrate the gravity force over the volume of the rotating body, and then the equations of translational motion will be:

$$\frac{d\vec{V}}{dt} = -\left(1 + \frac{V^2}{c^2} + 4\frac{\varphi_N}{c^2} + \frac{I\omega^2}{Mc^2}\right)\vec{\nabla}\varphi_N + 4\frac{\vec{V}}{c}\left(\frac{\vec{V}}{c}\cdot\vec{\nabla}\varphi_N\right) + \frac{3}{Mc^2}\int [\vec{\omega}\vec{r}]([\vec{\omega}\vec{r}]\cdot\vec{\nabla}\varphi_N)dm. \qquad (6.21)$$

Equation (6.21) shows that the orbital translational velocity $\vec{V}$ of the centre of mass of the body will have extra perturbations due to the rotation. The last term depends on the direction and value of the angular velocity $\vec{\omega}$ of rotation. Its order of magnitude is v_{rot}^2/c^2 and it should be possible to measure in laboratory experiments and astronomical observations using lunar laser ranging and timing of pulsars in binary systems (Baryshev 2002).

Gravitational Waves from Binary Systems The best test of the gravitational radiation formulae is offered by binary pulsars. For a binary system the loss of energy due to the *tensor* gravitational radiation is given by the quadrupole luminosity (the same in field gravity and general relativity):

$$\langle \dot{E} \rangle_{\{2\}} = \frac{32G^4 m_1^2 m_2^2 (m_1 + m_2)\left(1 + \frac{73}{24}e^2 + \frac{37}{96}e^4\right)}{5c^5 a^5 (1 - e^2)^{7/2}}. \tag{6.22}$$

Here m_1, m_2 are masses of the two bodies, a is the semimajor axis and e is the eccentricity of the relative orbit.

Within field gravity theory there is an additional loss of energy due to the *scalar* monopole radiation (which does not appear in general relativity), given by the relation (Baryshev 1982, 1995):

$$\langle \dot{E} \rangle_{\{0\}} = \frac{G^4 m_1^2 m_2^2 (m_1 + m_2)\left(e^2 + \frac{1}{4}e^4\right)}{4c^5 a^5 (1 - e^2)^{7/2}}. \tag{6.23}$$

Hence the ratio of the scalar to tensor luminosity is

$$\frac{\langle \dot{E} \rangle_{\{0\}}}{\langle \dot{E} \rangle_{\{2\}}} = \frac{5}{128} \cdot \frac{\left(e^2 + \frac{1}{4}e^4\right)}{\left(1 + \frac{73}{24}e^2 + \frac{37}{96}e^4\right)}. \tag{6.24}$$

The value of this ratio lies in the interval [0, 1.1%] and for a circular orbit equals zero. However, for a pulsating spherically symmetric body there is no quadrupole radiation and the scalar radiation becomes decisive. In particular it follows that in the field gravity theory it is impossible to have a "quiet" relativistic collapse of a spherical body.

The orbit of the binary pulsar PSR 1913+16 has an eccentricity $e = 0.6171309(6)$ (Damour and Taylor 1991), hence the scalar part (6.24) is

$$\Delta_{\text{scalar}} = 0.735\%.$$

Because the rate of change of the orbital period $\dot{P}$ is proportional to the total energy loss ($\langle \dot{P} \rangle = -\frac{3}{2}\frac{E}{P}\langle \dot{E} \rangle$), one expects a corresponding excess in the decrease of the orbital period due to scalar gravitational radiation.

The data by Weisberg and Taylor (2002) show that the excess of the orbital period decrease relative to the predicted quadrupole energy loss is

$$\Delta_{\text{obs}} = \Big((\text{obs}) - (\text{quadr})\Big) = +0.78\% \pm 0.06\%.$$

This is interestingly close to the expected value 0.735% for the additional energy loss by scalar gravitational radiation (Baryshev 1995).

It has been shown by Damour and Taylor (1991) that one must take into account the accelerations of the pulsar and the Sun in the Galaxy, and the proper motion of the pulsar. The distance d to the pulsar PSR 1913 + 16 is a critical parameter in the calculation of the Galactic effect. Unfortunately, the line of sight to the pulsar passes through a complex region of our Galaxy, and one must be very careful, when using known distances to other pulsars for a distance estimate to PSR 1913 + 16.

Damour and Taylor used indirect arguments to re-estimate the standard dispersion-measure distance of 5.2 kpc. With their new distance 8.3 kpc the Galactic effect is +0.69%, which could almost explain the observed excess. However, the arguments by Pynzar (1995), based on a study of the pulse structure of PSR 1913 + 16, lead to a distance of about 3 kpc. For such a short distance the Galactic effect is only +0.11%. In a new analysis, Weisberg et al. (2010) inferred that for a distance of 9.9 kpc the expected (general relativistic) and observed values would agree.

Indeed, the precise distance to the pulsar PSR 1913 + 16 may be viewed as a test of fundamental physics, related to the nature of gravitation. Also distances to other binary pulsars will be crucial for gravity physics.

The Double-Pulsar J0737-3039A/B The system J0737-3039A/B is expected to provide the most precise test of general relativity (Deller et al. 2009). As all classical relativistic effects are the same in GR and FG (Table 6.1), this system also tests field gravity. The extra scalar radiation in field gravity is in this case essentially smaller than the quadrupole gravitational radiation. This is because the scalar radiation is proportional to the eccentricity of the orbit (Eq. (6.24)). For the system J0737-3039A/B the orbit is quite round, $e = 0.08$. Thus the additional radiation would be only 0.03%.[2] Deller et al. (2009) determined the distance to this system by measuring its annual geometric parallax with the Australian Long Baseline Array of radio telescopes, and concluded that with another 10 years of pulsar-timing observations, PSR J0737-3039A/B will be able to test the validity of general relativity and other theories of gravitation with an accuracy of 0.01.

The BL Lac Object OJ287 as a Test Object Studies of binary supermassive black hole candidates in active galaxy nuclei can also yield promising tests of scalar and tensor gravitational radiation.[3] The BL Lac object (a very active quasar) OJ287 shows a quasi-periodic pattern of prominent outbursts; 11 well-identified outbursts and several probable outbursts are known since 1891 (Fig. 6.1). The outbursts seem to come in pairs separated by one to two years, and the pairs occur about 12 years apart (Sillanpää et al. 1996).

Valtonen and Lehto (1997) presented a model in which a secondary body (a black hole) pierces the accretion disk of the primary black hole and produces two sharp impact flashes per period, which may serve as time signals telling how the smaller body moves around the primary. It appeared that from the data for a sufficient number of outbursts it was possible to calculate definite post-Newtonian Kepler orbits for the smaller body, and this happened in the early 2007, just in time to predict the exact date for September 2007 when the second of the double peaks was due to appear.

It is interesting to write down the parameters of the calculated orbit: its eccentricity is 0.663, the precession rate of the major axis is 39.0 deg per orbit, and the mass

---

[2] In principle, there is also another small effect, related to measuring the shape of the orbit: the rotation of the orbiting body contributes to the equation of motion (Eq. (6.21)).

[3] Already Boris Komberg (1968) proposed a binary system as a quasar model.

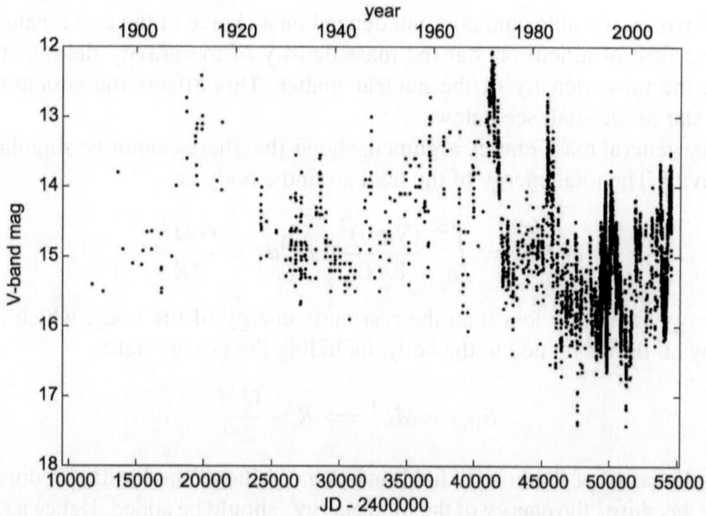

Fig. 6.1 Optical light curve for the quasar OJ287 (courtesy of Tuorla Observatory)

of the primary body is 18.0×10^9 solar masses. The model included the emission of gravitational radiation according to general relativity, and the predicted date for the event was September 13 plus or minus one day, as discussed in Valtonen et al. (2008a). The outburst happened as predicted. This is a test of general relativity (first proposed by Valtonen and Lehto in 1997) because if the system does not lose energy by emission of gravitational waves, then the orbit does not shrink, and the predicted outburst time is delayed by 20 days in this case.

Viewing OJ287 as a test of field gravity, the only difference from general relativity is the scalar gravitational radiation. Its contribution would be 1% of the total radiation, so that instead of 20 days, one would expect a delay 0.2d (5 hours) longer, which is now difficult to discern from the observations.

6.4.2 Strong Gravity and Compact Objects

For strong gravity the predictions diverge dramatically. In field gravity there are no black holes or singularities, and no such limit as the Oppenheimer-Volkoff mass. Hence compact massive objects in binary star systems and active galactic nuclei are major candidates for crucial tests.

No Black Holes in Field Gravity Now the mass density of the gravitational field around an object with mass M and radius R for a static weak field is

$$\rho_{(g)} = \frac{T^{00}_{(g)}}{c^2} = \frac{(\vec{\nabla}\varphi_N)^2}{8\pi G c^2} = 1.1 \times 10^{13} \left(\frac{M}{M_\odot}\right)^2 \left(\frac{10\text{ km}}{R}\right)^4 \text{ g/cm}^3. \tag{6.25}$$

It is positive, localizable, and does not depend on a choice of the coordinate system. On the surface of a neutron star the mass density of the gravity field is about the same as the mass density of the nuclear matter. This affects the structure of the neutron star as we shall see below.

A very general mass-energy argument shows that there cannot be singularities in field gravity. The total energy of the field around a body is

$$E_{(\mathrm{fg})} = \int_R^\infty \frac{(\vec{\nabla}\varphi_N)^2}{8\pi G} 4\pi r^2 dr = \frac{GM^2}{2R}. \tag{6.26}$$

This energy should be less than the rest mass energy of the body, which is determined by all things related to the body, including the gravity field.

$$E_{(\mathrm{fg})} < Mc^2 \Longrightarrow R > \frac{GM}{2c^2}. \tag{6.27}$$

The non-linearity of the gravity field makes the value of the limiting radius further increase, because "the energy of the field energy" should be added. Hence a safe estimate for the minimum radius of a massive body in the field gravity is $R_{\mathrm{lim}} > 0.5r_{\mathrm{g}}$.[4] Thus black holes and singularities are excluded by the existence of the positive energy density of the gravitational field.

Finite Gravity Force In the weak field approximation the field equation outside a mass M, surrounded by a positive field energy density (6.25), is

$$\Delta\varphi = +\frac{1}{c^2}(\nabla\varphi)^2, \tag{6.28}$$

which has the solution $\varphi = -c^2 \ln(1 + \frac{GM}{c^2 r})$. Hence the gravity force is

$$F_{\mathrm{g}} = m\frac{d\varphi}{dr} = \frac{GMm}{r^2}\frac{1}{(1 + GM/c^2 r)}. \tag{6.29}$$

For a maximally compact relativistic object having the radius $R_{\mathrm{M}} = GM/c^2$, the gravitational acceleration and the gravity force are restricted by

$$g_{\max} \le \frac{c^4}{GM} = \frac{c^2}{R_{\mathrm{M}}}, \quad \text{and} \quad F_{\mathrm{g}} \le \frac{mc^2}{R_{\mathrm{M}}}. \tag{6.30}$$

In general relativity the energy-density of the gravity field is negative (Sect. 5.2.3), hence the sign of the right side of (6.28) is opposite. In this case the solution is $\varphi = -c^2 \ln(1 - \frac{GM}{c^2 r})$, and the gravity force is $F_{\mathrm{g}} = (GMm/r^2)(1 - GM/c^2 r)^{-1}$, which is infinite, when $r \to GM/c^2$. This difference in the behaviour of the gravity force in general relativity and field gravity has important consequences for the structure of relativistic cosmic objects.

[4]This argument is a precise analogue to that of the classical radius of electron $R_{\mathrm{e}} = e^2/m_{\mathrm{e}}c^2$, following from the requirement that the electric field energy $E_{(\mathrm{fe})} = e^2/2R_0$ should be less than the electron's rest-mass energy $m_{\mathrm{e}}c^2$.

Hydrostatic Equilibrium Configurations In general relativity the equation of hydrostatic equilibrium (6.16) leads to a maximum mass of a neutron star, about $2\,M_{\odot}$, the Oppenheimer-Volkoff limit. Larger masses can exist only in the form of black holes.

In field gravity the equations of motion are contained in the conservation laws $T^{ik}_{(\Sigma),k} = 0$, where $T^{ik}_{(\Sigma)} = T^{00}_{(\text{gas})} + T^{00}_{(\text{int})} + T^{00}_{(\text{grav})}$ is the total EMT of considered system gas+gravity field in corresponding approximation. The post-Newtonian equation of hydrostatic equilibrium in FG (Baryshev 1988) depends on a particular choice of the interaction EMT and may be written

$$\frac{dp}{dr} = -\frac{G(\rho_0 + \delta\rho)(M_0^r + \delta M^r)}{r^2}, \tag{6.31}$$

where $\delta\rho = (e + p)/c^2 + 2\rho_0\Phi$, ρ_0 is the rest-mass density, $\Phi = \psi^{00}$, $M_0^r = \int_0^r 4\pi r^2 \rho_0 dr$, $\delta M^r = \int_0^r 4\pi r^2 [(e + 3p)/c^2 + 2\rho_0\Phi/c^2 + (d\Phi/dr)^2/8\pi G c^2] dr$.

The most important difference between Eqs. (6.31) and (6.16) is that within the field gravity theory the relativistic corrections lead to a decrease of the gravitating mass relative to the rest-mass, due to the negative value of the gravitational potential ($\Phi < 0$). According to Eq. (6.31) a hydrostatic equilibrium is possible for any large mass.

Tanyukhin (1995) constructed numerically the internal structure of neutron stars using the field gravity theory. He showed that such stars are more homogeneous than in general relativity, and with no upper limit on their mass.

Stability of Supermassive Stars Hoyle and Fowler (1963) suggested that a mass of the order of $10^8 M_{\odot}$ may condense in a galactic nucleus into a supermassive star (SMS), in which the nuclear energy generation take place. However, Fowler (1964) soon showed that in general relativity a SMS is unstable and will collapse to a black hole within a lifetime $\tau \approx 10(M/10^8 M_{\odot})^{-1}$ yr before the nuclear reactions begin. Hence in the standard GR only black holes can be the primary power sources of active galactic nuclei.

Within the field gravity theory the stability of supermassive stars was shown by Baryshev (1992a) who used the method developed by Fowler (1966) for considering the PN hydrostatic equilibrium and small adiabatic pulsations of a slowly rotating SMS. Its total equilibrium energy (excluding the constant term $\int \rho_0 c^2 = M_0 C^2$) in the form

$$E^{(\text{eq})}_{(\text{SMS})} = \int_V \left(e - 3p - \frac{1}{2}\varrho_0 v^2 \right) dV, \tag{6.32}$$

is a consequence of the relativistic virial theorem in the PN approximation. Here e is the thermal energy density, p is the pressure, $\frac{1}{2}\varrho_0 v^2$ is the density of the kinetic rotational energy, so that $\Delta E_{\text{rot}} = \int \frac{1}{2}\varrho_0 v^2 dV$. The first two terms can be expressed via the Newtonian potential energy E_{pot} plus the relativistic correction ΔE_{rel}, hence

$$E^{(\text{eq})}_{(\text{SMS})} = \frac{\beta}{2} E_{\text{pot}} - \Delta E_{\text{rel}} - \Delta E_{\text{rot}}, \tag{6.33}$$

where $\beta = p_{gas}/p_{tot} \ll 1$ is the gas to the total pressure ratio ($\beta \approx 10^{-3}$ for a mass $M \approx 10^8 M_\odot$) and the relativistic correction is

$$\Delta E_{rel} = K(n, \mathrm{gt}) \frac{G^2}{c^2} M_0^{7/3} \varrho_{0c}^{2/3}, \tag{6.34}$$

where M_0 is the rest-mass of the SMS, ϱ_{0c} is the central mass density and $K(n, \mathrm{gt})$ is a constant defined by the polytrope index n and the gravity theory. For the $n = 3$ polytrope the constant is[5]

$$K(3, FG) = +1.7349 \quad \text{while} \quad K(3, GR) = -0.9183. \tag{6.35}$$

The different signs of the corrections show that in general relativity we have a PN instability, while in field gravity a supermassive star is stable.

For small radial adiabatic pulsations of a SMS with radius R_0 and mass M_0, and $\delta r/r = (\delta R_0/R_0)\exp(-\omega_0 t)$, we get for the angular frequency

$$\omega_0^2 = \frac{1}{I}\left(-\frac{\beta}{2} E_{pot} + 2\Delta E_{rel} + 2\Delta E_{rot}\right), \tag{6.36}$$

where $I = \int r^2 \varrho_0 dV$ is the inertial moment of the SMS. Within the field gravity theory for the $n = 3$ polytrope, the oscillation period $P_0 = 2\pi/\omega_0$ of the supermassive star is

$$P_0 = \frac{2.11 R_0}{c\left[\frac{3}{8}\beta\frac{R_g}{R_0} + 4.78\frac{R_g^2}{R_0^2} + \frac{J^2}{(ckM_0R_0)^2}\right]^{1/2}}, \tag{6.37}$$

where R_g and J are the gravitational radius and the angular momentum of the SMS and k is a constant ≈ 1. This expression has no singularity contrary to the case of general relativity.

The PN stability in the field gravity theory radically changes the understanding of the evolution of the supermassive stars. In particular, in their last stages the main energy source will not be nuclear reactions with energy output of about 1%, but the gravitational binding energy of the order of $M_0 c^2$.

6.4.3 Relativistic Compact Objects

Observations of "black hole candidates" in X-ray binary stellar systems (masses about $10\,M_\odot$) and in galactic nuclei (10^6 to $10^9\,M_\odot$) provide tests of strong gravity effects. From the viewpoint of the FG theory the stellar-size objects and also the supermassive objects inferred to be common in centres of galaxies, are not black holes, but *relativistic compact objects* (RCO).

---

[5]Calculation of $K(3, FG)$ was done by A. Raikov (details in Oschepkov and Raikov 1995).

Probing the Existence of Black Holes Classical general relativity predicts black holes—objects having an event horizon at the Schwarzchild radius $2GM/c^2$ after which a one-way fall into the singularity is inevitable. There is no singularity and event horizon in the field gravity theory. As we explained above, the positive localizable energy density of the gravity field prevents the appearence of a singularity at the centre and also at the gravitational radius of a RCO. Instead of black holes, field gravity produces compact relativistic objects having radii $\approx GM/c^2$. Such objects have a highly redshifted surface and an intrinsic magnetic field. Another important prediction of field gravity is the existence of quark stars with masses of about 8 solar masses (Sokolov 1992c).

To prove observationally the existence of black holes means to prove the existence of the event horizon in relativistic compact objects. Crucial observational tests, which would convincingly reveal such a one-way sphere, have not yet been made, and the proof would be difficult, because many astrophysical processes are involved. Abramowicz et al. (2002) even argued that it is impossible to prove that an object has an event horizon.

ADAF as a Test of the Event Horizon It is hard for black hole models to understand the observed low luminosity of the accretion disc in a certain variability phase, when the accretion rate is still large. It was suggested (Narayan et al. 1997; Narayan and Quataert 2005) that this could be explained by the "advection dominated accretion flow" (ADAF) or "radiatively inefficient accretion flow"—it is assumed that protons and electrons are decoupled in the flow, so the kinetic energy is absorbed by the event horizon without an outward radiation. Bisnovatyi-Kogan and Lovelace (2000) and Binney (2003) argued that the magnetic field present in astrophysical plasmas of the accretion flow makes ADAF practically impossible.

Robertson and Leiter (2002, 2003, 2004, 2005) analysed data on black hole candidates in X-ray binary stars and active galactic nuclei and found evidence for intrinsic magnetic fields, in conflict with the black hole model. The low luminosity phase is naturally explained by the "propeller effect" of the magnetic field of the relativistic compact object. Robertson and Leitner's "magnetospheric eternally collapsing object" has no event horizon though its size is close to the Schwarzschild radius. Observations of the gravitationally lensed quasar Q0957 + 561A,B revealed the inner structure of the accretion disc, which demands an intrinsic magnetic field of the central RCO and may be well modeled as a MECO (Schild et al. 2006). The field gravity RCOs also can explain the magnetic fields in the "black holes".

Another unexpected finding is the very small radius of radiating matter in accretion discs. For instance, in the best studied accretion disc, around the central object of the Sy1 galaxy MCG-6-30-15, the inner radius of the orbiting matter is $r_{\text{inner}} = 0.615 R_g$. This is smaller than the Schwarzschild radius and was interpreted in terms of an extremely rotating Kerr black hole (Wilms et al. 2001). In comparison, within the field gravity theory the minimum radius of a relativistic compact object is about $0.5\ R_g$, from Eq. (6.27).

Crucial observational tests are difficult. Perhaps the most direct test of the black hole model was suggested by Falcke et al. (2000), who discussed VLBI observations of the black hole candidate in the Galactic centre with micro-arcsecond angular resolution. The profile of such an image can even distinguish between non-rotating and rotating black holes. Another direct test of the strong gravity effects would be the detection of a gravity wave signal from the relativistic collapse.

Relativistic Jets It is natural to expect matter accreting onto a rotating relativistic compact object and forming a disc around it. This process could be accompanied by relativistic jets which are observed in active galactic nuclei. The nature of this phenomenon is still debated and many observational facts have been gathered thus far. For example, an interesting effect, discovered by Babadzhanyants and Belokon' (1985, 1987) (also Belokon' 1991), is that the time of the birth of a superluminal radio component coincides with a strong optical flash. This effect was recently confirmed also for gamma-flashes in BLLac objects (Marscher et al. 2011).

Core-Collapse Supernovae, γ-Ray Bursts and Gravitational Waves The field gravity (no black holes) makes dramatic changes in the physics of supernova explosions. The collapse of the iron core of massive pre-SN stars will have a pulsational character and leads to gravitational signals of long duration (seconds), comparable with neutrino signals and gamma ray bursts.

The relation of the gamma-ray-burst (GRB) phenomenon to relativistic core-collapse supernovae has become a generally accepted interpretation of the GRBs (Paczynski 2001; Sokolov 2002). The compact GRB model suggested by Sokolov et al. (2006) predicts a correlation of the gamma and x-ray signals with gravitational bursts.

The gravitational antenna GEOGRAV observed a signal from SN1987A (Amaldi et al. 1987) together with the neutrino signal observed by the Mont Blanc Underground Neutrino Observatory (Aglietta et al. 1987). This could be a possible detection of the scalar gravitational radiation from the spherical core-collapse of the pre-supernova (Baryshev 1997). An observational strategy to make a distinction between scalar and tensor gravitational waves by using siderial time analysis was considered by Baryshev and Paturel (2001) and Paturel and Baryshev (2003a, 2003b). It is based on the difference in the statistics for the detections of tensor and scalar waves, the pulses assumed to arrive from burst sources in the Local Supergalaxy (the flat nearby galaxy system within 20 Mpc).[6]

6.4.4 The Hubble Law of Redshifts as a Key Observation

The Hubble law of redshifts starts immediately beyond the border of the Local Group. Such a law is also a part of the Friedmann model of the expanding universe, based on general relativity and homogeneous distribution of matter. We shall discuss Friedmann's model in the next chapter and focus here on the relation of the Hubble law to gravity theories.

[6]Controversial claims about possible detections of gravitational signals by *Nautilus* and *Explorer* antennas (Astone et al. 2002, 2006), if confirmed, require a new analysis of the potential sources of gravitational waves (Coccia et al. 2004).

The Hubble Law in Expanding Space In the 1930s Howard Robertson and Goeffrey Walker realized that there is an intimate link between the homogeneous matter distribution and the expansion law in the universe. General geometrical reasoning shows that if a uniform space (matter distribution) expands so that it remains uniform, then there is everywhere a linear velocity–distance relation (not observable directly). This regular expansion causes the redshift of light travelling in space and leads to the observed linear relation between redshift and distance in the local universe.

This connection between the supposed uniformity of matter distribution and the Hubble law was the first great success of the expanding world model. In addition to the linearity, the value of the slope (the Hubble constant) is closely linked to the age of the universe and its overall agreement with the determinations of the ages of old stars lends good support to the expansion picture. Further evidence for expansion is given by the surface brightness test and the time dilation versus redshift effect (Chap. 8).

General relativity does not permit a static matter distribution, and if the distribution is homogeneous, one necessarily expects regular expansion (or contraction). Taken together, the success of the Friedmann model to explain the Hubble law and many other astronomical observations, lend strong indirect support to the validity of general relativity even on the largest observed scales.

On the other hand, the lumpy distribution of visible matter has been seen as conflicting with the linear Hubble law, hence casting doubt on the expansion paradigm where the linearity of the velocity–distance law is linked with homogeneity. However, the everywhere uniform dark energy may help the linear Hubble law to appear even there where one would not expect it otherwise (Chap. 12).

The Hubble Law in Static Space The expanding space paradigm (and general relativity), explains well astronomical observations. At the same time, it is in spirit of science to test alternative ways to understand the cosmological redshift and the linear Hubble law. If one can show that known alternatives cannot offer explanations at the same level of success as the expanding space paradigm, the position of the latter, together with general relativity, is made stronger. And if some alternative remains reasonable, one should not close the case.

We note that a global Minkowski space is not automatically excluded or obsolete in cosmology; field gravity allows one to consider an infinite matter distribution in Minkowski space without the gravitational potential paradox. Thus assuming a static global Minkowski space (a part of the hypothesis of gravity to be tested), one may first ask if the redshift could be due to motions within space, together with the ordinary Doppler effect.

Davis and Lineweaver (2004) argued that the Hubble diagram for distant SNIa supernovae rules out the special relativistic Doppler interpretation at a high confidence level. One may add the problem of how to give galaxies their motions resulting in the linear Hubble law, without sacrificing the cosmological principle of "no-centre".[7] It is true that this has been discussed in Milne's kinematic relativity, based on special relativity with its length contraction and time dilation effects, but that interesting scheme (in fact without gravitation!) cannot be regarded as a modern contender even though Milne's Cosmological Principle has remained so central for modern cosmology.

The gravitational redshift has been tested in laboratory and appears both in field gravity and in general relativity. Then the critical question is: Is there some way of producing the cosmological redshift and the Hubble law using field gravity within a global Minkowski space? Although this may sound unlikely, we shall discuss in Chap. 9 a possible gravitational redshift effect related to the fractal distribution of matter. Such a possibility is tempting to study, because then the inhomogeneity, which may be problematic for the classical interpretation of the Hubble law, could be the very reason for the Hubble law within a basically different framework.

Concluding Remarks Our discussion of gravity physics emphasized that there are crucial experiments and astronomical observations for testing the nature of gravitation. In particular, the geometrical and field approaches are not equivalent experimentally, though the classical relativistic gravity effects in the weak field are identical in both theories. The following observations, if made, would be against the predictions of general relativity:

- the detection of scalar gravitational waves
- an extra deflection in the translational motion of rotating bodies
- too small sizes of compact objects in active galactic nuclei
- the presence of intrinsic magnetic fields in black hole candidates
- the non-expansion origin of cosmological redshifts

Similarly, evidence against such phenomena would be in direct conflict with the field approach to gravitation. Hence, astronomical observations around these subjects can potentially influence strongly cosmological physics.

In the next chapter we discuss the cosmological model constructed on the basis of the main stream geometrical gravity theory, general relativity.

[7] An expansion in Newtonian *absolute space* would imply one privileged point where the celestial body must remain at rest. Therefore, if one observes expansion of matter and has grounds to think that it is a universal phenomenon, one cannot simultaneously accept absolute space and the no-centre principle. In the modern paradigm the concept of no centre results from the overall expansion of space together with the uniform substance.

Chapter 7
The Friedmann Model

In early history of relativistic cosmology Alexander Friedmann (1923), in his book *The world as space and time*, modestly and wisely viewed cosmological models as "schematic and simplified, reminding one of the real world only to the extent that a dim reflection from a mirror of a schematic drawing of the cathedral of Cologne may be reminiscient of the cathedral itself". Since those pristine days, cosmology has grown into an ambitious project dealing with applications of modern physics to the description of the largest observable universe. Here we first take a brief look at Newtonian cosmology and the first world model by Einstein and then describe the Friedmann model, the main theoretical tool in the hands of today's cosmologists, which has developed into a many-component model containing ordinary matter, radiation, and, for the most part, dark unknown substances.

7.1 Newtonian Cosmology

Under "Newtonian cosmology" we mean the early application of Newton's theory to the static infinite universe, a good baseline for modern world models. It is based on classical mechanics and gravity theory together with the cosmological principle of a uniform star distribution. Newtonian universe is only apparently simple, and its paradoxes have been discussed up to our days. Interesting historical details may be found in North (1965), Harrison (1981), Baryshev and Teerikorpi (2002), and Teerikorpi et al. (2009).

7.1.1 Newtonian Cosmological Model

In our laboratories it is natural to regard the space outside the Earth as an empty cosmos. Physicists may even forget that there is a huge universe around the lab. For our Solar System the other stars are far enough so that the empty cosmos begins just outside it.

Y. Baryshev, P. Teerikorpi, *Fundamental Questions of Practical Cosmology*, Astrophysics and Space Science Library 383,
DOI 10.1007/978-94-007-2379-5_7, © Springer Science+Business Media B.V. 2012

Uniform Stellar World A totally new situation is encountered in cosmology. Contrary to the ancient Stoic view, still entertained in the 19th century, we think that there is no place outside the matter in the universe, and "far enough from a gravitating mass" carries little sense in cosmology. However, this does not mean that there are no inertial frames in a universe filled by matter (cf. Narlikar 1993, p. 29). Indeed, let us consider a star within a homogeneous isotropic matter distribution. Then all forces from the other stars are mutually compensated and the net cosmological gravity force vanishes. This means that in accordance with its definition, an inertial frame exists for each such star, when the sum of all forces equals zero.

Newton viewed the fact that we see all those unmoving stars in the sky, instead of one lump of stars, as evidence for an infinite uniform distribution of stars in space. Fluctuations of the gravity force from nearest neighbours are expected to produce small inhomogeneities in the star distribution.

Holtsmark's Distribution A mathematical proof of Newton's guess about the possibility of an infinite homogeneous universe came from Norwegian physicist Jan Holtsmark (1919). He considered a statistically uniform distribution of particles (Poisson's law) in Euclidean space. Actually he had charged particles interacting with each other via Coulomb's force, but this problem is identical to Newton's one.

Holtsmark found that in an infinite universe of such interacting particles there is only a finite average force acting on any particle. The strength of the force is determined by the nearest neighbours. Holtsmark's probability distribution for the amplitude of the cosmological force F is

$$P_{\rm H}(F) = \frac{2}{\pi F}\int_0^\infty \exp(-ax^{3/2}/F^{3/2})x \sin x dx, \tag{7.1}$$

where $a = (4/15)(2\pi G)^{3/2}m^{3/2}n_0$, m is the mass of a particle, and n_0 is the number of particles per unite volume. This Holtsmark's distribution has an infinite dispersion because of close encounters between point-like particles. Irina Petrovskaya (1986) generalized the distribution for finite-sized stars; then the dispersion becomes finite. This means that in the infinite homogeneous Newtonian universe there is no gravitational *force* paradox. The net result of the action of an infinite number of masses is a finite force with its amplitude essentially determined by the nearest neighbours. Similar arguments were presented already by Svante Arrhenius (1908) in his article about the infinity of the world.[1]

[1] In this same study Arrhenius explains the absence of spiral nebulae close to the band of the Milky Way as due to the extinction of light by absorbing material. At that time and before, the odd distribution of nebulae was often regarded as evidence for their status as constituents of our Milky Way instead of being remote "Island Universes".

7.1.2 Paradoxes of the Newtonian Cosmology

Although the force paradox may be avoided as noted above, several other interesting paradoxes appear within the Newtonian cosmological model having a homogeneous distribution of eternal stars in infinite Euclidean space.

The Paradox of Gravitational Potential At the level of the concept of the gravitational potential the Newtonian cosmology has a problem known as the Neumann-Seeliger *gravitational potential paradox*: as the volume of a matter distribution of a finite density tends to infinity the Newtonian gravitational potential at any point increases without limit. Indeed, the solution of the Poisson equation inside the homogenous ball (Eq. (5.3)) gives the gravitational potential in the centre of a ball of radius R as

$$\varphi_N(0) = -2\pi G\rho_0 R^2 \rightarrow -\infty \quad \text{for } R \rightarrow \infty, \tag{7.2}$$

i.e. it has an infinite value for an infinitely large ball, for any density $\rho_0 > 0$. Another way to see the problem is: considering that in a uniform matter distribution also the potential should be constant, then only the solution $\rho_0 = 0$ (no matter!) satisfies Poisson's equation. This is why it has been said that the Newtonian gravity theory can not be applied to the cosmological problem (e.g., Landau and Lifshitz 1971, p. 333).

The Photometric Paradox Two other problems of the Newtonian cosmology are the *paradox of dark sky* (Cheseaux, Olbers) and the *thermodynamic paradox* (Boltzmann).

If the stars are uniformly distributed in infinite space then the night sky should be blazing as a typical star surface. Let n_0 be the mean number density of stars in space, hence the sky fraction $f(r)$ covered by stars is:

$$f(r) = \frac{1}{4\pi}\int_0^r \frac{A}{r^2} n_0 4\pi r^2 dr, \tag{7.3}$$

where $A = \pi R_*^2$ is the area of a star's cross section and $\Omega_* = A/r^2$ is the solid angle of the star at distance r. Therefore the sky will be completely covered ($f(r) = 1$) if the stars are distributed up to the radius

$$r_{\mathrm{Olb}} = (An_0)^{-1} \approx \frac{1}{\sigma n_0}, \tag{7.4}$$

where the last term is like the length of typical free motion between collisions of gas particles having the cross section σ and the number density n_0. In terms of the typical distance L between the particles and the particle size d, the radius of "Olbers's sphere" r_{Olb} can also be expressed as:

$$r_{\mathrm{Olb}} = \left(\frac{L}{d}\right)^2 L. \tag{7.5}$$

The total energy density of the background radiation at the Earth is

$$\rho_{\rm BR}(r) = \frac{1}{c}\int_0^r \frac{L_*}{4\pi r^2} n_0 4\pi r^2 dr = \frac{L_* n_0 r}{c}, \tag{7.6}$$

where L_* is the typical luminosity of a star. For $r = r_{\rm Olb}$, the sky background is as bright as the surface of a star.

The Thermodynamic Paradox This means the thermal death of Boltzman's universe filled with elastic interacting molecules. The second law of thermodynamics tells that such a system inevitably drives towards an asymptotic homogeneous state with thermodynamic equilibrium and small fluctuations. On the contrary, we observe around us a highly structured universe. This has the interesting implication that the age of this complex world should be finite, otherwise all structures should have been destroyed.

There is another more modern aspect of the thermodynamic paradox, related to energy. The energy sources in the universe are restricted by the finite value of the mass-energy of the matter mc^2. In particular, the nuclear reactions in stars will eventually stop, and stars cannot shine forever.

7.1.3 Suggestions to Resolve the Paradoxes

These paradoxes have been of great importance for cosmological research, stimulating active searches for self-consistent world models.

Change the Poisson Equation The first solution of the gravitational paradox was proposed by von Seeliger (1895) and Neumann (1896) as a modification of one initial postulate, the theory of gravitation. Perhaps gravity falls off at large distances faster than the inverse-square law so that the gravitational potential has the Newtonian expression multiplied by the extra factor $e^{-\alpha r}$, where α is small enough to be consistent with the usual theory for short distances. This would make an infinite quasi-Newtonian universe possible.

An analogous idea was used by Einstein (1917), who added the Λ-term into general relativity. He replaced Poisson's equation $\Delta\phi = 4\pi G\rho_0$ by

$$\Delta\phi - \Lambda\phi = 4\pi G\rho_0, \tag{7.7}$$

where Λ is a universal constant. The solution of Eq. (7.7) is:

$$\phi = -\frac{4\pi G}{\Lambda}\rho_0 = \text{const.} \tag{7.8}$$

Now the potential can be constant even for a finite matter density. Local inhomogeneities in finite regions of space can be considered as causing an additional local potential which is Newtonian if sufficiently small relative to c^2.

Change the Mass Distribution It was argued by Fournier d'Albe (1907) and Charlier (1908, 1922) that the Newtonian universe may be constructed without gravitational and night sky paradoxes if there is an unlimited clustering hierarchy (see Chap. 10). In the model the typical value of the mass within the distance r from a hierarchy element scales as $M(r) \propto r^1$. The size and the mass of the universe are arbitrarily large, but the mean density $M(r)/r^3 \sim r^{-2}$ converges to zero and Olbers's paradox is avoided. Though the modern counterpart to hierarchy, the idea of fractality, is no longer needed to solve Newtonian paradoxes, it has special significance to cosmology in studies of large-scale structure formation.

Let Light Be Redshifted In his pioneering discussion of the cosmological significance of the paucity of radiation, Bondi (1952) dropped the assumption of a static universe and suggested that the redshift of light arising in large-scale expansion could decrease the background radiation to the observed low level. In fact, he considered that one might choose either the redshift or a young universe as alternative explanations of the dark sky.

Allow a Finite Age Stars have not existed always, but were switched on some finite time ago: when we look far in space, we also look back in time, at the ancient era when stars were not yet shining. This idea was pointed out by J.H. Mädler (1873) in his book on the history of astronomy and in fact earlier mentioned by Edgar Allan Poe, the American father of modern detective fiction. In 1848, he visioned in his cosmological essay *Eureka* that the distance of the background where the stellar discs are blended into one is "so immense that no ray from it has yet been able to reach us at all".

William Thomson, better known as Lord Kelvin (after the river flowing near his university in Glasgow), noted that if the stars have been shining no longer than 100 million years, then the radius of the visible universe is at most 100 million light years. Ages of this order appeared in a popular theory, which ascribed the hotness of the Sun to the energy released when it gradually contracts under its own gravity. Essentially such a solution for Olbers's paradox is offered by the 14 milliard-year-old big bang cosmos (Harrison 1987).

As an illustration, let us calculate the radius of Olbers's sphere r_{Olb} for a world filled by stars similar to our Sun. Assume that the number density n_0 is 1 star per cubic parsec. Then $1/n_0 = \mathrm{pc}^3 \approx 3 \times 10^{40}\ \mathrm{km}^3$ and $A = \pi(7 \times 10^5)^2\ \mathrm{km}^2$. Equation (7.4) gives $r_{\mathrm{Olb}} \approx 2 \times 10^{28}\ \mathrm{km} \approx 10^{15}$ pc. In megaparsecs this formidable length is 10^9 Mpc which corresponds to $0.3 \times 10^6 R_{\mathrm{H}}$ where R_{H} is the Hubble distance. In other words, light would need about 300000 ages of the (Friedmann) universe in order to travel from the edge of Olbers's sphere to us!

Abandon Static Space The new idea of expanding space came from the geometrical revolution in gravitation theory, general relativity. After the static world model of Einstein (1917), non-static solutions of Einstein's equations (Friedmann 1922, 1924) showed that the universe can expand or contract. The new model had a finite age, the light within it was redshifted, and it did not suffer from the paradox of dark sky.

7.2 The Friedmann Cosmological Model

The Friedmann model (or the Friedmann-Lemaître-Robertson-Walker model), is the main-stream framework for interpreting astrophysical observations and the basis of the standard cosmological model. Here general relativity works together with Einstein's cosmological principle of homogeneity.

7.2.1 Basic Equations

As we discussed above, general relativity has been successfully tested in the weak gravity conditions (of the Solar System and binary neutron stars), and the next natural step is to ask whether it can be applied to the Universe as a whole. Assuming that it can, we arrive at the Friedmann model.

The Friedmann model is described by the exact non-linear Einstein equations for the special case of homogeneous and isotropic matter distribution. Current versions of standard cosmology make a distinction between ordinary matter (with positive pressure) and dark energy (negative pressure). So it is convinient to write Einstein's equations in the two-fluid form:

$$\Re^{ik} - \frac{1}{2} g^{ik} \Re = \frac{8\pi G}{c^4} \left(T^{ik}_{(\mathrm{m})} + T^{ik}_{(\mathrm{de})} \right), \tag{7.9}$$

where $T^{ik}_{(\mathrm{m})}$ is the energy-momentum tensor for ordinary matter (the 1st fluid), and $T^{ik}_{(\mathrm{de})}$ is the dark energy component (the 2nd fluid; e.g., the famous cosmological constant, Carroll et al. 1992).

From the Bianchi identities for the Riemann tensor $\Re_{iklm}$ (from which the "geometric" left side of Eq. (7.9), or the Einstein tensor) is constructed, one gets the continuity equation for the right side:

$$T^{i}_{k\,;i} = (T^{i}_{(\mathrm{m})k} + T^{i}_{(\mathrm{de})k})_{;i} = 0. \tag{7.10}$$

T^{i}_{k} is the total EMT of matter and dark energy. For non-interacting matter and dark energy the covariant divergence of each EMT equals zero separately. The general case of interaction, with energy transfer between matter and dark energy was studied by Gromov et al. (2004).

7.2.2 Einstein's Cosmological Principle

Einstein's Cosmological Principle states that the universe is spatially homogeneous and isotropic on "large scales". This term is required, because the universe certainly is inhomogeneous on scales of galaxies and their clusters.

Einstein's Static World Model In 1917, Albert Einstein applied general relativity, soon after he had invented it, to the cosmological problem. If the matter distribution is uniform, the curvature of space is the same everywhere. Einstein showed that in the case of a static spherical space with radius of curvature R his Eq. (7.9), when applied to ordinary matter of finite density $\rho > 0$ without pressure (dust model) do not have a solution. It is instructive to write down the resulting two independent equations:

$$-\frac{3}{R^2} = -\frac{8\pi G}{c^2}\rho, \qquad -\frac{1}{R^2} = 0. \tag{7.11}$$

In order to secure a solution, Einstein added to the left "geometry" side of his field equations the constant term $g_{ij}\lambda$, leading to a new pair of equations:

$$\lambda - \frac{3}{R^2} = -\frac{8\pi G}{c^2}\rho, \qquad \lambda - \frac{1}{R^2} = 0. \tag{7.12}$$

Now the solution exists linking the cosmological constant λ, the constant density ρ and the radius of curvature of the spherical universe R:

$$\lambda = \frac{4\pi G\rho}{c^2} = \frac{1}{R^2}. \tag{7.13}$$

This remarkable equation in early practical cosmology tells that by measuring the matter density ρ in a local (but representative) region, one can infer the size of the universe (if it indeed has the static spherical geometry).

Extension to Time-Dependent Universes After Friedmann's work on non-static models, Einstein's hypothesis of homogeneity and isotropy can be taken to mean that on all scales $r > r_{\rm hom}$ the total energy density $\varepsilon = \rho c^2$ and the total pressure p are functions of time only:

$$\varepsilon(\vec{r}, t) = \varepsilon(t), \qquad p(\vec{r}, t) = p(t). \tag{7.14}$$

Here the total energy density is the sum of the energy density of ordinary matter ($\varepsilon_{\rm m}$) and dark energy ($\varepsilon_{\rm de}$), and similarly for the total pressure:

$$\varepsilon = \varepsilon_{\rm m} + \varepsilon_{\rm de}, \qquad p = p_{\rm m} + p_{\rm de}. \tag{7.15}$$

In models with one "effective" fluid the equation of state is usually suggested to have the ideal form:

$$p = \gamma\varepsilon. \tag{7.16}$$

To emphasize the difference between usual matter (with positive pressure) and dark energy (with negative pressure) one can use separate notations for the equations of state of the partial components:

$$p_{\rm m} = \beta\varepsilon_{\rm m}, \quad 0 \le \beta \le 1, \qquad p_{\rm de} = w\varepsilon_{\rm de}, \quad -1 \le w < 0. \tag{7.17}$$

Values $w < -1$ have also been considered for a hypothetical substance nicknamed phantom energy (Caldwell et al. 2003).

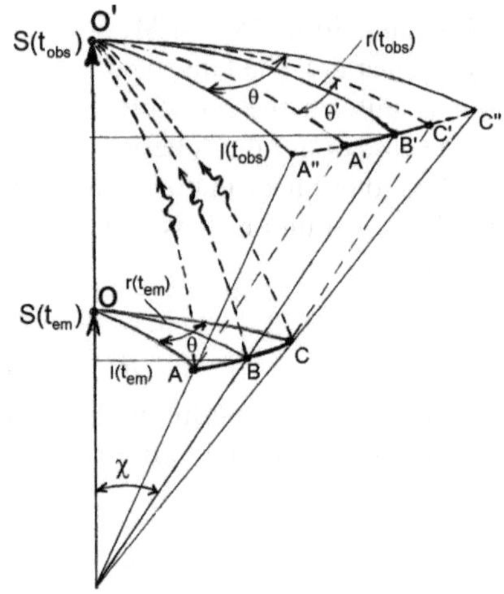

Fig. 7.1 Properties of the universe at the times of emission t_{em} and observation t_{obs} of light from a distant galaxy having the linear size ABC. Note that $l(t_{em})$ and $r(t_{em})$ are the external and internal metric distances, respectively, when the scale factor is $S(t_{em})$

7.2.3 Space Expansion Paradigm

The standard cosmological model includes the expanding space paradigm, so that the metric of a constant curvature space depends on time, and distances between galaxies may increase (or decrease) with time.

The Robertson-Walker Metric Cosmological models deal with 4D space-time geometry. An important consequence of homogeneity and isotropy of such spaces is that their metric g^{ik} (the line element $ds^2 = g_{ik}dx^i dx^k$) may be presented in the form found by Robertson and Walker. As we discussed in Chap. 2, the line elements of curved spaces may be given in different systems of coordinates, useful for certain purposes. Expressed in the "spherical" comoving space coordinates χ, θ, ϕ the RW line element is

$$ds^2 = c^2dt^2 - S(t)^2[d\chi^2 + I_k(\chi)^2(d\theta^2 + \sin^2\theta d\phi^2)], \tag{7.18}$$

where t is the cosmic time, $I_k(\chi) = \sin(\chi), \chi, \sinh(\chi)$ correspond to the curvature constants $k = +1, 0, -1$, respectively, and $S(t)$ is the scale factor. The comoving coordinates of a galaxy do not change during the expansion of the universe, while the scale factor expresses how the distance between the observer (at $\chi = 0$) and the galaxy (at χ, θ, ϕ) changes with time.

In the expanding space paradigm the *proper* (internal) metric distance r of a body with a fixed comoving coordinate χ is given by

$$r = S(t) \cdot \chi. \tag{7.19}$$

It is *internal* in the sense that it is measured wholly within the curved space. It changes with time t as the scale factor $S(t)$. In other words, r and χ give the Eulerian and Lagrangian representations of the comoving distance. The dimension of the metric distance $[r] = \text{cm}$, hence if $[S] = \text{cm}$, then the comoving coordinate distance χ labeling the galaxy is dimensionless (see Fig. 7.1).

In fact, χ is the spherical angle and $S(t)$ is the radius of the (pseudo)sphere embedded in the 4D Euclidean space. This means that the "cm" (the measuring rod) itself is defined as an unchangeable unit in the embedding Euclidean space. Hence the distance r measured in cm may be viewed as the "internal" proper distance on the 3D hypersurface of the embedding space.

When "cylindrical" comoving coordinates μ, θ, ϕ are used, the line element is

$$ds^2 = c^2dt^2 - S(t)^2\left[\frac{d\mu^2}{1-k\mu^2} + \mu^2(d\theta^2 + \sin^2\theta d\phi^2)\right]. \tag{7.20}$$

The meaning of the coordinate μ is given by the fact that the *external distance* l is expressed as

$$l = S(t)\cdot\mu. \tag{7.21}$$

In analogy with the 2D space within the 3D space (Chap. 2), this is the distance from the z-axis in the embedding Euclidean 4D space. To avoid confusion, it is important to use different designations for the distances defined by the intervals (7.18) and (7.20). Note that the proper metric distance r is additive, while the external distance l is not. The line element (7.20) is convenient, when one calculates the angular size of a rigid rod as a function of distance, and in fact, $S(t)\mu$ is what is called the angular size distance! It is also expressed in cm, but in physical meaning it differs from the metric distance r. The relation between the two distances is

$$r = S(t)I_k^{-1}(l/S), \qquad l = S(t)I_k(r/S), \tag{7.22}$$

where I_k^{-1} is the inverse function for I_k.

Expanding homogeneous space implies that at any given moment the recession velocity is proportional to distance. The *recession velocity* V_{exp} of a body with fixed χ, due to "space expansion", is accurately defined as the *rate of increase of the metric distance* r. Its simple and exact relativistic expression follows from (7.19):

$$V_{\text{exp}} = \frac{dr}{dt} = \frac{dS}{dt}\chi = \frac{dS}{dt}\frac{r}{S} = H(t)r = c\frac{r}{R_{\text{H}}}. \tag{7.23}$$

Here $H(t) = \dot{S}/S$ is the Hubble constant (also a function of time) and $R_{\text{H}} = c/H(t)$ is the Hubble distance at the time t. ("dot" means the time derivative d/dt). This also signifies that the linear velocity-distance relation $V = Hr$, identified to be behind the observed Hubble law, is a consequence of the uniform matter distribution.

7.2.4 Friedmann's Equations

Up to now we have considered the "kinematics" of expansion. Now we discuss the "dynamics", i.e. how the scale factor S changes with time.

In comoving coordinates the total EMT of the cosmological fluid is

$$T_i^k = \mathrm{diag}\,(\varepsilon, -p, -p, -p)\,. \tag{7.24}$$

For an unbounded homogeneous matter distribution (7.14) Einstein's equations (7.9) are reduced to Friedmann's equations. From the initial set of 16 equations we have only two independent equations for the (0, 0) and (1, 1) components, and these may be written as

$$\frac{\dot{S}^2}{S^2} + \frac{kc^2}{S^2} = \frac{8\pi G}{3c^2}\varepsilon, \tag{7.25}$$

$$2\frac{\ddot{S}}{S} + \frac{\dot{S}^2}{S^2} + \frac{kc^2}{S^2} = -\frac{8\pi G}{c^2}p. \tag{7.26}$$

The Bianchi identity (7.10) implies the continuity equation

$$3\frac{\dot{S}}{S} = -\frac{\dot{\varepsilon}}{\varepsilon + p}, \tag{7.27}$$

which complements (7.25) and (7.26) as a consistency condition.

Using the Hubble "constant" $H = \dot{S}/S$, we rewrite (7.25) and (7.26) as

$$H^2 - \frac{8\pi G}{3}\rho = -\frac{kc^2}{S^2}, \tag{7.28}$$

$$\ddot{S} = -\frac{4\pi G}{3}\left(\rho + \frac{3p}{c^2}\right)S. \tag{7.29}$$

In terms of the critical density $\rho_{\mathrm{crit}} = 3H^2/8\pi G$, the total density parameter $\Omega = \rho/\rho_{\mathrm{crit}}$, the curvature density parameter $\Omega_k = kc^2/S^2H^2$, and the deceleration parameter $q = -\ddot{S}S/\dot{S}^2$, these equations have also the forms

$$\Omega - 1 = \Omega_k, \tag{7.30}$$

$$q \equiv -\frac{\ddot{S}S}{\dot{S}^2} = \frac{1}{2}\Omega\left(1 + \frac{3p}{\rho c^2}\right). \tag{7.31}$$

Here Ω, p, and ρ are total quantities, i.e. sums of the corresponding components for matter and dark energy.

Friedmann's equation (7.31) determines how the scale factor $S(t)$ or the distance $r(t)$ depend on time. For a cosmological fluid with the equation of state (7.16) and zero curvature ($k = 0$) the scale factor behaves as

$$S(t) \propto t^{\frac{2}{3(1+\gamma)}}, \tag{7.32}$$

where $\gamma > -1$. For $\gamma = 0$ (dust) the scale factor is $S(t) \propto t^{2/3}$, for $\gamma = 1/3$ (radiation) $S(t) \propto t^{1/2}$, and for $\gamma = -1$ (cosmological vacuum) $S(t) \propto \mathrm{e}^{\alpha t}$.

Equation (7.31) links together the deceleration q and the density parameter Ω, both of which may be studied observationally. Thus it could be seen as a test of general relativity on very large scales. In fact, it would work nicely for a simple world with dust-like matter ($p = 0$). However, the beauty and simplicity of the test is deteriorated by the introduction of several components with differing physics; as $\rho = \sum \rho_i$ and $p = \sum p_i$ the model contains many free parameters. This makes the Friedmann model very flexible when confronted with new observational data and, at the same time, lessens the value of (7.31) as a promising large-scale test of general relativity.

The Main Parameters of Friedmann Models The critical matter density ρ_{crit}, the Hubble parameter H, the deceleration parameter q, and the density parameter Ω are the main physical parameters that define a Friedmann model. Such quantities as the metric distance, the redshift, the time since the Big Bang, and relations between them are also used when the Friedmann models are confronted with observations.

A special value of the total density determines the curvature of space:

$$\rho_{\text{crit}} = \frac{3H^2}{8\pi G} = 1.88h^2 \times 10^{-29}\ \text{g/cm}^3 = 2.8h^2 \times 10^{11}\ M_\odot/\text{Mpc}^3. \tag{7.33}$$

This *critical density* can be derived from Eq. (7.28), by putting the curvature equal to zero ($k = 0$).

All models with the density $\rho > \rho_{\text{crit}}$ or, equivalently, the density parameter $\Omega = \rho/\rho_{\text{crit}} > 1$ have a positive curvature. When exactly $\Omega = 1$ the expanding space is flat, and for a low density universe $\Omega < 1$ the curvature is negative. Here the density ρ includes fields and matter, with positive or negative pressure, but does not include the density of the gravity field.

Modern Friedmann models contain at least two important components of substance (matter, dark energy), and their relative contributions to the cosmic density are characterized by the ratio

$$\alpha(S) = \frac{\rho_{\text{de}}(S)}{\rho_{\text{m}}(S)}. \tag{7.34}$$

This is generally time dependent. It is also affected by a possible interaction (energy transfer) between cosmic components, in particular between dark energy and dark matter (e.g., Peebles and Ratra 2003). In an interesting special case the relative contributions remain constant ($\alpha = \text{const}$, a "coherent" Friedmann model; Gromov et al. 2004).[2] Due to the flexibility of the Friedmann model, one may test on a phenomenological level even the possible interaction between dark matter and dark energy, and thus their reality and nature (Teerikorpi et al. 2003).

[2]The parameters of the partial equations of state (7.17) and the associated "effective" one-fluid parameter (7.16) are related by $\gamma = (w\alpha + \beta)/(\alpha + 1)$. For a coherent model with constant equation of state parameters for matter (β) and dark energy (w), the associated one-fluid model also has $\gamma =$ constant in $p = \gamma\varepsilon$.

7.3 Redshift, Distance, and Recession Velocity

The Friedmann model is rich with theoretical cosmological laws, which we cannot observe directly; these are expressed as relations between quantities, one or both of which are not observable from our limited vantage point. For instance, we can observe the redshift (which together with the magnitude of a standard candle leads to the Hubble law), but we cannot observe the recession velocity of the object in question, nor its present metric distance.

7.3.1 Scale-Factor and Redshift

As a direct consequence of the continuity equation (7.27) there is a relation between the density and the scale factor. For the ideal equation of state (7.16) this theoretical cosmological law of changing density has a simple form independent of the value of the curvature of space:

$$\rho \propto S^{-3(1+\gamma)}. \tag{7.35}$$

The density may change because the scale-factor is a function of time $S(t)$.

Lemaître's Redshift–Scale Factor Relation In standard cosmology the cosmological redshift z is a new phenomenon of geometric physics, caused by space expansion making the wavelength of a travelling photon stretch according to the equation first derived by Lemaître (1927, 1931):

$$(1+z) = \frac{\lambda_0}{\lambda_1} = \frac{S_0}{S_1}. \tag{7.36}$$

Here λ_1 and λ_0 are the wavelengths at emission and reception, respectively, and S_1 and S_0 are the corresponding values of the scale factor. This Lemaître's relation is a consequence of the radial null-geodesics ($ds = 0$, $d\theta = 0$, $d\phi = 0$) of the RW line element.

In order to derive Lemaître's law we ask what the observed time interval δt_o is for two light signals initially separated by the interval δt_e at emission, after they propagate in expanding space? On its trajectory from a galaxy with comoving radial coordinate χ to the observer (us) at $\chi = 0$ for light is valid: $0 = cdt - S(t)d\chi$, and hence

$$\text{constant} = \int_0^r dr = c\int_{t_{em}}^{t_o} \frac{dt}{S(t)} = c\int_{t_{em}+\delta t_e}^{t_o+\delta t_o} \frac{dt}{S(t)}, \tag{7.37}$$

where t_{em} and t_o are the times of emission and observation. Using the identity between the last two integrals above and referring to Fig. 7.2 it is easy to show that

$$\int_{t_{em}}^{t_{em}+\delta t_{em}} \frac{dt}{S(t)} = \int_{t_o}^{t_o+\delta t_o} \frac{dt}{S(t)}, \tag{7.38}$$

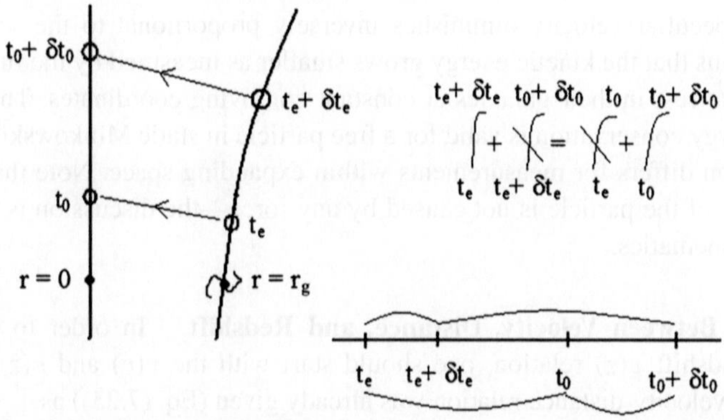

Fig. 7.2 Lemaître's redshift law. The integral from the time of emission to the time of detection over $dt/S(t)$ is a constant for the galaxy in question. The interval δ_e between the emissions of two signals is changed into the longer interval δ_0 at detection. The two integrals can be written as the sums of two integrals as shown at the bottom

hence

$$\frac{\delta t_{\rm em}}{S(t_{\rm em})} = \frac{\delta t_{\rm o}}{S(t_{\rm o})}. \tag{7.39}$$

Thus the time interval stretches along with the scale factor. As the wavelength is proportional to 1/frequency (= period), it also is proportional to the scale factor, which is Lemaître's law.

Another Effect: Cooling-Down of Peculiar Velocities A photon moving in expanding space loses its energy (is redshifted). Also a moving non-zero mass particle loses its (kinetic) energy or slows down in expanding space. Let us look closer at this interesting phenomenon.

A galaxy having a non-zero peculiar velocity does not remain at rest, but its comoving coordinate changes. Assume that the origin of the coordinate system is at another galaxy which is genuinely at rest. Then the change of rate of distance $V(t)$ between these galaxies can be written as a sum of two terms: the regular expansion plus the peculiar velocity $v(t)$

$$V(t) = \frac{d[S(t)\chi_{\rm g}]}{dt} = v_{\rm exp}(t) + v(t) = \frac{dS}{dt} \times \chi_{\rm g} + S(t) \times \frac{d\chi_{\rm g}}{dt}. \tag{7.40}$$

During the interval dt the galaxy experiences a change of $v(t)dt$ in its comoving coordinate and at the same time the space is expanding with the rate of $H(t)v(t)dt = (\dot{S}/S)v(t)dt$, hence what remains at $t + dt$ is the peculiar velocity $v(t + dt) = v(t) - (\dot{S}/S)v(t)dt$. Finally we get the equation for the time evolution of $v(t)$: $\dot{v}/v = -\dot{S}/S$. Its solution is simply

$$v(t) \propto 1/S(t). \tag{7.41}$$

Thus the peculiar velocity diminishes inversely proportional to the scale factor, which means that the kinetic energy grows smaller as measured by the fundamental observers at rest in their galaxies at constant comoving coordinates. The ordinary law of energy conservation is valid for a free particle in static Minkowski space, but the situation differs for measurements within expanding space. Note that the "deceleration" of the particle is not caused by any force—the discussion is entirely in terms of kinematics.

Relations Between Velocity, Distance, and Redshift In order to obtain the velocity-redshift $v(z)$ relation, one should start with the $v(r)$ and $r(z)$ relations. The exact velocity-distance relation was already given (Eq. (7.23)) as

$$v_{\text{exp}}(r) = Hr. \tag{7.42}$$

Here $H(t) = \dot{S}/S$ is the Hubble parameter at the moment t. Its value at the present epoch t_0 is called the Hubble constant $H_0 = H(t_0)$ and it determines the characteristic time scale $1/H_0 = r(t_0)/v_{\text{exp}}(t_0)$.

The observer on the Earth lives at the epoch $t = t_0$, and the observed redshift serves as a measure of distance at this epoch. For photons moving along the coordinate χ, $cdt = \pm S d\chi$, and we may write

$$S_0 d\chi = \frac{cdz}{H(z)}. \tag{7.43}$$

Integration over the redshift gives the *metric distance-redshift relation*

$$r(t_0, z) = r(z) = \int_0^z \frac{cdz'}{H(z')}. \tag{7.44}$$

In the practical case of redshifts $z < 1000$, with non-interacting cold dark matter (dm), radiation (rad) and dark energy (de) making the current total density parameter $\Omega^0_{\text{tot}} = \rho^0_{\text{tot}}/\rho^0_{\text{crit}} = \Omega^0_{\text{dm}} + \Omega^0_{\text{rad}} + \Omega^0_{\text{de}}$, the Hubble ratio may be given, from Eqs. (7.28) and (7.35), as

$$\begin{aligned} H(z) = H_0[&\Omega^0_{\text{dm}}(1+z)^3 + \Omega^0_{\text{rad}}(1+z)^4 + \Omega^0_{\text{de}}(1+z)^{3(1+w)} \\ &+ (1 - \Omega^0_{\text{tot}})(1+z)^2]^{1/2}. \end{aligned} \tag{7.45}$$

Using the $r(z)$ relation one gets the *exact velocity-redshift relation*

$$v_{\text{exp}}(z) = r(z) H_0 = c\frac{r(z)}{R_{H_0}}, \tag{7.46}$$

where the expansion velocity is for an object (with the redshift z) observed at the time $t = t_0$, and R_{H_0} is the present value of the Hubble radius.

After All, why the Linear Distance–Redshift Law in Near Space? Consider a nearby galaxy and the change of the scale factor during the small time interval dt when the light travels from this galaxy to the observer (us):

$$S(t_{\mathrm{obs}}) \approx S(t_{\mathrm{em}}) + \frac{dS}{dt} \times dt. \tag{7.47}$$

When the distance is short and the expansion speed much less than the speed of light, then the distance R measured by the astronomer using any of the available ways is practically equal to the metric distance $S(t)\chi$ and the time interval can be approximated as $dt \approx S\chi/c = R/c$. From this follows that $S(t_{\mathrm{obs}})/S(t_{\mathrm{em}}) \equiv 1+z = 1+(\dot{S}/S(t_{\mathrm{em}}))R/c$ and finally

$$cz = HR \quad \text{for small distances.} \tag{7.48}$$

This is also how Lemaître predicted the linear Hubble law.

7.3.2 *Measuring Distance and Time in Friedmann Cosmology*

Any observation of a distant object involves both distance and time, because to reach us the light had to travel a finite time. As McVittie (1974) says, distance is a degree of remoteness; in some sense or another, faint galaxies must be remote. Only a cosmological model gives the exact recipe for calculating from the observed properties of an object its distance.

The Fundamental Meter Stick Distance Amongst the variety of distances, one would like to think that there exists a fundamental one, corresponding to meter sticks put one after another from the Sun to the centre of a galaxy. Somewhat paradoxically, such a distance which one may find "natural" to think about, is the one which cannot be directly measured.

In fact, when McVittie (1974) critically discussed various kinds of cosmic distances, he did not prefer any of them as the most natural choice for expressing very large distances. However, *within the Friedmann model*, the metric distance is unquestionably unique and real, even though not directly measurable.[3] It is a primary quantity of Riemannian geometry, while other distances (luminosity etc.) are extremely useful workhorses of practical cosmology, but hardly distances at all! They are not additive and their values in "cm" have no direct physical sense.

Even when a layman asks the distance to a high-redshift quasar, it may be a good idea to give the present metric distance. If the metric distance of a quasar is 800 Mpc, this can be characterized by saying that it is about $800/0.8 = 1000$ times more distant than the Andromeda galaxy, giving a vivid impression of its remoteness.

[3] Because of the indirect, non-local measurement of distance, cosmological (metric) distances will always be tied to a cosmological model and tend have some unphysical aura around them (Samuel 2005).

In the 1977 Tallinn conference, Jõeveer and Einasto proposed the cell structure of the galaxy universe. If one assumes, for the sake of argument, that there were such a periodic structure with a metric size of 100 Mpc, then a quasar at the present distance of 3000 Mpc, could be said to lie behind about 30 "cell walls", illustrating the usefulness of metric distance within the Friedmann model.

The Universal Time The question of the universal time and simultaneity appeared already in Einstein's static model, where he divided the global space-time strictly into curved space and uncurved time, giving "a kind of absolute time distinct from space" (Eddington 1923a). In view of the special theory of relativity, rejecting any preferred time (or space), the global cosmic time in the Friedmann model may sound strange. And the universal time does not come alone, as Lehti (2002) has emphasized, but together with the preferred spatial reference frame, where the observers are at rest relative to the large-scale matter and the background radiation.

When one constructs the Friedmann model, one puts into the world the homogeneous substratum and defines the time coordinate so that the density only depends on it. Similarly, when we apply the model, we accept (usually silently) that the structure of the concrete world happens to be so regular from the beginning (the initial conditions) that it can be described by the universal time together with the momentary space where the metric distance has reality. Lehti (2002) illustrated the situation: "The physicist looks at space from the perspective of general theories, and the astronomer from the perspective of our unique cosmos". That the universal time is tightly linked to the structure of the universe (in the Friedmann model) is shown by the fact that observers everywhere could use the temperature of the background radiation as a clock showing the cosmic time (Lehti 2000).

7.3.3 Ages and Horizons

We observe distant events as they occurred when the photon now received was emitted at the event. However, we usually would like to have a map of events as they occur *now* all over the universe, or more generally, a series of such maps for different moments. British cosmologist E. Arthur Milne (1896–1950) thus classified cosmological models into two kinds.

World Picture and World Map A World Map (WM) portrays the universe as it would look if the speed of light were infinite: the model shows all parts of the universe at the same (present) cosmic time. A World Picture (WP) gives an observer's view of the universe: distant events are seen as they were the light-travel time ago.

A World Picture is familiar from physics, where the past-light cone diagram is usually conveniently drawn in space-time with two spatial and one time coordinate, so being a visualizable surface in the 3D space-time. Actually, the past-light cone is a 3D hypersurface in the 4D space-time. The Friedmann model may be viewed as a World Map, since it contains the universal time parameter whose fixed value defines a homogeneous 3D "hypersurface". We can see only a small patch of the hypersurface around us. In order to compare observations with theory predictions, one must transform this theoretical WM into a WP-like representation.

Age-Redshift Relation In the Friedmann model there is the origin of the time coordinate, $t = 0$, and the age of the universe. This was pointed out by Friedmann himself in his first cosmological article. The connection between time and scale factor differentials, $cdt = cdS/(dS/dt)$, may be written as $cdt = -cdz/(H(z)(1+z))$, giving the age of the universe at redshift z as

$$t(z) = \int_z^\infty \frac{dz'}{(1+z')H(z')}. \tag{7.49}$$

For $z = 0$ this integral gives the present age of the universe t_0. For a flat ($\Omega = 1$) Einstein-de Sitter model ($\Omega_{\mathrm{m}} = 1$), $t_0 = (2/3)H_0^{-1}$, while for an empty $\Omega = 0$ universe $t_0 = H_0^{-1}$, where the characteristic Hubble time is

$$t_{H_o} = H_0^{-1} = (3.24h_{100} \times 10^{-18}\ \mathrm{sec}^{-1})^{-1} = 0.98h_{100}^{-1} \times 10^{10}\ \mathrm{yrs}. \tag{7.50}$$

Table 8.1 shows ages and look-back times for different Friedmann models.

The Existence of Horizons The *particle horizon* χ_{part} is defined as the largest comoving distance from which the light emitted at $t = 0$ could have reached us by now. More distant objects have not been able to affect us causally. It also means the largest distance visible to us now, in principle at least. From $cdt = S(t)d\chi$ we obtain

$$\chi_{\mathrm{part}} = c\int_0^{t_0} \frac{dt}{S(t)} \tag{7.51}$$

The thus defined comoving distance χ_{part} increases with time, hence parts of the universe which earlier were beyond the particle horizon, enter later the horizon (become visible). The metric distance to the particle horizon is $r_{\mathrm{part}} = S_0\chi_{\mathrm{part}}$. As a simple example, for a flat dust model $S(t) = S_0(t/t_0)^{2/3}$ with its age $t_0 = (2/3)H_0^{-1}$, the integration gives $r_{\mathrm{part}} = 2c/H_0$.

The *event horizon* χ_{event} is the most distant *present* event ($t = t_0$) from which the signal could reach us at some finite future time.[4] The events beyond the event horizon will be forever hidden for us. Allowing an infinite future, it is valid that

$$\chi_{\mathrm{event}} = c\int_{t_0}^{\infty} \frac{dt}{S(t)}. \tag{7.52}$$

---

[4]One may find it difficult to remember the difference between the two kinds of horizon. Note that here a particle refers especially to a photon which we observe (and which left its origin sometimes *in the past* $0 \to t_0$), while "events" refer to phenomena all over the universe (and which will be observed or not sometimes *in the future* $t_0 \to \infty$).

It is easy to show that for the scale factor behaving as $S(t) \propto t^p$ the event horizon is at a finite distance only if $p > 1$, meaning an accelerating universe ($d^2S(t)/dt^2 \propto p(p-1)t^{p-2} > 0$). For example, the flat dust model does not have a finite event horizon, or $\chi_{\mathrm{event}} = \infty$, so eventually all events just now happening in that infinite universe will be seen. An interesting example is the exponentially accelerating de Sitter universe with $S(t) = e^{Ht}$ (H is a constant): the metric distance to the event horizon is now finite and approaches zero in the distant future: $\chi_{\mathrm{event}} = (c/H)e^{-Ht}$. At each moment there are an infinite number of events beyond the event horizon, which will never be observed.

7.4 Basic Observable Quantities: Angle, Flux, Surface Brightness

The metric distance r of an object having the cosmological redshift z is

$$r(z) = R_{H0} \int_0^z \frac{dz'}{h(z')}, \tag{7.53}$$

where $R_{H0} = cH_0^{-1}$ is the Hubble radius $0.93h_{100}^{-1} \times 10^{28}$ cm $= 3000h_{100}^{-1}$ Mpc.

7.4.1 Mattig's Relations

Figure 7.1 shows the difference between the momentary distance $r(t_{\mathrm{obs}})$ at the moment of reception of a photon and the distance $r(t_{\mathrm{emit}})$ at the moment of emission of the photon by a distant galaxy. The internal metric distance to an object with redshift z at the moment of observation t_{obs}, the $r(z)$ distance, may be analytically expressed only in some simple cases.

The External Metric Distance $l(z)$ For the zero-pressure dust universe this relation was first derived by Mattig (1958)[5] as:

$$l(z) = \frac{c}{H_0} \frac{zq_0 + (q_0 - 1)((2q_0 z + 1)^{1/2} - 1)}{q_0^2(1+z)}, \tag{7.54}$$

where $q_0 = \Omega_0/2$ is the deceleration parameter at $t = t_0$ (Eq. (7.31)). In terms of the internal metric distance it has the form

$$r(z) = S_0 I_k^{-1}\left[\sqrt{\frac{2q_0 - 1}{k}} \frac{H_0}{c} l(z)\right], \tag{7.55}$$

[5] Wolfgang Mattig has worked predominantly in the field of solar physics, and, as he mentioned in a letter to us, he never worked in extragalactic research and cosmology was his hobby. He derived the famous Mattig's equation when he had to deliver a lecture on cosmology in connection with his doctoral thesis.

where $I_k[\chi] = \sin\chi, \chi, \sinh\chi$ for $k = 1, 0, -1$ and the scale factor $S(t_0)$ is

$$S_0 = \frac{c}{H_0}\sqrt{\frac{k}{\Omega_0 - 1}}. \tag{7.56}$$

For the non-interacting two-fluid dust-vacuum Friedmann model

$$r(z) = \frac{c}{H_0}\int_0^z \frac{dz'}{(\Omega^0_{\rm vac} + \Omega^0_{\rm m}(1+z')^3 - \Omega^0_k(1+z')^2)^{1/2}}. \tag{7.57}$$

The corresponding $l(z)$ relations for $k = -1$, 0, and 1, respectively, are

$$l(z) = \frac{c}{H_0}\frac{1}{(-\Omega^0_k)^{1/2}}\sinh\left(\int_{\frac{1}{1+z}}^1 \frac{(-\Omega^0_k)^{1/2}dy}{y(\Omega^0_{\rm m}/y - \Omega^0_k + \Omega^0_{\rm vac}y^2)^{1/2}}\right), \tag{7.58}$$

$$l(z) = \frac{c}{H_0}\int_{\frac{1}{1+z}}^1 \frac{dy}{y(\Omega^0_{\rm m}/y + \Omega^0_{\rm vac}y^2)^{1/2}}, \tag{7.59}$$

$$l(z) = \frac{c}{H_0}\frac{1}{(\Omega^0_k)^{1/2}}\sin\left(\int_{\frac{1}{1+z}}^1 \frac{(\Omega^0_k)^{1/2}dy}{y(\Omega^0_{\rm m}/y - \Omega^0_k + \Omega^0_{\rm vac}y^2)^{1/2}}\right). \tag{7.60}$$

Gromov et al. (2004) give relations for interacting cosmological fluids.[6]

7.4.2 *The Angular Size-Redshift Relation*

Before going to the angular size-redshift relation in Friedmann models let us first extend the discussion (in Chap. 2) of curved spherical 2D space in 3D Euclidean space to the 4D case of expanding curved spaces. Expansion complicates the interpretation of the measured properties of distant objects.

2D Expanding Sphere Consider a "galaxy" with the linear size d and which was at the distance $u(t_{\rm e})$ when the now-observed light was emitted and the world radius was $S(t_{\rm e})$ (see Fig. 7.1). What is its observed angular size now?

An emitted photon continues its flight along the one and same great circle, i.e. the geodetic line leading to the observer at P. The photons emitted at the moment $t = t_{\rm e}$ from the end points A and B travel along the great circles PA and PB that form the angle $\theta(t_{\rm e})$. So from Eq. (2.35)

$$\theta(t_{\rm e}) = \frac{d}{S(t_{\rm e})\sin\frac{u(t_{\rm e})}{S(t_{\rm e})}}. \tag{7.61}$$

---

[6] We discuss here Friedmann models, but note that any developed non-Friedmann world model must also have its own redshift-distance (z–r) and redshift-time (z–t) relations and also the rules which relate luminosity and angular size distances to the metric distance r, in order to be able to predict observable effects and test these predictions (Chap. 8).

These stretching great circles have by the moment of observation transformed into the circles $P'A'$ and $P'B'$. These form the angle $\theta(t_{\rm obs})$ that is exactly the same as $\theta(t_{\rm e})$. Hence the observed angular size expressed in terms of the radius of the sphere $S(t_{\rm obs}) = (1+z)S(t_{\rm e})$ at the time $t_{\rm obs}$ is

$$\theta(t_{\rm obs}) = \frac{d(1+z)}{S(t_{\rm obs}) \sin \frac{u(t_{\rm obs})}{S(t_{\rm obs})}}. \qquad (7.62)$$

Note that the ratio u/S does not change during expansion.

Friedmann Models The desired relation comes from the line intervals in 4-space (Eqs. (7.18) or (7.20)). The latter one gives simpler expressions (the relation between the angle and the distance for the *external* metric is Euclidean). Thus consider a rod with a linear size dl, perpendicular to the line-of-sight, with the comoving radial coordinate μ. The distance element between the ends of the rod at the moment of emission t_1 ($dt = 0$, $d\mu = 0$) is

$$ds^2 = -dl^2 = -\mu^2 S^2(t_1)(d\theta)^2. \qquad (7.63)$$

The coordinates were so chosen that $d\phi = 0$. Then the angular size becomes

$$\theta_1 = \frac{d}{\mu_1 S(t_1)} = \frac{d(1+z)}{\mu_1 S(t_0)}, \qquad (7.64)$$

where the relation $S(t_1) = S(t_0)/(1+z)$ was used. Hence the angular size-redshift relation in the Friedmann model has the form

$$\theta(z) = \frac{d(1+z)}{l(z)} = \frac{d}{l_{\rm ang}(z)}, \qquad (7.65)$$

and the angular size distance $l_{\rm ang}(z)$ is

$$l_{\rm ang}(z) = \frac{l(z)}{1+z}. \qquad (7.66)$$

Here appears the external metric distance $l(z) = S(t_0)\mu$. It may be expressed by the Mattig equation (7.54) for the dust universe or by the Eqs. (7.58)–(7.60) for the two-fluid Friedmann models. If space is flat ($k = 0$), then the external and proper metric distances are equal ($l(z) = r(z)$).[7]

We see that a consequence of space expansion is that the angular size of a distant object is $(1+z)$ times larger than the size that the object would have at its present metric distance in a static Euclidean universe.

Generally the distance $l_{\rm ang}(z)$ has a maximum (or the angle has a minimum) at some z whose value depends on the Friedmann model. If a true standard rod can be found, then this test is not only parametric (telling about the Friedmann model), but it also tests the reality of space expansion.

[7] A useful expression for the angular size distance for different dust-vacuum models was given by Demianski et al. (2003), accurate to 1.5% in the z range from 0 to 10.

7.4.3 The Magnitude-Redshift Relation

Consider a distant galaxy at the comoving radial coordinate μ in the Friedmann universe. The galaxy emits light isotropically around it with the rate L [energy/time], the bolometric (total) luminosity of the source.

Draw a sphere with the galaxy in the centre and the observer (we) at the surface at the moment of reception (now, $S = S(t_0)$). The area of the sphere is $4\pi(S(t_0)\mu)^2$, so we measure the flux

$$f = \frac{L}{4\pi(S(t_0)\mu)^2(1+z)^2}. \tag{7.67}$$

Here one factor $1+z$ comes from the redshift of each photon and the other one is due to the slowed-down reception rate. We see that the luminosity distance $l_{\text{lum}}(z) = S(t_0)\mu(1+z) = l(z)(1+z)$. Since $l_{\text{ang}}(z) = l(z)/(1+z)$, we may summarize the relations between external metric, angular size and luminosity distances:

$$l(z) = l_{\text{ang}}(z)(1+z) = \frac{l_{\text{lum}}(z)}{1+z}. \tag{7.68}$$

In terms of magnitude the bolometric magnitude-redshift relation is

$$m_{\text{bol}}(z) = 5\log(l(z)(1+z)) + 25 + M_{\text{bol}} + A_{\text{bol}}, \tag{7.69}$$

where $l(z)$ is the external metric distance in Mpc, M_{bol} is the bolometric absolute magnitude of the source, and A_{bol} is the extinction correction.

7.4.4 Surface Brightness

Using the above results on fluxes and angular sizes, we can now derive how the surface brightness depends on the redshift in the Friedmann model. Let us consider a shining sphere (luminosity L, diameter D) at redshift z. Then the flux f_{obs} arriving from the sphere and its size θ in the sky are

$$f_{\text{obs}} = \frac{L}{4\pi[(1+z)l(z)]^2}, \qquad \theta = \frac{D(1+z)}{l}.$$

Hence the surface of the sphere shines with the brightness B

$$B \propto \frac{f_{\text{obs}}}{\theta^2} \propto (1+z)^{-4}. \tag{7.70}$$

The surface gets remarkably dim at high redshifts; at $z = 1$ the surface brightness is only $1/16$ of the value at a nearby distance! If in a static space the light is for some "tired light effect" redshifted, then the surface brightness at redshift z is merely lowered by the factor $1+z$, as each photon loses this much of its energy. The large expected effect makes the search for the redshift dependence of surface brightness a very important test of the reality of expansion (Chap. 8).

7.4.5 Spatial Volumes and Cosmological Malmquist Bias

In some cosmological tests, such as the number–redshift relation (Chap. 8), and also in a proper analysis of the magnitude–redshift relation (when one takes into account the cosmological Malmquist bias), one has to know how to calculate the spatial volume $V(z)$ as a function of redshift.

Volumes in a Friedmann World From the expression for the RW-metrics (7.19), (7.21), the comoving volume element of the expanding space is

$$dV = S^3(t)I_k^2(\chi)d\chi d^2\omega = S^2 I_k^2(r/S)drd^2\omega = l^2 \frac{dl}{\sqrt{1-(l/S)}} d^2\omega, \qquad (7.71)$$

where $d^2\omega = \sin\theta d\theta d\phi$ is the solid angle element (4π for the whole sphere), and $r = S\chi$, $dr = Sd\chi$, $I_k(\chi) = I_k(r/S)$, $l = S\mu$, $dl = Sd\mu$, $l = SI_k(r/S)$.

Hence the volume $V(r)$ of the sphere with the radius equal to the metric distance r, or the volume-distance relation, is

$$V(r) = \int_0^r dV = \frac{4\pi}{3} S^3(t)\chi^3\sigma_k(\chi) = \frac{4\pi}{3} r^3 \sigma_k(r/S), \qquad (7.72)$$

where $\sigma_k(\chi) = 3\chi^{-3}\int_0^\chi I_k^2(y)dy$ is $3\chi^{-3}(\frac{\chi}{2} - \frac{\sin 2\chi}{4})$ for $k = +1$; is 1 for $k = 0$; and is $3\chi^{-3}(\frac{\sinh 2\chi}{4} - \frac{\chi}{2})$ for $k = -1$. Then, the known relation $r(z)$ leads to the corresponding volume-redshift $V(z)$ relation in Friedmann models.

For the popular case $k = 0$, the volume is simply $V(z) = \frac{4\pi}{3} r(z)^3$, where the present metric distance $r(z)$ is obtained from Eq. (7.57) with $\Omega_k^0 = 0$. For example, in the Einstein-de Sitter dust model, the co-moving volume $V(z)$ has the especially simple expression (where $A = (32\pi/3)(c/H_0)^3$)

$$V(z) = A\left[1 - 1/(1+z)^{1/2}\right]^3. \qquad (7.73)$$

Classical vs. Cosmological Malmquist Bias As we discussed in Chap. 3, one may inspect the Hubble diagram in two ways: as $\log z$ against m or m against $\log z$. In the 2nd case, the magnitude limit distorts the average Hubble relation at increasing redshifts in an easily visualizable manner (type 2 bias). In the 1st case, the magnitude limit is not important, but at higher redshifts we have the rather intricate cosmological Malmquist bias, an analogue of the classical Malmquist (type 1) bias (Teerikorpi 1998).

In the treatment of the classical Malmquist bias (Chap. 3), the space is assumed to be Euclidean, static, and transparent. The standard candles do not change with lookback time nor have any K-effect caused by the redshift of the spectrum across a finite wavelength band. In modern cosmology, some of these assumptions are not valid when one observes deep space. Here we briefly explain why the cosmological bias differs from the classical one.

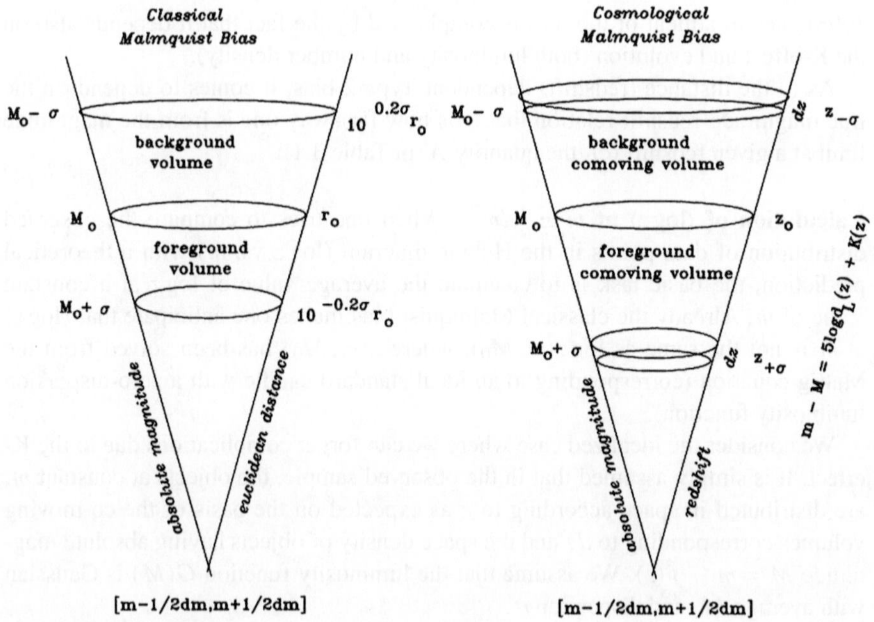

Fig. 7.3 (*left*) Classical Malmquist Bias: Symmetrical parts of the Gaussian luminosity function, observed through the magnitude window $[m - 1/2dm, m + 1/2dm]$, originate from different foreground and background volumes as determined by the r^3-law of Euclidean volumes and the r^{-2}-law of fluxes. (*right*) Cosmological Malmquist Bias in Friedmann models: Symmetrical parts of the Gaussian luminosity function, observed through the magnitude window $[m - 1/2dm, m + 1/2dm]$, originate from different comoving volumes. Fundamental theory gives the change of M with redshift z and the value of comoving volume, allowing one to calculate the average value of $\log z$ for standard candles with apparent magnitude m. (From Teerikorpi 1998; reproduced with permission © ESO)

In classical space the flux of light is proportional to r^{-2}, while the differential volume at the distance r increases as $r^2 dr$ (and similarly the number of sources if they are uniformly distributed). In a more general cosmological case, where it is convenient to replace r by the redshift z and consider comoving volumes, both of these proportionalities usually differ from the classical ones. This changes Type 1 Malmquist bias: when looking the universe through a narrow magnitude window, the shift in the average absolute magnitude of a standard candle is no longer the classical one. Even for a homogeneous spatial distribution the shift generally is not constant, as classically, but depends on the derived distance modulus.

The behaviour of the cosmological bias as a function of magnitude, as compared with the constant classical bias, depends on how the luminosity distance and the comoving volume increase with the redshift (Fig. 7.3). For example, for the Einstein-de Sitter model $\Omega = \Omega_{\mathrm{m}} = 1$, the volume increases at such a slow rate that the bias is progressively smaller than the classical $1.382\sigma^2$. If one ignores this bias or believes that it is constant, one might conclude that $\Omega < 1$ (see examples in Teerikorpi

1998). The treatment of the bias is complicated by the fact that it depends also on the K-effect and evolution (both luminosity and number density).

As to the distance (redshift) dependent Type 2 bias, it comes to depend on the true magnitude-redshift relation that tells how far away one is from the magnitude limit at a given redshift (cf. the quantity A' in Table 3.1).

Calculation of $\langle \log z \rangle$ at $m \pm \frac{1}{2}dm$ When one tries to compare the observed distribution of data points in the Hubble diagram ($\log z$ vs. m) with a theoretical prediction, the basic task is to calculate the average value of $\log z$ at a constant value of m. Already the classical Malmquist bias makes one anticipate that $\langle \log z \rangle$ at m is not the same as $\log z(m, M_0)$, where $z(m, M_0)$ has been solved from the Mattig equation (corresponding to an ideal standard candle with a zero-dispersion luminosity function).

We consider the idealized case where we can forget complications due to the K-effect. It is simply assumed that in the observed sample, the objects at constant m, are distributed in space according to z as expected on the basis of the co-moving volumes corresponding to dz and the space density of objects having absolute magnitude $M = m - f(z)$. We assume that the luminosity function $G(M)$ is Gaussian with average M_0 and dispersion σ.

The average value of $\log z$ for a fixed m can then be calculated as

$$\langle \log z_m \rangle = \frac{\int_0^\infty \log z dV(z) G(m - f(z))}{\int_0^\infty dV(z) G(m - f(z))}, \tag{7.74}$$

where $dV(z) = (dV(z)/dz)dz$. The derivative $dV(z)/dz$ can be obtained from expressions giving the comoving volume $V(z)$ for the considered Friedmann model. The function $f(z)$ comes from the magnitude–redshift relation (Eq. (7.69)). For a detailed discussion and applications, see Teerikorpi (1998).

7.5 The Hot Big Bang Scenario

The standard cosmological model includes Friedmann's model of expanding space as a general framework and the Hot Big Bang scenario according to which the universe began from a high-temperature high-density initial state (as reviewed, e.g., by Padmanabhan 2005).

At the present epoch the causally connected regions on the cosmic background radiation sky have angular sizes of a couple of degrees (as can be inferred from the particle horizon, Eq. (7.51), at the epoch of recombination $z \approx 1500$) giving rise to the "horizon problem" within the classical Friedmann model. The modern version of the big bang scenario includes the initial "inflationary" state of the universe with an exponential expansion rate (Guth 1981). This has the advantage that at the beginning $t = 0$ the radius of the universe was finite, and the horizon problem is solved. The second plus of the inflation is that the density of the universe exactly equals the critical value so that $\Omega_{tot} = 1$ and the spatial geometry is Euclidean. A third consequence is the quantum production of small initial density perturbations from which the large-scale structure of the universe then grew.

At high redshifts the universe is radiation dominated and the scale factor grows as $S(t) \propto t^{1/2}$. While space expands, the photon gas and matter are cooling down and the temperature decreases as $T \propto t^{-1/2}$. The temperature of the cosmic background radiation at redshift z is predicted to be $T(z) = T_0(1+z)$, where $T_0 = 2.726$ K is the present temperature.

After the first three minutes the primordial nucleosynthesis was completed, resulting in a helium-to-hydrogen mass ratio $^4\mathrm{He/H} \approx 0.25$. Modern observations of the CBR fluctuations are interpreted to give the baryon fraction in the universe $\Omega_{bar} h^2 = 0.024$. This means that the matter contents of the universe are mostly non-baryonic, and the composition of the dominating component is still unknown.

As the universe expanded, a time came when the matter density equaled the radiation density. At redshifts $z > 1500$ the CBR temperature was $T > 4000$ K and the intergalactic gas was ionized plasma. After the epoch of recombination $z < 1000$ the intergalactic gas was in a neutral state, and the first structures began to grow. The initial density-temperature fluctuation were imprinted into the "surface" of the last scattering, and these may be observed as an anisotropy in the background radiation.

Because the primordial nucleosynthesis predicts a low baryonic matter density $\Omega_{bar} \approx 0.05$, this led to the revised standard model dominated by non-baryonic dark matter. This new "cold dark matter" component was also needed for producing the observed strong clustering and at the same time preserving the low level of anisotropy of the background radiation (Chernin 1981).

The small primordial density fluctuations can be amplified by the gravitational instability leading to large-scale structure formation. The modern version of the standard model includes as its main mass components the dark energy $\Omega_{de} \approx 0.7$ and the non-baryonic cold dark matter $\Omega_{cdm} \approx 0.3$. The baryonic component is less than 5% and does not play an important role in the large-scale structure formation. The actually observed luminous matter contribution is $\Omega_{lum} \approx 0.005$, which also means that 90% of baryonic matter has an unknown chemical composition and hence the abundance of light elements can not be considered as a strong test of the big bang model.

Chapter 8
Classical Cosmological Tests

In science, an experiment is usually a way of investigating "cause-and-effect" processes in Nature by creating a special situation where we can vary the physical conditions and see how this affects the outcome of the process. Obviously, in large-scale physics such operations are quite limited and an experiment rather means a carefully planned set of observations directed to test a theoretical prediction. Modern physics views the observable universe as a place where the physical laws may be studied on the largest available scales. The cosmic laboratory has many features which complicate the work, including non-locality of observations and selection effects always putting their finger on observed relations.

8.1 Cosmological Tests

A world model can use information from nearby space as well as from the farthest depths of the cosmos. Even the Solar System or the Local Group of galaxies may serve as test fields of cosmological physics and be sources of fundamental discoveries having impact on the study of much larger scales.

8.1.1 From Low to High Redshifts

The local universe may be defined as a sphere around us where the redshifts of galaxies are less than about 0.1. The intermediate universe refers to the range $0.1 < z < 1$, and the deep universe only starts around $z = 1$.

The Local Universe In this important region, the cosmic distance scale and the Hubble law are established from well-studied galaxy samples. Galaxies of different types, their masses, luminosities, chemical compositions and stellar contents, as well as their spatial distribution, can be studied without relying heavily on specific world models. Euclidean geometry rules and evolutionary effects are still small.

Y. Baryshev, P. Teerikorpi, *Fundamental Questions of Practical Cosmology*, Astrophysics and Space Science Library 383,
DOI 10.1007/978-94-007-2379-5_8, © Springer Science+Business Media B.V. 2012

This "local calibration" of cosmic properties makes possible comparisons at larger distances and look-back times.

A special role is left for the *very local universe*, within 10 Mpc from the Milky Way, where the Hubble law emerges in an environment like a miniature copy of the structures on large scales, with groups, walls and voids. Kraan-Korteweg and Tammann (1979) speak about "The Local Volume".

The Intermediate Universe In the redshift range 0.1–1 detailed observations of galaxies become harder, while new types of rare, but very luminous objects start to appear: quasars and γ-ray bursts. The physics sufficient for describing the local universe requires corrections, and cosmological tests are expected to reveal first signs of the curvature of a Friedmann universe. The assumed homogeneity of the universe should be established here.

The Deep Universe The redshift ≈ 1, where deepness starts, roughly corresponds to the Hubble distance $c/H_0 \approx 3000h^{-1}$ Mpc. Strong cosmological influences on observations are expected in this promising territory for testing world models. Dramatically, the Friedmann model predicts that the angular sizes of identical galaxies first decrease with increasing distance (redshift) and after $z \approx 1$–2 start to grow. The look-back times are a large fraction of the age of the (big bang) world and evolutionary effects should be visible.

Understanding deep-space observations requires an adequate relativistic cosmological model. This is needed even for the practical tasks of calculating distances, sizes, and luminosities of celestial bodies.

8.1.2 Classical and Crucial Cosmological Tests

A cosmological model should give definite mathematical descriptions for observable quantities (redshift, flux density, angular size etc.) and predict relations between them. In fact, apart from its logical consistency and reasonable physics, it is only through such quantities that the model is linked to reality. To illustrate this we show, along with the Friedmann predictions, relations for the classical steady-state and some other models.

One may speak about two kinds of cosmological tests. *Crucial* tests serve to probe the validity of the basic assumptions of a world model, while *parametric* tests are used to estimate the model parameters.

Parametric Tests The classical tests were considered by Sandage (1961) as a program for the 5-m Hale telescope to derive the parameters of the Friedmann model by observations in the galaxy universe. Among them were

- the angular size–redshift relation,
- the magnitude–redshift relation, and
- the number count–magnitude relation.

Such classical parametric tests have been summarized in a pedagogic manner by Sandage et al. (1995) in *The Deep Universe*. For instance, by comparing the observations of "standard candles" at small and large distances with theoretical predictions one may estimate values of the model parameters.

Modern cosmology is actively engaged in the derivation of cosmological (Friedmann model) parameters from the anisotropies in the cosmic background radiation (Hu and Dodelson 2002; Spergel et al. 2007). Such data from $z \approx 1000$ valuably complement the results of the classical tests at redshifts $z \sim 1$. All cosmologies, also such which do not relate their parameters to the CBR fluctuations, may be tested using observations within the galaxy universe.

Crucial Tests Newton made his *Experimentum Crucis* when he studied the nature of light and colour with glass prisms at his home at Woolsthorpe, Linconlshire, during the years 1665–66 when the University of Cambridge was closed because of the plague. "Crucis" refers to the cross, or an experiment of crossroads, where one excludes an alternative and goes further.

We view as crucial those tests that concern the validity of the important initial hypotheses of cosmological models or their fundamental predictions. Therefore, crucial tests for the standard cosmology include at least:

- testing the validity of general relativity
- determining the matter distribution in space
- testing the reality of space expansion
- measuring the temperature of the background radiation at different redshifts
- determining the ages of the oldest objects

Even parametric tests, considered as a whole, may be crucial: there may be no combination of model parameters that satisfy them. Crucial tests are interesting, but generally difficult to perform: concerned with the frontiers they require the best technology of the time. The measurement of distances to spiral nebulae in the days of the Island Universe debate is an example.

8.2 A Résumé of Selection and Distortion Effects

Astronomical selection effects may be divided into two types: observer-related and physical (intrinsic and intervening) effects. The observer-related effects may be fundamental limitations or technical effects. An important limitation comes from our position in space and time. It gives rise, among other things, to various forms of the Malmquist bias (Chap. 3). Purely technical selection effects are caused by the limited observing capabilities.

8.2.1 K-Correction, Absorption and Evolution Effects

The K-effect appears when objects with different redshifts are observed with fixed filters, as we discussed in Sect. 4.1.1. Absorption of light by intervening medium makes the object fainter. It also contributes to selection: in a magnitude-limited sample the directions with less absorption are preferred.

When one constructs a volume-limited sample from a magnitude-limited one, one should note that the K- and A-effects change the effective magnitude limit. For instance, if the K-effect makes a galaxy fainter ($K_B(z) > 0$), the effective B-limit for such galaxies at the redshift z is brighter by this same amount. The K-effect can be quite large for E-galaxies in optical wavelengths (Fig. 4.1), a few magnitudes around $z \sim 2$. Therefore, unevolved galaxies similar to the local ones may be much fainter in the sky than expected from the distance alone, in the same band.

Evolution If a global cosmic time exists and galaxies were formed almost simultaneously in the past, as in the big bang model, then nearby galaxies around us are older than those that we see at high redshifts. "Standard candles" as a class may actually change in cosmic time and we must then add evolution to the theoretical prediction.

The K-correction depends only on the actual spectral distribution, while the evolution correction $E_i(z)$ is derived from phenomenological models of the evolution of the spectral energy distribution for different Hubble types (e.g., Poggianti 1997; Bruzual and Charlot 2003).

Evolution has often been invoked to explain differences between theory and observations. However, if one defines evolution in this way, one should have independent proof that the model itself is sufficiently close to reality—otherwise there is the risk of a circular argument.

8.2.2 Other Distortion Effects

The K-effect is an observer-related technical effect and evolution can be seen as a poorly known physical distortion effect. They interfere in a complicated manner with the Malmquist bias, both Type 1 and Type 2. Also other selection effects tend to distort relations between observable quantities.

Surface Brightness One factor influencing the detection of galaxies is their surface brightness, together with size. Very distant compact galaxies will look like a star on a photograph, while galaxies of "normal" luminosity but with low surface brightness have much of their surface below the sky background so they look small and are hard to detect. These effects limit our ability to have an unbiased view of the whole galaxy population (Arp 1965; Disney 1976) and can make one underestimate the number of low surface brightness (LSB) galaxies even in local surveys (McGaugh 1994).

How much of the galaxy universe has been missed because of the low surface brightness galaxies? Impey and Bothun (1997) stated that these "make up a significant amount of the luminosity density of the local universe. They contribute substantial but poorly determined amounts to the census of baryons and dark matter."

Now we have quantitative estimates. From a blind neutral hydrogen survey, Minchin et al. (2004) found that LSB galaxies contribute over half of the number density of galaxies, and they may make around 20% of the total mass density, but only 7% of the luminosity. These numbers may be still lower limits, but already suggest that the low surface brightness galaxies do not contribute radically to the cosmic mass.

Furthermore, the LSB galaxies are similarly clustered as the galaxies in general on scales larger than 5 Mpc, while there are differences on scales from 2 to 5 Mpc—they appear to favour the inner rims of filaments as defined by the other galaxies (Rosenbaum and Bomans 2004). Earlier surveys (e.g., Binggeli et al. 1990) already showed that dwarf and low surface brightness galaxies fall into the structures delineated by the luminous ones and that there is no evidence that these galaxies fill voids.

Anisotropic Radiation The relativistic beaming is an intrinsic physical effect, but it also gives rise to an observer-related selection effect, because we cannot move around the object. It appears for active galaxies that contain high-speed plasma jets. The effect has a strong influence on measured fluxes and angular sizes, and on number counts of objects (e.g., Padovani and Urry 1992). An analytical method for calculation of the probability distribution of the observed quantities in the double radio sources was developed by Baryshev and Teerikorpi (1995) who also discussed an age selection effect.

Gravitational Lensing This useful astronomical tool based on the deflection of light by gravitating masses, is also an intervening selection effect which distorts fluxes and angular sizes (e.g., Schneider et al. 1992; Refsdal and Surdej 1994). For instance, it has been suggested that the amplification caused by gravitational lensing may contribute to the Arp effect, i.e. the enhanced number of quasars around lower-redshift galaxies (Chap. 9).

To Conclude Cosmological tests require great care because one must take into account *all* essential selection effects, and all are not necessarily yet known. The various effects work together hiding true relations between cosmological quantities. This may produce strange behaviour of the quantities, perhaps interpreted as revealing some new physics, or at least may result in erroneous numerical values; the Hubble constant is a classical example.

8.3 The Angular Size-Redshift Relation

In an early study Tolman (1930) discussed angular sizes and light fluxes of distant nebulae in non-static space and considered their relation to redshift as a test of expansion. Hoyle (1959) had a lecture on the Steady State model at a radio astronomy meeting and suggested to study the angular size–redshift relation as a way to distinguish between alternative cosmologies. The idea simply requires a celestial body with a constant size and observable at various distances. It is also interesting to calculate the angular size–redshift relation for a co-moving structure, expanding together with the universe.

8.3.1 Angular Size in the Friedmann Model

As was told in Chap. 7, the angular size distance $l_{\mathrm{ang}}(z)$ is obtained from the "external" metric distance $l = S(t)\mu$ as $l_{\mathrm{ang}}(z) = l(z)/(1+z)$, while l is related to the usual metric distance r as $l = S(t)I_k(r/S)$ (Eq. (7.22)). The $l(z)$ relation (Eqs. (7.58)–(7.60)) for a two-fluid model and Mattig's Eq. (7.54) for a simpler pressureless dust universe) contains the cosmological parameters. In the test using "standard rods" at different redshifts, H_0 drops out.

Generally the angular size distance $l_{\mathrm{ang}}(z)$ has a maximum (or the angular size has a minimum) at some z whose value depends on the specific Friedmann model. For the Einstein-de Sitter model ($\Omega_{\mathrm{m}} = 1$, $\Omega_\Lambda = 0$), the angular size of a rod with the linear size d is

$$\theta_{\mathrm{E-deS}} = \frac{d}{R_{H_0}} \frac{(1+z)^{3/2}}{2[(1+z)^{1/2} - 1]}, \tag{8.1}$$

which has the minimum at $z = 1.25$ (the expression contains the Hubble distance $R_{H_0} = c/H_0$). For the current standard model with $\Omega_{\mathrm{m}} = 0.3$ and $\Omega_\Lambda = 0.7$, the minimum is expected to be around $z = 1.61$.

For structures expanding together with space one must divide the observed angle by $1+z$ (these were at redshift z smaller by a factor $1+z$).

8.3.2 Notes on Other Models

Some non-Friedmann models that have been discussed in the literature do not have any minimum in the angular size at a finite redshift.

Classical Steady-State Model The parameters of this famous model correspond to those of the zero-curvature Friedmann model with an exponentially growing scale factor $S(t) = S_0 e^{Ht}$ ($H =$ const.). The metric distance is $l(z) = R_H z$ (Narlikar 1993). Similarly as in Friedmann models, due to space expansion, the angular size distance is $(1+z)$ times smaller, and the angular size-redshift relation for a standard rod is

$$\theta_{\mathrm{ss}}(z) = \frac{d}{R_{H_0}} \frac{1+z}{z}. \tag{8.2}$$

For structures expanding with space the formula becomes simply $(d/R_{H_0})/z$.

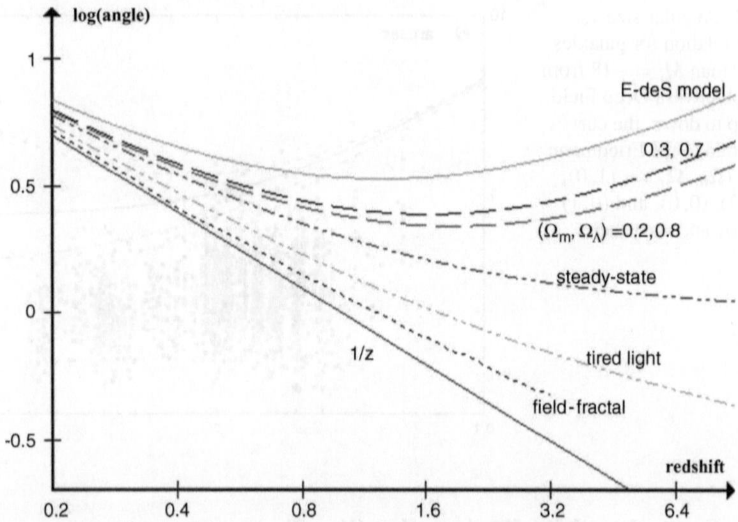

Fig. 8.1 Predicted (log) angular size vs. redshift relations for different models

Static Models In the usual *tired light model* (LaViolette 1986), a photon traversing a distance r suffers an energy loss given by $E(r) = E_0 e^{-\beta r}$. The redshift is then related to the distance as $z(r) = e^{(\beta r - 1)}$. In static Euclidean space, $r_{\text{ang}} = r$ and the angular size θ decreases with increasing redshift as

$$\theta_{\text{tl}} = \frac{d}{R_{H_0}} \frac{1}{\ln(1+z)}. \tag{8.3}$$

In the *field-fractal model* (Sect. 9.5.1) the cosmological redshift is due to a global gravitational effect of dark matter with a stochastic self-similar distribution (with the fractal dimension $D = 2$) up to the Hubble distance. The metric distance is given by $r(z) = R_{H_g} Y(z)$, where $R_{H_g} = c/H_g$. The function $Y(z) = W^{-1}(z)$, and $W(z)$ is defined by Eq. (9.41). This static Euclidean space should have $r_{\text{ang}} = r$, leading to the angular size–redshift relation

$$\theta_{\text{ff}}(z) = \frac{d}{R_{H_g}} \frac{1}{Y(z)}. \tag{8.4}$$

Figure 8.1 shows that the static models (with no "extra" $1+z$-term) predict a quickly decreasing angular size. In particular, the field-fractal solution leads to a relation not so far from the "$1/z$-law".

Similar "Euclidean" $\theta \sim 1/z$ behaviour (or not far from it) has been inferred in some cosmological constructions, where it is assumed that the standard rods are expanding along with space (e.g., Taganov 2008; Suntola 2011). As was noted above, this is valid also for the steady state model.

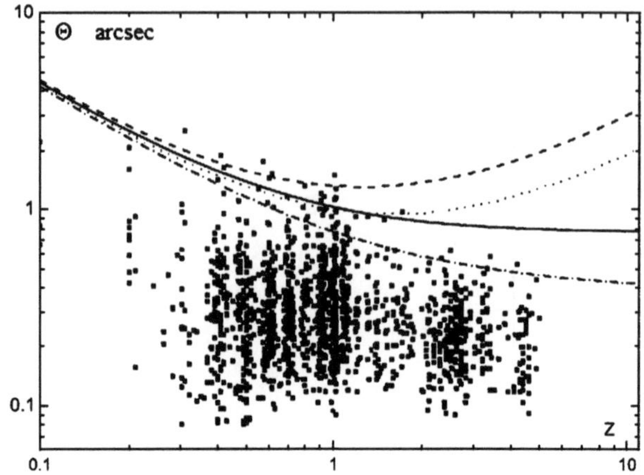

Fig. 8.2 Angular size vs. redshift relation for galaxies brighter than $M_i = -18$ from the Hubble Ultra Deep Field. From up to down, the curves correspond to the Friedmann models $(\Omega_{\mathrm{m}}, \Omega_\Lambda) = (1, 0)$, $(0.3, 0.7)$, $(0, 0)$, and $(0, 1)$ (Nabokov and Baryshev 2008a)

8.3.3 Optical and Radio Angular Size Test

The main problem for the angular size test is how to identify a true "standard rod" having the same size at different redshifts. There is also always some scatter, because the "rod" is generally viewed from different angles.

If size depends on luminosity, so that more compact objects are brighter (e.g., Nilsson et al. 1993), this may cause, because of an observational flux limit, a selection biasing the angular size-redshift relation. A similar effect appears, if the flux is relativistically amplified due to outward motions.

Optical Galaxy Sizes The early history of the angular size-redshift test, including the problematic definition of sizes for galaxies with diffuse edges, has been described by Sandage (1995a). Djorgovski and Spinrad (1981) constructed the $\theta(z)$ relation for a sample of elliptical galaxies with $z < 1$. Their data showed a behaviour not far from z^{-1} (Chap. 4).

Using data on the angular sizes of galaxies from the Hubble Ultra Deep Field and other deep surveys, Bouwens et al. (2004) and Ferguson et al. (2004) looked for the evolution of linear sizes within the standard cosmological model ($\Omega_\Lambda = 0.7$, $\Omega_m = 0.3$). Nabokov and Baryshev (2008a) considered different cosmological models for the UDF in the range $0.1 < z < 3.5$. Not surprisingly, the formally required evolution of galaxy sizes (Full Width at Half Maximum sizes) much depends on the choice of the model (Fig. 8.2).

These studies suggest that (within the standard cosmology) high-redshift galaxies were smaller than now. The trend with redshift (size $\sim H^{-1}(z)$; Ferguson et al. 2004) roughly agrees with the popular scenario where galactic disks are formed within dark-matter halos (Fall and Efstathiou 1980).

Sizes of Radio Sources The angular size $\theta(z)$ test in radio is uncertain due to selection effects in radio samples and possible size evolution of the radio com-

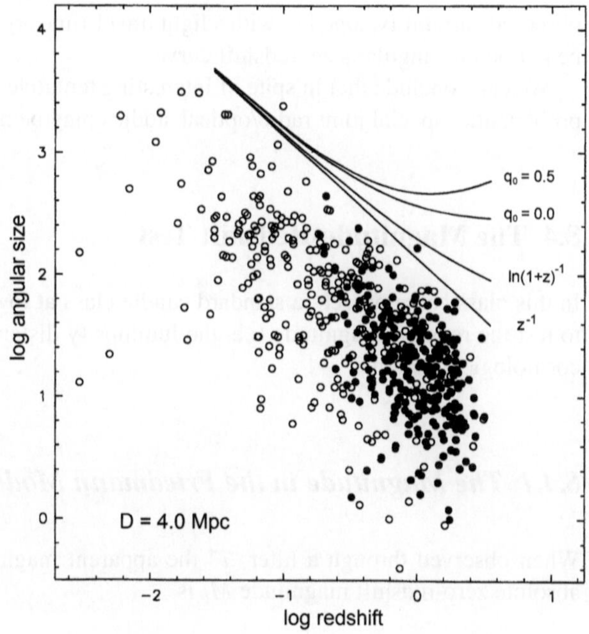

Fig. 8.3 Angular size vs. redshift relation for double radio sources in radio galaxies (*open circles*) and quasars (*dots*) (Courtesy of K. Nilsson)

ponents. For classical double radio sources Kapahi (1987) found an approximate $\theta \propto z^{-1}$ behaviour up to $z \approx 2$. Nilsson et al. (1993) confirmed that the outer double components in radio galaxies and quasars have this enigmatic $\theta(z)$ relation without minimum (Fig. 8.3). They interpreted this conflict with Friedmann models as a selection effect reflecting a radio power-size anticorrelation (rather than as a real size evolution).[1] For double-lobed quasars in the VLA FIRST survey, Buchalter et al. (1998) constructed a subsample with no (apparent) size evolution, and derived a $\theta(z)$ relation roughly consistent with a wide range of Friedmann models.

A compact VLBI radio source (with a typical size of about 50 pc, in comparison with the 500 kpc wide double radio sources) is produced by a relativistic jet from the central engine of a quasar. For such $\sim$ milliarcsecond structures Kellermann (1993) found an angular size-redshift relation in agreement with the Einstein–de Sitter model ($q_0 = 1/2$), whereas Jackson and Jannetta (2006) derived the quite different estimate $q_0 \approx -0.7$ from analogous data on ultra-compact radio sources. For $\Omega = 1$, the latter value corresponds to $\Omega_\Lambda \approx 0.8$ in the two-component Friedmann model, because $q = 0.5\Omega_m - \Omega_\Lambda$ (from Eq. (7.31)).

Wiik and Valtaoja (2001) used "knots" (shock fronts) appearing in high resolution radio maps. The size of a knot (of the order of 5 pc) may be derived from the

[1] Crawford (1995) defended the possibility that the deviation is due to a static cosmology. In this case there would be no evolution and no power-size anticorrelation so the selection effect discussed by Nilsson et al. (1993) would be of minor importance.

observed variability, together with a light travel-time argument, and the sources may be put on one angular size–redshift curve.

We can conclude that in spite of interesting tentative results the θ–z test remains problematic; special joint radio/optical studies may be needed.

8.4 The Magnitude-Redshift Test

In this classical approach, a standard candle class at low and high redshifts is used to test the redshift-magnitude (i.e. the luminosity distance) predictions of different cosmological models.

8.4.1 The Magnitude in the Friedmann Model

When observed through a filter "i" the apparent magnitude of an object with the absolute zero-redshift magnitude M_i is

$$m_i(z) = 5\log(l(z)(1+z)) + C_i(z), \tag{8.5}$$

where $C_i(z) = 25 + M_i + K_i(z) + A_i + E_i(z)$. If the K-, extinction, and evolution corrections are known for a standard candle class, Eq. (8.5) can be used to derive the redshift-luminosity distance relation $l_{\text{lum}}(z) = l(z)(1+z)$.

Special Cases For pure dust matter Friedmann model there is a small-z approximation often seen in early "pre-Mattig" studies

$$m_{\text{bol}} \approx 5\log z + 1.086(1 - q_0)z + C_i(z). \tag{8.6}$$

The first term in the right side of (8.6) gives the slope 5 of the usual Hubble relation. We see that only for the deceleration $q_0 = 1$ ($\Omega = \Omega_{\text{m}} = 2$), the m–z relation has the exact linear form $m \propto 5\log z$. If the universe is presently accelerating ($q_0 < 0$), the observed magnitude of a standard candle tends to be fainter than for a decelerating universe ($q_0 > 0$).

The "pure vacuum" flat model ($\Omega = \Omega_\Lambda = 1$) has the linear relation $l(z) = r(z) = R_{\text{H}_0} z$ (the same as in the steady state model!), hence the magnitude-redshift relation is $m_{\text{bol}} = 5\log(z(1+z)) + C_i(z)$.

8.4.2 Notes on Other Models

Similarly as for the angular size, it is of interest to see how the $m(z)$ relations predicted by unconventional models differ from that of the standard model.

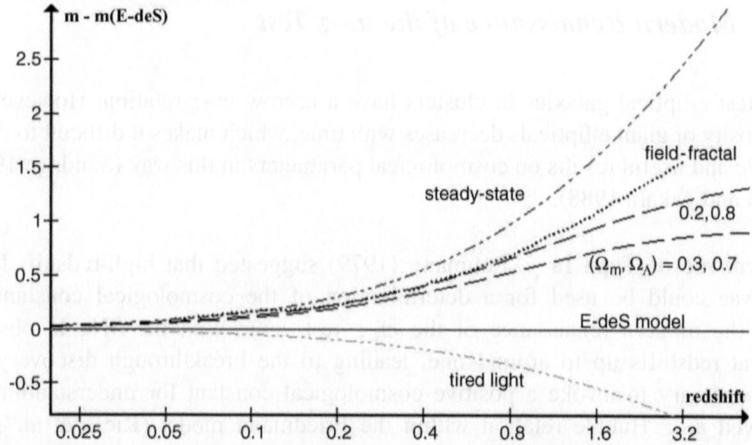

Fig. 8.4 Predicted magnitude vs. redshift relations for different models, presented as residuals from the Einstein-de Sitter model (the *horizontal line* at $m - m(\mathrm{EdeS}) = 0$)

The Classical Steady State The luminosity and metric distances are related similarly as in flat Friedmann models: $r(z) = l_{\mathrm{lum}}(z)/(1+z)$. Because $r(z) = R_{\mathrm{H}}z$, the magnitude-redshift relation is

$$m_i(z) = 5\log(R_{\mathrm{H}}z(1+z)) + C_i(z). \tag{8.7}$$

Static Models In the simplest tired light model with Euclidean static space (for instance, Jaakkola 1993), the magnitude of a standard candle depends on the redshift as follows:

$$m_i(z) = 5\log(R_{\mathrm{H}}\ln(1+z)) + 2.5\log(1+z) + C_i(z). \tag{8.8}$$

In the field-fractal model, the luminosity and metric distances are related as $r_{\mathrm{lum}}(z) = r(z)(1+z)$. This result includes the lost energy of individual photons and their diminished arrival rate due to gravitational time dilation. The magnitude-redshift relation then becomes

$$m_i(z) = 5\log(R_{\mathrm{H_g}}Y(z)(1+z)) + C_i(z). \tag{8.9}$$

Figure 8.4 shows how the predictions of different models differ from the Einstein-de Sitter model. Note that the static field-fractal model predicts a magnitude–redshift relation which is not so far from the Λ-dominated models (except for $z > 1.5$ where the theoretical relation probably breaks down), whereas the static tired-light model deviates strongly.

8.4.3 Modern Renaissance of the *m–z* Test

Brightest elliptical galaxies in clusters have a narrow m–z relation. However, the luminosity of giant ellipticals decreases with time, which makes it difficult to derive reliable and useful results on cosmological parameters in this way (Sandage 1995b; Yoshii and Takara 1988).

Supernovae of Type Ia Tammann (1979) suggested that high-redshift Ia supernovae could be used for a determination of the cosmological constant. Indeed, the modern renaissance of the m–z test came with the SNe Ia observations at redshifts up to around one, leading to the breakthrough discovery that it is necessary to invoke a positive cosmological constant for understanding the observed m–z Hubble relation within the Friedmann model (Riess et al. 1998; Perlmutter et al. 1999): around $z = 1$ the supernovae are about 0.7 magnitudes fainter than expected from the Einstein-de Sitter model (cf. Figs. 1.3 and 8.4).

The data are best described with a flat model containing about 70% dark energy and 30% of gravitating (dark and luminous) matter. Within the standard model the flatness condition comes, apart from inflation arguments, from the angular spectrum of the fluctuations in the thermal background radiation, where the location of the first acoustic peak suggests $\Omega \approx 1$.

After these pioneering efforts, several teams have provided a large increase in the total number of measured SNe Ia. At high redshifts ($z > 1$), the Hubble Space Telescope has been used for high-precision optical and infrared follow-up of supernovae discovered from the ground and to carry out both search and follow-up from space. Extensive ground-based projects have been adding hundreds of low-redshift ($z < 0.3$) supernovae to the Hubble diagram. The number of well-measured objects beyond $z \approx 1$ is smaller (around twenty in 2010). Kowalski et al. (2008) provided a framework to analyse the supernova datasets in a homogeneous manner and created the "Union" SNe Ia compilation of the world's published SNIa data (for an analysis of the updated "Union2" compilation, see Amanullah et al. 2010).

8.4.4 Other Attempts to Interpret the SNIa Data

Even though the Lambda term is now standard in cosmological equations, it is no wonder—in view of its unknown physical nature[2]—that also other ways have been searched for in order to explain the magnitude–redshift relation for the distant supernovae. And even if minor, such effects should be taken into account in accurate studies of dark energy and its time dependence.

[2] The vacuum and "antigravitating" dark energy have a quantum nature. At present they enter general relativity on a phenomenological level only. As parts of any new quantum gravity theory these entities will likely affect our understanding of dark matter and large-scale structure formation.

We restrict ourselves to explanations based on possible problems in the standard candles. Quite another kind of approach admits that the Ia supernovae are good distance indicators, but says that it is problematic to take the magnitude (luminosity distance) vs. redshift relation at its face value and to interpret it in terms of the ideal homogeneous world model. The "acceleration" might be just a consequence of the assumed homogeneity (e.g., Mattsson 2010; Célérier et al. 2010).

Selection Effects? The supernovae Ia have some scatter even after the correction for the dimming rate. As is the case with Cepheid stars whose variation amplitude enhances the usual magnitude selection effect (Chap. 3), the required detection of a supernova when it is still brightening causes pressure towards brighter than average objects and increasingly so at higher redshifts. The correction for this selection would mean making the observed magnitude still fainter by some amount and one does not expect to explain in this way the "too" faint magnitudes of high-z supernovae. Minor effects due to this Type 2 selection effect and gravitational lensing have been discussed, e.g., by Kowalski et al. (2008) and Sarkar et al. (2008).

Intergalactic Dust? Extinction due to intergalactic dust has been proposed to be the agent dimming the Ia supernovae fluxes beyond $z = 0.5$. This alternative can be observationally constrained since it should be accompanied by reddening, unless the dust is unusual ("grey": extinction independent on wavelength) and suitably distributed at different redshifts (Aguirre 1999). In its original form, where the density of the dusty medium follows that of non-relativistic matter ($\sim (1+z)^3$), this model is contradicted by SNIa data beyond $z \approx 1.0$ where the supernovae are too bright (Riess et al. 2004). In the grey dust model of Robaina and Cepa (2007) the density follows the stellar formation rate density evolution (with some delay due to the injection of the dust into intergalactic space). Such a "replenishing grey dust" model is flexible enough for explaining the run of the Hubble relation. Independent evidence would be needed for the presence of the intergalactic grey dust. Minor effects from reddening-producing intergalactic dust have been discussed by Ménard et al. (2010).

Problems with the Candles? Could the average properties of Ia supernovae be changing over cosmological times? After all, their use is a purely empirical method to determine the luminosity distance. Hence, along with collecting SNe Ia events, it is important to continue studying their physical nature as thermonuclear explosions of carbon-oxygen white dwarfs (Hoyle and Fowler 1960) which have in some way gained mass from their surroundings. Theoretically, these explosions are not yet fully understood.

In fact, the Ia supernovae do not seem to form a uniform class. One speaks about two components in their population: the "prompt" one proportional to the instantaneous host galaxy star formation rate (in S galaxies) and the "delayed" component, delayed by several Gyr and proportional to the total stellar mass (in E galaxies). Thus some of the photometric properties which are used to derive their luminosity distance depend on the host galaxy type (Hamuy et al. 1995). The more luminous prompt population with broader light curves and younger progenitor stellar population dominates at high redshifts where the star formation rate was higher. The resulting small redshift-dependent shift (~ 0.05 mag) in the Hubble diagram, even if not fully controlled (Howell et al. 2007), cannot change the general run of the Hubble relation.

Another sign of non-uniformity among the Ia supernovae came from the different expansion speeds of the ejected material (Branch 1987; Benetti et al. 2005). Maeda et al. (2010) suggested that this diversity is not intrinsic, but caused by asymmetric explosions viewed from random directions. The asymmetry would result from the ignition occurring at an offset from the centre of the white dwarf progenitor. This result also suggests that the expansion speed diversity does not undermine the use of SNIae as standards.

Confirmation from Other Objects? Of course, one would like to have other types of standard objects observable from low to high redshifts and based on other physical mechanisms. Already before the supernova breakthrough, Jackson and Dogdson (1997) used 256 ultracompact radiosources in the redshift range from 0.5 to 3.8 to construct the redshift–angular size diagram (first done by Kellermann 1993 using 79 sources). Their analysis excluded the Einstein-de Sitter model and it was stated that the best-fit flat model has $\Omega_\Lambda = 0.8$ (for more recent results, see Jackson and Jannetta 2006). Also, it is interesting that the suggested class of the optically most luminous quasars (Teerikorpi 1981b, 2000) appeared best when the magnitudes were calculated using the flat Friedmann model with $\Omega_\Lambda \approx 0.7$, as analysed using the cosmological Malmquist bias approach (Teerikorpi 2003).

8.5 Galaxy Counts and the Background Radiation

The counts of stars as a function of their apparent magnitude is a classical way to study our Milky Way. Since the 1920s, the counts of galaxies have been recognized as a way to study the structure of the galaxy universe. They can also be compared with the optical background radiation originating from all galaxies. However, if the galaxy population evolves in a complicated manner over cosmic times (as it does according to the standard view), the counts are a problematic way to determine the cosmological model. The number-magnitude relation, especially for faint magnitudes, is rather a test of the evolutionary processes within a given cosmological model, than a test of the model itself.

Apart from evolution, the counts of faint galaxies can be much affected by selection. In a Friedmann world the surface brightness rapidly gets dimmer with increasing redshift and many high-z galaxies remain undetected below the threshold surface brightness in a galaxy survey (Yoshii 1993). In order to study all such problems, one needs counts from bright to faint magnitudes in several photometric bands and separately for different galaxy types (cf. a case study by Cohen et al. 2003). Here we describe briefly how to predict the counts and the background radiation in the ideal situation of non-evolving and uniformly distributed light sources.

8.5.1 The Number-Redshift Relation

If the objects fill the space uniformly, their numbers directly reflect the volume that they occupy and one may derive the growth of the cosmological volume with distance (redshift). The number-redshift relation is also a step in the derivation of the number-magnitude relation.

For different geometries, the known relations $r(z)$ lead to corresponding volume-redshift $V(z)$ relations (Sect. 7.4.5).

Friedmann Models If the current number density of objects is n_0, the number-redshift relation can be found by using the known $r(z)$ relation. After integration over the redshift z and over the whole sky (the solid angle $= 4\pi$) one derives

$$N(z) = 4\pi n_0 V(r(z)). \tag{8.10}$$

Here the definition of the comoving distance r automatically accounts for space expansion, so the number density is taken for the present epoch.

For some particular cases there are simple analytical formulas, e.g., for $(\Omega_m, \Omega_\Lambda)$ equal to $(0,0)$, $(1,0)$, $(0,1)$.

Other Models In the steady-state model with its constant density n the number–redshift relation for a uniform distribution is simply (Narlikar 1993):

$$N(z) = 4\pi n \left(\frac{c}{H_0}\right)^3 \left[\ln(1+z) - \frac{3z^2+2z}{2(1+z)^2}\right]. \tag{8.11}$$

In the field-fractal model the space is flat and static, and the metric distance is $r(z) = R_{H_0} Y(z)$. Hence the volume and the number behave as

$$V(z) = \frac{4\pi}{3} R_{H_0}^3 [Y(z)]^3, \qquad N(z) = \frac{4\pi}{D} R_{H_0}^3 n_0 \left(\frac{r_0}{R_{H_0}}\right)^{3-D} [Y(z)]^D. \tag{8.12}$$

For the $N(z)$ relation, the number density is taken to be $n(r) = n_0 (r/r_0)^{D-3}$.

Prospects for the $N(z)$ Test For small redshifts the number–distance relation $N(r)$ is sensitive to the spatial distribution, but not to the cosmological parameters. Systematic redshift surveys give now the possibility to measure the number–redshift relation. However, even the extensive SDSS survey (Chap. 11) is still affected by large fluctuations due to the presence of huge structures around us, like the Sloan Great Wall (Sylos Labini et al. 2009b).

8.5.2 The Number-Magnitude Test

The conclusion by Sandage (1988a) about the classical $N(m)$ test is still relevant: one may expect a gross agreement of the slope at bright magnitudes, but the uncertain evolution corrections, the lack of detailed homogeneity in the spatial distribution and the relative insensitivity to the curvature undermines its use as a check of the standard model.

Friedmann Models It is assumed that there are objects with a known luminosity function uniformly distributed in space. Let $dA(m, z)dmdz$ be the number of objects with redshift $[z, z + dz]$ contributing to the counts per magnitude bin around m_i, i.e.

$$dA(m_i, z)dmdz = \Phi(M_i)\frac{dV}{dz}dm_i dz, \tag{8.13}$$

where $\Phi(M_i)$ is the luminosity function and M_i is the i-filter absolute magnitude ($M_i = m_i - 5\log(l(z)(1 + z)R_{H_0}(\mathrm{Mpc})) - 25 - K_i(z)$, no luminosity evolution, space fully transparent). The differential number counts (per one steradian) in the i band magnitude bin dm_i are obtained by integrating (8.13) over z:

$$N_d(m_i) = \int_0^{z_{\max}} dA(m_i, z)dmdz, \tag{8.14}$$

where $z_{\max}$ is the maximum redshift for the objects.

Other Models Even though the galaxy counts may not deliver the parameters of the Friedmann model, it may be useful to study the predictions of non-Friedmann models in order to see how large differences would be expected. For the steady-state model (with no evolution correction) and the static field-fractal model (where the question of evolution is open) the $N(m)$ predictions are also derived by using the $N(z)$ and $m(z)$ relations in the same way as for the Friedmann model.

Modern Data on Faint Galaxy Counts At small and intermediate redshifts ($z < 0.2$) the galaxy and space-time evolution can be neglected. In principle it should be possible to verify experimentally the relation $\alpha = D/5$ for $N(m) \propto 10^{\alpha m}$. To do this properly, one has to analyse luminosity properties together with correlation properties in volume-limited subsamples, in order to avoid any bias due to observational selection effects (Sylos Labini et al. 1998; Gabrielli et al. 2005).

It has been found (e.g., Tyson 1988; Lilly et al. 1991) that for the B-magnitude the slope of the counts is $\alpha \approx 0.6$ in the range $15 < m < 18$ and $\alpha \approx 0.45$ for $m \sim 19$ up to $m \sim 27$. The fact that $\alpha \approx 0.6$ at intermediate magnitudes has been seen as a proof of homogeneity of matter distribution (Peebles 1993), while the subsequent change of slope has been interpreted as an effect of galaxy evolution, or a combined effect of galaxy and space-time evolution (see Yoshii 1993).

In the infrared K-band $\alpha \sim 0.67$ for $12 < K < 16$ and the slope changes at $K \sim 18$ to $\alpha \approx 0.26$, in contrast with the B-band counts (e.g., Gardner et al. 1993).

Several sources of bias and selection should be taken into account in the study of the observations of faint galaxies (Yoshii 1993; McGaugh 1994), including the Eddington bias in the counts (Teerikorpi 2004). If one defines the evolution as the deviation from the Euclidean behaviour ($\alpha = 0.6$) there is the a priori assumption that the distribution becomes homogenous and that locations are not correlated with luminosities; this should be tested independently.

8.5.3 The Total Background Radiation Due to Galaxies

The sum of the photons arriving from different distances (redshifts), contains information about all populations of objects during the whole history of the universe. The background radiation was first detected in the microwave, radio, and X-ray bands. Its observation in the optical wavelength region (and in the UV and IR wavebands) has been much more challenging (the extragalactic background light, EBL). Also, in this range the evolution of galaxies tends to mask the predicted differences due to different cosmological models. We illustrate the calculation of the background intensity for the Friedmann model.

Integrated Radiation Consider the flux density from an object with a power-law spectrum $L_{\nu_{\rm em}} \sim \nu_{\rm em}^{-\alpha}$ at the observed frequency $\nu_{\rm obs}$ (cf. Eq. (7.66))

$$F_{\nu_{\rm obs}} = \frac{L(\nu_{\rm obs}(1+z))}{(1+z)^3 4\pi l_{\rm m}^2} = \frac{L(\nu_{\rm obs})}{(1+z)^{1+\alpha} 4\pi l_{\rm m}^2}. \qquad (8.15)$$

For a uniform spatial distribution the number of sources per steradian in a spherical shell dr is $dN = nl^2 dr$ and the background intensity $I(\nu_0)$ is

$$I(\nu_0) = \int F(\nu_0) dN = \frac{L(\nu_0) n_0}{4\pi} \int_0^\infty (1+z)^{-(1+\alpha)} dr. \qquad (8.16)$$

Now we can use the expression for the $dr - dz$ relation in the form $dr = cdz/H(z) = cdz/[H_0(\Omega_0 z + 1)^{1/2}(1+z)]$, where the last equality is for the simple dust universe (Eq. (7.45)). In this case

$$I(\nu_0) = \frac{c}{H_0} \frac{L(\nu_0) n_0}{4\pi} \int_0^\infty \frac{dz}{(\Omega_0 z + 1)^{1/2} (1+z)^{2+\alpha}}. \qquad (8.17)$$

In Friedmann models the integral in (8.17) converges for all $\alpha > -1.5$, which is the case for any realistic spectrum. To an order of magnitude the background intensity is that originating within the Hubble radius:

$$I(\nu_0) \approx \frac{c}{H_0} \frac{L(\nu_0) n_0}{4\pi}. \qquad (8.18)$$

In Newtonian cosmology without redshift the integral is infinite (Olbers!):

$$I(\nu_0) = \frac{L(\nu_0) n_0}{4\pi} \int_0^\infty dr. \qquad (8.19)$$

Analogous calculations can be done within other cosmological models.

Detection The measurement of the optical background light is very difficult (Mattila 1990, 2003; Bernstein et al. 2002, 2005), because its expected intensity ($\sim 10^{-9}$ erg cm^{-1} s^{-1} sr^{-1} Å^{-1}) is a tiny fraction ($\sim 1\%$) of the total sky brightness due to other sources of diffuse radiation, such as the zodiacal light, the airglow, the integrated galactic starlight, and diffuse galactic light. Up to now there is no generally accepted measurement of the EBL. The upper limits obtained do not deviate much from the expectation (8.18), which suggests that astronomers have not missed a very large fraction of luminous galaxies in their surveys, in agreement with the studies of the low surface brightness galaxies (Sect. 8.2.2).

8.6 Classical Crucial Tests for the Friedmann Model

Crucial tests on the reality of the expansion of space were formulated years ago, but possibilities to perform some of them have emerged only recently.

One may ask why such tests are so relevant, amid a lot of evidence, when taken together, pointing at expansion. For instance, the Hubble expansion time $T_H \approx 15 \times 10^9$ yrs is not far from the ages of the oldest globular clusters and the age derived from chemical elements. The thermal background radiation is natural to understand as a result of the compressed and hot past state.

Nevertheless, the paradigm of the big bang cosmology is basically that of expansion—expanding space (and the Hubble law) follow elegantly from the general relativity-based Friedmann model. Furthermore, the new concept of expanding space is related to the general relativistic view of space, time and gravitation, hence its study is fundamental. To prove that space is really expanding, would mean strong support for general relativity (or its various modifications; Clifton et al. 2011). In Sandage's (1995a) list of 23 astronomy problems for the next decades, "Is the expansion real?" was the first one in cosmology.

Here we describe the surface brightness-redshift test, the redshift-time dependence test, the time dilation test, and the temperature-redshift test. Also the ages of high-z galaxies are crucial for finite-age cosmologies.

8.6.1 The Tolman-Hubble Surface Brightness-Redshift Test

Hubble and Tolman (1935) proposed a test of the nature of the cosmological redshift by measuring the surface brightness of near and distant galaxies. After attempts with ground-based observations, it was applied to distant (up to $z = 0.92$) elliptical galaxies measured by the HST by Sandage and Lubin in 2001 and by Sandage (2010) who gives references to earlier work.

The surface brightness B is expected to change with the redshift in the following ways in a few well-known cases:

- classical space without redshift: $B \propto (1+z)^{-0}$,
- tired light effect: $B \propto (1+z)^{-1}$,
- expanding space: $B \propto (1+z)^{-4}$.

The expected effect in expanding space, even though strong, is not easy to detect. Sandage (1995b) gives a good description of the problems involved. Indeed, there are no extended surfaces (galaxies, their systems) which can be regarded as equally bright intrinsically at different redshifts and whose surface brightness can be simply defined and measured. In galaxies the surface brightness decreases towards the edge. One always measures some average brightness and the procedure should be such that it catches equivalent regions both for large nearby objects and for the remote ones which are much smaller in angular size and whose faint outer parts have fallen below the detection limit and innermost parts are beyond the angular resolution.

In Chap. 4 we mentioned the Petrosian size measure for elliptical galaxies. Based on the surface brightness profile, this method can, in principle, pinpoint intrinsically same linear radii for identical elliptical galaxies, nearby and distant. The surface brightness dimmening does not influence the resulting size. Using this formalism Sandage has searched for the expected dimmening in giant elliptical galaxies at redshifts up to about 0.9.

There are a few further problems in this test (Sandage 1995b, 2010). Elliptical galaxies are not identical in luminosity and surface brightness; bigger ones have fainter surfaces within some fixed diameter, which must be taken into account when one compares galaxies with similar Petrosian sizes. The brightness of the elliptical galaxies evolves over the cosmic time and one must include the evolution correction in the test.

Lubin and Sandage (2001) and Sandage (2010) have derived a behaviour close to that expected for expanding space. Some uncertainty still remains because of the mentioned problems; one would also like to see the test for different types of objects. The result by Sandage (2010) may be given using the exponent n in the $B(z)$ relation $(1+z)^{-n}$. They derived $n = 2.80 \pm 0.25$ (filter R) and $n = 3.48 \pm 0.14$ (filter I). An evolution of the surface brightness is needed in order to obtain $n = 4$ from these data. It is indeed regarded that the giant E galaxies were brighter in the past, due to the passive evolution of the stellar population, roughly as needed. Sandage's (2010) conclusion was: "The Tolman prediction is verified. The expansion would seem to be real."

It should be noted the calculation also involves a dependence on the Friedmann model, and Moles et al. (1998) have emphasized that both the size calculations and evolution give uncomfortable extra freedom in the interpretion of the observations. It would be interesting to repeat the analysis using some other cosmological models. For instance, the static field-fractal model would seem to predict $n = 2$ (without evolution), but in that model a fixed angular size corresponds to a larger linear size than in the Friedmann models.

8.6.2 Maximum Age–Redshift Test

Friedmann models possess an exact relation between the time since the beginning $T(z)$ and the redshift z. The age of an object, now observed to have the redshift z and born at $z_{\mathrm{form}} > z$, is given by $t_{\mathrm{obj}} = T(z) - t(z_{\mathrm{form}})$.

Table 8.1 Ages in Myr as a function of redshift for $\Omega = 1$ universes with ($\Omega_\Lambda = 0.7$) and without dark energy. $H_0 = 70$ km sec^{-1}/Mpc

z	$T^{\mathrm{de}}_{\mathrm{back}}$	$T^{\mathrm{de}}(z)$	T_{back}	$T(z)$
0	0	13470	0	9310
0.5	5040	8430	4240	5070
1	7720	5750	6020	3290
3	11360	2110	8150	1160
5	12320	1150	8680	630
7	12720	750	8900	410
10	13000	470	9060	250
12	13100	370	9110	200

Table 8.1 gives the "look-back time" T_{back} and the "elapsed from the big bang time" $T(z)$ for a few redshifts, in the flat model with $\Omega = \Omega_{\mathrm{m}} = 1$ using $H_0 = 70$ km sec^{-1}/Mpc, and similarly for an accelerating universe with the Lambda-term: $\Omega_{\mathrm{m}} = 0.3$, $\Omega_\Lambda = 0.7$. In these models, the present age of the universe is 9.3 and 13.5 milliard years, respectively. To adjust these ages to another value of the Hubble constant, multiply them by $70/H$.

The age paradox has always followed the expanding world model: there obviously cannot be objects older than the universe itself. In the 1930s the Hubble time was estimated to be less than 2 milliard years, shorter than the age of the earth. Later globular clusters became critical objects with high ages calculated from stellar evolution. These ages have now come down below about 14 milliard years, admissible for the Friedmann model. Nowadays, when the highest measured redshifts have exceeded 8 (and a case has been made for an object with $z \sim 10$ by Bouwens et al. 2011), a related question concerns the high-z galaxies, their stellar populations, and also the supermassive objects in the centres of active galaxies. Has there been sufficient time for their formation?

8.6.3 *Sandage's Redshift-Time Dependence Test*

Sandage (1962) considered how the redshift changes with time, when a distant galaxy is observed. A general formula for the very slow change expected in an expanding universe was derived by McVittie in the Appendix of Sandage (1962) for a time interval short compared with the age of the universe. Nowadays this effect has been discussed in the context of Friedmann models with dark energy (Loeb 1998; Gudmundsson and Björnsson 2002).

The redshift change. The radiation from a source arrives at times t_0 and $t_0 + \Delta t_0$, having left the source at t and $t + \Delta t$. The observed redshifts are

$$z = \frac{S_0}{S} - 1, \qquad z + \Delta z = \frac{S(t_0 + \Delta t_0)}{S(t + \Delta t)} - 1. \tag{8.20}$$

Hence, to the first order of magnitude, taking into account that $S(t+\Delta t) = S(t) + \dot{S}(t)\Delta t$, $\Delta t/\Delta t_0 = S/S_0$, and $\Delta t_0 \to 0$, we get

$$\left(\frac{dz}{dt}\right)_0 = (1+z)H_0 - H(z), \tag{8.21}$$

where $H(z)$ is given by Eq. (7.45).

For a two-component universe ($\Omega = \Omega_m + \Omega_\Lambda$)

$$\left(\frac{dz}{dt}\right)_0 = -(1+z)H_0\left[(\Omega_m(1+z) + (1-\Omega_{tot}) + \Omega_\Lambda(1+z)^{-1})^{1/2} - 1\right]. \tag{8.22}$$

For a dust universe ($\Omega_\Lambda = 0$), all redshifts are decreasing at the present time. If $\Omega_\Lambda > 0$, the redshifts may be either increasing or decreasing, depending on the value of z (nearby objects: z increasing; distant objects: z decreasing). Only now the time is ripening for this difficult dz/dt-test. In terms of radial velocity, the predicted change dv/dt is ~ 1 cm s^{-1}/yr. This may be within the reach of the future 42 m E-ELT telescope, though even then requiring some 4000 hours of observing time over a period of 20 years (Liske et al. 2008).

In the steady state model the redshift increases with time: $c\dot{z} = +cH_0 z$. For the rare case of an empty universe $\dot{z} = 0$; now there is no deceleration and for any two objects the ratio of scale factors remains constant during the expansion. Because $1 + z = S_0/S_{em}$, the redshift is constant, too. Also for static models the prediction is $\dot{z} = 0$, making this test a powerful tool to distinguish between fundamentally different world models.

8.6.4 Wilson's Supernova Time Dilation Test

Wilson (1939) suggested supernovae as a test of the nature of the cosmological redshift: in an expanding universe the light-curve of a supernova occurring in a distant galaxy should appear to be expanded along the time axes in the ratio $(1 + z) : 1$ with respect to the standard local light-curve. This time delay test was also discussed by Rust (1974) and Teerikorpi (1981c). Recent observations of the supernovae Ia have finally given the possibility to perform the test (Leibundgut 2001; Goldhaber et al. 2001).

The observed width τ_{obs} of the supernova light-curve can be written as

$$\tau_{obs}(z) = \tau_{em}(1+z)^p, \tag{8.23}$$

where $p = 1$ for the Doppler, gravitational and space expansion Lemaître effects, while $p = 0$ for the tired light. For the Friedmann and steady-state models $p = 1$

because of space expansion, and also for models based on gravitational redshift $p = 1$.

The light-curves for 35 Type Ia supernovae with redshifts up to $z \approx 1$ were analysed by Goldhaber et al. (2001). They derived the dilation parameter $p = 1.0 \pm 0.1$. Another study by Blondin et al. (2008) measured the spectral ages in the supernova rest frame. Comparison with the observed time led again to the $(1+z)^{-1}$ factor expected for expanding space.

The time dilation test provides good evidence against the tired-light hypothesis, but it can not distinguish between expanding space models and models involving cosmological gravitational redshift. However, it is interesting to mention that Hawkins (2010) could not see the time dilation in quasar light curves, though he notes that this result may tell more about the physics of quasars than about the cosmological effect searched for.

8.6.5 Background Radiation Temperature vs. Redshift

An important cosmological test is the measurement of the cosmic background radiation temperature at different redshifts. In the Friedmann universes with a high initial temperature of matter and radiation, the predicted temperature depends linearly on redshift, i.e. $T(z) = T_0(1+z)$. In the steady-state model the temperature does not evolve. Within static models the evolution of the background radiation is not known.

The Bahcall-Wolf CBR Temperature-Redshift Test Bahcall and Wolf (1968) suggested to use fine structure transitions for measuring physical properties of the interstellar and intergalactic media. In particular the temperature of the photon exitation may be measured at high redshifts in the absorption line systems of quasars, and Bahcall and Wolf viewed this as a possible test for the existence at large redshifts of the cosmic black-body radiation.

Using high-resolution spectroscopy at large telescopes, the exitation temperature of the thermal background radiation has been estimated from atomic fine-structure transitions in cool absorption-line systems along the line of sight to high-redshift quasars (e.g., LoSecco et al. 2001; Molaro et al. 2002). The first results gave the value $T_{\mathrm{CBR}} = 11 \pm 1.6$ K at redshift $z = 3.025$, while the predicted value is $4.025 \times 2.726 = 11.0$ K.

A major problem for such estimates is how to take into account the different competing local excitation processes. Furthermore, in this method the radiation of the cold dust can not be separated from the background radiation having a similar temperature.

CBR Temperature-Redshift Test by the SZ Effect The determination of the temperature of the cosmic background radiation from the Sunyaev-Zeldovich effect,

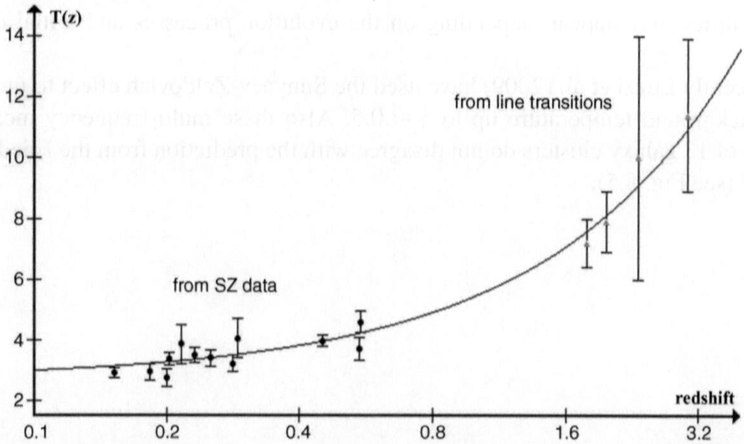

Fig. 8.5 Measurements of the background radiation temperature at high redshifts (based on the data in Luzzi et al. 2009). The data points come from the SZ effect ($z < 1$) and from the line transition observations. The *curve* gives the relation $T = T_0(1+z)$

which involves the scattering of the CBR photons from the hot gas inside a cluster of galaxies (Zeldovich and Sunyaev 1969a, 1969b), was suggested by Fabbri et al. (1978) and Rephaeli (1980) who saw it as a test of the universality of the radiation. Its use as a crucial cosmological test at high redshifts was emphasized by Baryshev (1992b) and discussed by Horellou et al. (2005) having in view the multifrequency SZ observations to be made by the Planck satellite.

The basic idea comes from the fact that the intensity of the observed extra radiation changes its sign at a frequency ν_0 that can be expressed in the form of parameter x as

$$x_0 = \frac{h\nu}{k_B T_{\mathrm{CBR}}} = 3.830. \tag{8.24}$$

If the temperature of the CBR changes with redshift as

$$T_{\mathrm{CBR}}(z) = T^0_{\mathrm{CBR}} \times (1+z)^{\delta}, \tag{8.25}$$

where $T^0_{\mathrm{CBR}} = 2.726 \pm 0.002$ is the present epoch temperature, then the crossover frequency will change as

$$x_0(z) = \frac{h\nu_0(1+z)}{k_{\mathrm{B}} T^0_{\mathrm{CBR}}(1+z)^{\delta}} = 3.830 \times (1+z)^{1-\delta}. \tag{8.26}$$

For Friedmann models $\delta = 1$ and there is no change of the frequency $x_0 = 3.83$. For the steady-state model $\delta = 0$, hence the characteristic zero intensity frequency will increase with the redshift as $x_0 = 3.83(1+z)$. Within static models different possibilities may appear, depending on the evolution processes and initial conditions.

Recently Luzzi et al. (2009) have used the Sunyaev-Zeldovich effect to measure the background temperature up to $z \approx 0.5$. Also these multi-frequency measurements of 13 galaxy clusters do not disagree with the prediction from the Friedmann model (see Fig. 8.5).

Chapter 9
Constructing Universes: A Gallery of Ideas

Amidst the success story of standard cosmology one should not lose sight of healthy reminders of why also off-the-mainstream ideas can be useful and have the right to exist in contemporary science. *First*, the finite observable part of the possibly infinite universe does not allow one to test directly the initial hypotheses on the universe as a whole. The possibility of a major reform is not excluded. *Second*, even the known phenomena may have different interpretations, each corresponding to a specific choice of the basic framework able to explain key observations. *Third*, theoretical physics is a developing subject and "new physics" may offer a variety of cosmological applications. *Fourth*, observations and theoretical understanding are always limited, hence even a quite credible world model has its limitations, too (in current cosmology the nature of 95% of the substance is unknown).

These points emphasize the importance of crucial observational tests as the only way to decide between alternative cosmological ideas; the different theories should not only explain the known empirical facts, but they should also make predictions which can be tested by new observations.

9.1 Territories of Cosmological Ideas: Classifying World Models

It is curious to see in history that there tend to be "incorrect" ideas, which are then surprisingly transformed into main stream paradigms (so, the spiral nebulae were generally viewed as part of the Milky Way, before they became Island Universes in the 1920s). In the shadow of the main paradigm, other ideas are easily ignored or little studied and one cannot really know how well these might explain the data. It is wise to keep in view a variety of thoughts and to seek ways to test their validity observationally.

A variety of cosmological ideas have been proposed, many of them only rudimentary. Some may be obscure and hard to understand, except perhaps by their authors. In actual world models one can discern three cornerstones, highlighting the basic structure: *Observation*, *Theory*, and *Cosmological Principle*. One may characterize a model by asking:

Y. Baryshev, P. Teerikorpi, *Fundamental Questions of Practical Cosmology*,
Astrophysics and Space Science Library 383,
DOI 10.1007/978-94-007-2379-5_9, © Springer Science+Business Media B.V. 2012

- What empirical facts are viewed as cosmologically important or relevant?
- What physical theories, in particular for gravitation, are used?
- What assumptions, beyond our finite experience, are the initial principles?

In brief, theory is the physics known from laboratories, and extended by the aid of principles into the expanse of the cosmos, where it is used to understand important observations. One should note that two models may see the same observations as important, but may explain them differently. For instance, the usual idea about the background radiation is, of course, its origin in the hot early universe. A quite different approach has tried to see it as an integrated glow from radiation sources at a variety of redshifts.

Examples In order to see Observation, Theory, and Principle in action, consider first Ptolemy's model. It purported to explain the important observed celestial movements, the most basic being the daily rotation of the sky. Aristotelian physics and the epicyclic theory described on the basis of Euclidean geometry formed the physical framework. The cosmological principle consisted of the uniform circular motion around the centre of the universe.

For Newton's cosmology the unchanging starry sky was a key observation (proper motions of stars were still unknown through most of Newton's lifetime), while the planetary motions were explained by Newton's mechanics and gravitation. In cosmology, this theoretical apparatus was extended to the whole infinite universe uniformly filled by gravitating stars.

Modern cosmology is based on key observations in the galaxy universe, where we see the cosmological redshift and the Hubble law, and which is bathed in the thermal background radiation. General relativity and Friedmann's expanding model form the theoretical framework, together with the standard particle physics. According to the Cosmological Principle all main matter components have a homogeneous and isotropic spatial distribution.

Positive Roles of Other Ideas The Friedmann model is strong in explaining key observations. There are some alternative approaches, but usually they can explain only one or two empirical facts on which they focus attention (see, e.g., López-Corredoira 2003, 2010). Nevertheless, rather than being just a "noisy minority", they can be valuable. For example, they may contain a part of a future world model, as a novel inferred cosmological process, or they can define such territories of theoretical ideas that push scientists to devise new cosmological tests.

Hubble and Tolman (1935) suggested the number counts and the surface brightness as ways to test alternative explanations of the cosmological redshift. Another good example is Hoyle's (1959) angular size vs. redshift test, which he hoped could support the classical steady-state model, some aspects of which remain fascinating. The continuous creation of matter when space expands was a controversial element of that model. Now the continuous appearance of dark energy is a part of the new cosmological physics.

9.2 Cosmological Principles

In order to bridge the known and the unknown we postulate something which as if extends our knowledge to regions from where we can never obtain information. Some principles hide in the shadows, like the belief that ordinary analytical mathematics adequately describes reality or the epistemological assumption that the knowledge about the whole universe is accessible to us. A general idea, since Giordano Bruno, is that the physical laws found on Earth are valid everywhere, and the builder of world models usually assumes that (1) physical laws are the same in all space and time, (2) fundamental physical constants are true constants, and (3) particular properties, including measuring standards, are the same in all space and time.

When cosmology advances, these statements may require adjustments in the light of new observations and theoretical ideas. For example, multi-dimensional theories predict variations of physical constants. The requirement of the same physical laws does not make the large-scale world just a blown-up copy of our neighbourhood; it also permits new laws of cosmological physics which appear only on very large spatial, temporal and mass scales.

9.2.1 The Perfect Cosmological Principle: Steady State

The ordinary Cosmological Principle makes the universe spatially uniform and isotropic at each cosmic time (but its density may change). In the Steady State model (Bondi and Gold 1948; Hoyle 1948; see also Narlikar 1993 and the book *A Different Approach to Cosmology* by Hoyle et al. 2000), the Perfect Cosmological Principle makes the universe globally similar in space and time. One might think that here "perfectness" has some aesthetic appeal, but Bondi and Gold emphasized that the philosophical motivation was the desire to use terrestrial physics unambiguously in cosmological conditions. Especially, the creation process, instead of being in the deep past, was "brought within the scope of physical inquire" (Bondi 1952); it was assumed to happen all the time over the entire cosmos, even though very slowly.

General Conclusions In a world with no beginning and no overall evolution, the mean matter density does not change with time on scales large enough. Regarding the cosmic kinematics, Bondi and Gold (1948) argue that the Perfect Cosmological Principle predicts regular expansion from the observed absence of thermodynamic equilibrium (which an infinitely old static universe would have reached). So the reason behind the cosmological redshift is space expansion: the spatial volume increases with time. To compensate for the decreasing matter density, the creation of matter, along with space, has to be assumed.

The spatial homogeneity and isotropy lead to the Robertson-Walker line element characterized by the space curvature $k = 0$ and the deceleration parameter $q = -1$. Hence the scale factor increases exponentially: $S(t) = S_0 e^{Ht}$, where the Hubble constant H is a true *constant*. The cosmic density has the critical value

$\rho = 3H^2/8\pi G$. The theory includes a modified form of general relativity, i.e. gravity is geometrical in nature.[1]

Burbidge (2006) paid attention to the fact that within the steady-state model Hoyle and Sandage (1956) predicted the acceleration of the universe. A main property of the current cosmology was hidden in a minor alternative!

The discovery of the thermal background in the 1960s quickly made the big bang idea the ruling paradigm. Indeed, the main difficulty for the steady state is the accurate thermal spectrum of the background radiation. Hoyle (1982) noted that the nuclear reactions going on inside stars during their life yield the correct order of magnitude for the energy density of the microwave background. In this ordinary physical process, a fraction of baryonic energy is converted into electromagnetic radiation—the big problem is how to thermalize it. Also, the first attempts to measure the temperature of the background radiation at high redshifts ($z > 1$), though affected by uncertainties due to unknown local excitation radiation, favour the changing temperature predicted by the Friedmann model (Chap. 8).

Note the difference between steady and static states. For instance, Toivo Jaakkola (1993) sketched "Equilibrium cosmology", the universe in steady state but static (the redshift having some other reason than expansion). This has a historical antecedent in the speculations of William MacMillan (1871–1948) before the galaxy universe was found. For MacMillan cosmology was "the study of transformations of energy throughout the cosmos, the study of the origins being of no more interest than the study of dissolutions". The flow of energy would be such that "the singular points [stars] may change their positions and their brilliancy, but it is not necessary to suppose that the universe as a whole has ever been or ever will be essentially different from what it is today".

9.2.2 Einstein's Cosmological Principle

When Einstein (1917) applied his general relativity to cosmology, not yet knowing about galaxies, he imagined a world filled with stars and argued that the stars have a natural uniform distribution: matter lumps around any preferred centre should with time evaporate and disperse all over space (somewhat similarly the atomist Epicurus argued against the Stoic world of one big island of matter within infinite space: "If space were infinite and the bodies were limited in number, these could not stay in some one place, but would be moved into infinity, they would be dispersed without any assistance or propulsion other than collisions.")

The hypothesis of the large-scale homogeneity was called Einstein's Cosmological Principle by Edward Milne who analysed the foundations of cosmology in the 1930s. The Copernican principle "all places in the Universe are alike" is naturally fulfilled in a homogeneous world. (The many faces of the Copernican principle have been interestingly discussed by Rudnicki (1995).) Besides the absence of a centre,

[1]In Hoyle-Narlikar's approach a scale-invariant gravitation theory is used making possible the creation of matter by means of a C-field. The C-field has negative energy and negative stresses and is conserved together with usual matter.

another advantage of uniformity was a simplification of equations, which permitted Einstein to derive the static world model. Soon Friedmann (1922) liberated the universe from this stiff state, allowing the uniformly distributed matter and space to expand.

In these early years observational evidence for a homogeneous universe was meagre at best and it was mainly theoretical reasoning that guided the cosmologist.

Derivation of Uniformity from Local Isotropy Walker (1944), a British mathematician who worked closely with Milne, proved that uniformity follows from his hypothesis of "Local Spherical Symmetry" which supposes that isotropy exists locally about each point of a Riemannian manifold.

A simple reasoning leading to homogeneity when there is isotropy around each point, may be found in the *First Three Minutes* by Weinberg (1977). He shows how one can go from any one point to another arbitrary place along circle arcs on which the density remains the same. Hence the density is the same on every point. Strictly speaking this conclusion requires a hidden mathematical assumption of regularity, i.e. a smooth density around each point, only then *Local Isotropy* + *No Centre* + *Regularity* ⇒ *Uniformity*. "Local isotropy plus no centre" means that all points are equivalent and around each point the density law does not depend on the direction (it may depend on the distance from this point). The "regular" matter distribution is described by continuous, smooth mathematical functions—the *smooth fluid approximation* as explicitly mentioned by Narlikar (1993) when discussing simplifying assumptions of cosmology. It means going over from a discrete distribution of particles to a continuum distribution so that one may use the concept of mass density at each point of space. It is thus the union of local isotropy, no centre, and smoothness which gives homogeneity.

9.2.3 Mandelbrot's Cosmological Principle

The lumpy distribution of luminous matter makes it worthwhile to bear in mind that Einstein's homogeneity and Mandelbrot's fractality might both be useful approximations to reality, but relate to different spatial scales and different substances. In his *Fractals: form, chance, and dimension*, Benoit Mandelbrot (1977) foresaw that galaxies are fractal-like distributed and described the properties of such a distribution. He recalls how around 1965, his ambition was to implement the law of decreasing density with a model where there is "no centre of the universe" or "the centre is everywhere".

Mandelbrot views the fractal galaxy distribution as a major conceptual step in the description of the cosmological matter distribution. It is a kind of synthesis of hierarchical structures ("thesis") and homogeneity ("antithesis"), essentially based on randomness. Indeed, there is an essential difference between true random fractals and stiff hierarchical protofractals. Into protofractals the hierarchy is injected "ex-nihilo", by defining explicitly its levels. But fractals internally contain a scale invariance (self-similarity) and the impression of a hierarchy follows as an unavoidable consequence.

Interestingly, fractality (to be discussed in Chap. 10) carries within itself also a trace of uniformity. Within a fixed radius, i.e. for a fixed scale, every observer counts the same number of elements, *on average*. But upon changing the radius, a "new uniformity" is found with a new mean number density. Furthermore, there is no centre for random fractals—another "relic" from homogeneity.

Thus Mandelbrot made the first step for fractals in cosmology, generalizing Einstein's cosmological principle ($D = 3$), now allowing a non-uniform galaxy distribution with $D < 3$. His "Conditional Cosmographic Principle" states that all observers see similar cosmic landscapes around them, under the condition that they make observations from a structure element, galaxy.

Mandelbrot's cosmological principle of fractality—the observers attached to the structure elements are equivalent—satisfies Milne's formulation ("the whole world-picture as seen by one observer (attached to a fundamental particle or galaxy) is similar to the world-picture seen by any other observer"). It also automatically makes what Igor Karachentsev (1974) called "the ecological correction to the Copernican principle": the real observer can *live* only on or close to a celestial body.[2] In this sense the Copernican principle is contained by Mandelbrot's principle which may be seen as a generalization of the principle of homogeneity. Hence there should not be fears about an "unprincipled" fractal universe (Coles 1998; Wu et al. 1999).

Does Isotropy Always Imply Uniformity? The proof of uniformity is based on the density being smooth around all points; this is not valid for fractals. It is smoothness which wipes out fractality and makes uniformity. Actually local isotropy and no centre suggest a fractal structure, of which homogeneity is a special case with $D = 3$ (see Sylos Labini 1994).

Of course, there is never an *exact* local isotropy around every observer, not even in a uniform world, and still less inside a fractal. Instead one may speak of a statistical isotropy, so that the sky observed from any galaxy "looks much the same". It is natural to conjecture that for distributions made of discrete points, there is a generalization of the above chain of reasoning: *Statistical Isotropy + No Centre ⇒ Fractality*.

As will be considered in Chap. 10, the projection of a fractal structure with fractal dimension $D \geq 2$ could contribute to an apparent isotropic celestial distribution of galaxies. According to the theorem on fractal projections the resulting distribution will have the fractal dimension $D = 2$, which means homogeneity on a 2D plane or isotropy on the celestial sphere.

Two other factors contribute to celestial isotropy: lacunarity and the luminosity function. The patchiness on the sky depends on the fractal dimension, but also on

[2] This is usually called the Weak Anthropic Principle. The strong variant as formulated by Brandon Carter (1974) states that the universe and the fundamental parameters on which it depends must be such as to admit the creation of observers at some stage.

the lacunarity, a measure of how frequent large voids are. Numerical simulations show that fractals with a small lacunarity can have rather smooth projections on the sky. The second factor is the wide range of the luminosities of celestial bodies. As a result two objects with equal apparent brightness actually may have widely different distances. The mixing of nearby and distant objects hides clusters and fills in holes. For example, this decreases the celestial anisotropy for distant radio sources.

Towards Einstein–Mandelbrot Concordance? We do know of uniform cosmic components: the photon gas of the background radiation, the ocean of possible low-mass neutrinos, the hypothetical gravitons, and in particular, the suggested physical vacuum or dark energy. These relativistic ingredients cannot form structures. As the average density of the fractal matter decreases with increasing volume, there will eventually be a scale beyond which the uniform component is denser than the fractal one. Hence one may regard, after all, the universe as homogeneous on such scales. However, this is not due to the galaxies, but because of the uniform relativistic matter!

"Fractal universe" is sometimes linked with an infinite fractal, with zero average density. True, there are no scale limits to a pure mathematical fractal. But real physical objects usually have lower and upper cutoffs between which fractality is observed. Thus it is expected that the fractal galaxy distribution appears only within a finite interval of scales. Mandelbrot allowed for the possibility that the distribution is uniform on large scales, while fractal on smaller scales. With any uniform matter component, such as the photon-gas or the cosmological vacuum, the universe becomes homogeneous on a sufficiently large scale and it would have a non-zero average density. Thus the intuitions of both Einstein and Mandelbrot appear to have grasped fundamental features of the universe.

9.3 Fractality in Cosmological Physics

In the modern Friedmann model the presumedly uniform dark matter and dark energy determine the dynamics of the universe. On the other hand, the spatial distribution of luminous matter (galaxies) appears to be described by a fractal law (Chap. 11). Also, there is evidence for dark matter being about similarly distributed as luminous matter. Then, could fractality have some deep-seated significance for cosmology? Let us take a historical viewpoint.

9.3.1 The Einstein–Selety Correspondence

In his paper on relativistic cosmology Einstein (1917) emphasized that small stellar speeds spoke against large potential differences and gave support to large–scale smoothness. He viewed it important that the Poisson equation modified by the λ

term has a solution for the constant density. So a static, homogeneous matter distribution is possible, which also explains the small stellar velocities. Einstein then extended this result to general relativity, obtaining the spherical world model with a finite radius of curvature.

One may wonder why Einstein assumed that the matter is uniformly distributed. He really did not look at the non-uniform starry sky, but he viewed cosmology on the same level as Newton and Seeliger had done when they pondered whether the simplest situation, homogeneity, is on cosmological scales consistent with the local physics in the Solar System. When Einstein could thus show that his equations lead to a static world model, this must have further enhanced his positive attitude to the idea of homogeneity.

The Austrian scientist Franz Selety was aware of both Einstein's homogeneous and Charlier's hierarchical models (which will be discussed in Chap. 10). In *Annalen der Physik*, Selety (1922) argued that it is possible to construct hierarchical worlds which fulfil simultaneously the following conditions: (1) Infinite space & infinite total mass, (2) mass filling space so that locally there is everywhere a finite density, (3) zero average density of the mass in the whole world, and (4) non-existence of a unique middle point or middle region of the world.

Selety realized that the cosmological principle of "no centre" may also be valid for hierarchies: for an observer in a "molecular-hierarchic" system the universe appears everywhere basically similar. He also raised the question of Mach's principle in such universes and argued that it can be fulfilled. He pointed out that in such models a zero average density for the whole world exists together with its infinite total mass.

Einstein (1922) quickly replied. He expressed his opinion that Mach's principle is not fulfilled in a zero-density universe. Selety (1923) did not agree and once more discussed the crucial points of his model. Summarizing the arguments raised by Einstein and Selety, we see as main objections to hierarchical models in the 1920s (when the observations were scarce, too):

- Mach's principle invalid for a hierarchic world model with zero global density.
- Large potential differences → an excessively high velocity dispersion for stars.
- A hierarchic system evaporates & stars fill up the voids → homogeneity.
- A hierarchic world contains a preferred middle point.

A Retrospective View Mach attempted to link the inertial mass of a body to the large-scale mass distribution in the universe. One can define the acceleration of a particle only relative to distant masses. In fact, the nature of inertial mass is still a challenge for modern physics, including general relativity, where the rest mass of a particle is regarded as a relativistic invariant, independent of the large-scale mass around the particle. Mach's principle no longer defines admissible world models. Other items related to the old Einstein–Selety "debate" still have relevance to cosmology.

The small speeds of stars are now known to be due to their motion in our Galaxy and not related to the entire universe. In fact, the problem of velocities has moved from stars to galaxies. Astronomers have been asking why the velocity dispersion

Table 9.1 Observational research topics related to large-scale fractality

Subject	References
conditional density new methods of data analysis 2-point conditional column density	Pietronero 1987; Coleman and Pietronero 1992; Gabrielli et al. 2005; Bharadwaj et al. 1999; Best 2000; Martinez and Saar 2002; Baryshev and Bukhmastova 2004; Vasilyev 2004
fractal dimension of galaxy distribution scales of fractality	Coleman et al. 1988; Klypin et al. 1989; Lemson and Sanders 1991; Jones et al. 1988; Martinez and Jones 1990; Jones et al. 1992; Sylos Labini et al. 1998
local radial galaxy distribution number counts, normalization luminosity function, multifractals fractality of dark matter	Sandage 1995a, 1995b; Teerikorpi et al. 1998; Teerikorpi 2004; Baryshev 1981; Joyce and Sylos Labini 2001; Courtois et al. 2004; Sylos Labini and Pietronero 1996; Baryshev 1981; Durrer and Sylos Labini 1998; Massey et al. 2007a
dependence of correlation function on depth, luminosity type of object peculiar velocities	Einasto et al. 1986; Calzetti et al. 1987; Davis et al. 1988; Norberg et al. 2001, 2002; Klypin and Kopylov 1983; Bahcall 1988; Bahcall et al. 2003; Zehavi et al. 2002, 2005; Hawkins et al. 2003
local fractal dimension linearity and coldness of the local Hubble flow	Tikhonov et al. 2000; Tikhonov and Makarov 2003; Sandage et al. 1972; Sandage 1986–1987; Karachentsev and Makarov 1996; Ekholm et al. 2001; Whiting 2004; Karachentsev et al. 2009
local tests of cosmological vacuum and dark energy within fractals	Chernin 2001; Baryshev et al. 2001; Axenides and Perivolaropoulos 2002; Chernin et al. 2006; Macciò et al. 2005; Teerikorpi et al. 2008
statistical mechanics of self-gravitating fractal gas, $D = 2$	Perdang 1990; De Vega et al. 1996, 1998; Combes 1999; Huber and Pfenniger 2002

around the local Hubble flow is so small within the very lumpy galaxy distribution (Sandage et al. 1972; Sandage 1999) Einstein's Λ-term turns up here, too: it might contribute to the smooth Hubble flow as was first pointed out by Chernin (2001); see Chap. 12.

The stability of hierarchical (fractal) structures of gravitating particles is an open topic in modern gravithermodynamics. Perdang (1990) and De Vega et al. (1996) concluded that a statistical equilibrium may be possible for structures with $D \approx 2$.

The question of the central point is related to the cosmological principle. Fractals preserve important properties of the old hierarchical systems but are more realistic models of the galaxy distribution (Mandelbrot 1989; Pietronero et al. 1997). In particular, a stochastic fractal structure does not contain a privileged centre.

The arguments of Einstein and Selety are no longer reasons for rejecting inhomogeneous world models, but they continue to inspire questions in cosmology. In particular, the nature of the large-scale structure has prompted astronomers and physicists to study different aspects of self-similar systems. Tables 9.1 and 9.2 illus-

Table 9.2 Theoretical research topics related to large-scale fractality

Subject	References
protofractal (hierarchical) models relativistic fractal models cosmological tests, fractal universe	Wertz 1971; Wesson 1978; Soneira and Peebles 1977, 1978; Bonnor 1972; Ruffini et al. 1988; Ribeiro 1993; Gromov et al. 2001; Fang et al. 1991; Baryshev et al. 1994a, 1994b; Joyce et al. 2000
cosmological gravitational redshift	de Sitter 1917; Bondi 1947; Baryshev 1981, 1994
N-body simulation, local structure initial conditions and discreteness stability, velocity, force	Governato et al. 1997; Moore et al. 2001; Klypin et al. 2003; Macciò et al. 2005; Bottaccio et al. 2002; Baertschiger et al. 2002; Baertschiger and Sylos Labini 2004; Gabrielli et al. 2005
origin and evolution of large scale fractals	Haggerty 1971; Peebles 1974; Alfvén 1989; Lerner 1986; Ostriker and Cowie 1981; Schulman and Seiden 1986; Szalay and Schramm 1985; Maddox 1987; Luo and Schramm 1992
cosmological principle and fractality, isotropy and homogeneity fractal holography	Mandelbrot 1975, 1977; Pietronero and Sylos Labini 1995; Rudnicki 1995; Wu et al. 1999; Baryshev and Teerikorpi 2002; Mandelbrot 1989; Sylos Labini 1994; Mureika 2007

trate the wide range of observational and theoretical items touched by the fractal-like large-scale structure.

9.3.2 Fractal Sources for Gravity Field

In our simple classification of world models, the standard model has general relativity as Theory and homogeneous matter distribution as Cosmological Principle. If one assumes a fractal matter distribution instead of the homogeneous one, this leads to an unsolved problem of how to describe non-analytical fractal sources of the cosmological gravity field.

One has attempted to bypass this problem by using the global mass-radius relation $M(r)$, the main characteristics of the fractal matter distribution which determines the fractal density field $\rho(r)$. Taking Mandelbrot's cosmological principle as a generalization of Einstein's principle for the case of cosmological models with isotropic fractal structures, one may write

$$\rho(\vec{r},t) = \rho(r,t), \qquad p(\vec{r},t) = p(r,t). \tag{9.1}$$

Here the variable r is the radius of a sphere around each point of a structure. Can fractal cosmological models be based on solutions of gravity field equations with the sources described by the fractal density law?

Lemaître-Tolman-Bondi Models The LTB models are exact solutions of Einstein's equations for (1) spherical symmetry, (2) pressureless matter (dust) and (3) motion with no particle layers intersecting. The last point means that one views the matter expanding (or contracting) smoothly within spheres which retain their identity during the process (similarly as the comoving coordinates of galaxies keep fixed during the Friedmann expansion). These models are the simplest generalization of the Friedmann models with a non-zero density gradient and "the assumption of spherical symmetry supplies us with a model which lies between the completely homogeneous models of cosmology and the actual universe with its irregularities" (Bondi 1947).

The LTB model has helped us to understand the kinematics of galaxies around *individual* mass concentrations. For example, Teerikorpi et al. (1992) and Ekholm et al. (1999b) could see the expected behaviour around the Virgo cluster: (1) Hubble law at large distances, (2) retardation at smaller distances, (3) zero-velocity border at about half way from us to Virgo, and (4) collapsing galaxies at still smaller distances. Hanski et al. (2001) studied the Perseus-Pisces supercluster using the LTB model with the Λ-term.

Usually one considers the behaviour of a spherical layer of matter at distance r from the centre of symmetry. In Newtonian terms, the dynamics of a shell is described by two constants: (1) the mass inside the radius of the shell $M(r)$ and (2) the energy per mass of the shell

$$E(r) = \frac{1}{2}\left(\frac{dr}{dt}\right)^2 - \frac{GM(r)}{r} - \frac{\Lambda c^2 r^2}{6}. \tag{9.2}$$

We remind that the energy density of cosmological vacuum (or dark energy) corresponding to the Λ-term is $\rho_{\mathrm{DE}} = \Lambda c^2/(8\pi G)$ and the last term becomes $-(4\pi/3)G\rho_{\mathrm{DE}}r^2$. In practice, for a given $M(r)$ one can calculate the expected present radial velocity $V(r)$ at the distance r, assuming that the expansion of the system has began T years ago (often taken to be as the age of the large-scale Friedmann model). For $\Lambda = 0$, there is a parameterized solution (e.g., Olson and Silk 1979); for $\Lambda > 0$, one uses numerical integration (e.g., Hanski et al. 2001).

At the zero-velocity radius the peculiar velocity towards the mass concentration equals the cosmological Hubble velocity for the same distance. This distance (about 1 Mpc for the Local Group and 10 Mpc for the Virgo cluster) has been often used for estimating the total mass M_0 of a gravitationally bound system (Lynden-Bell 1981; Sandage 1986). The spherical model with $\Lambda = 0$ leads to the estimator

$$M_0 = (\pi^2/8G)t_0^{-2}R_0^3 = 2.74\,10^{12} M_\odot \left(\frac{t_0}{10^{10}\ \mathrm{yrs}}\right)^{-2}\left(\frac{R_0}{1\ \mathrm{Mpc}}\right)^3. \tag{9.3}$$

Here t_0 is the age of the universe, or more generally, the "bang time" when the system started expanding from the concentration. In fact, this formula is like Kepler's Third Law for a very eccentric orbit of a test particle with R_0 equal to two times the semi-major axis and t_0 one half of the period.

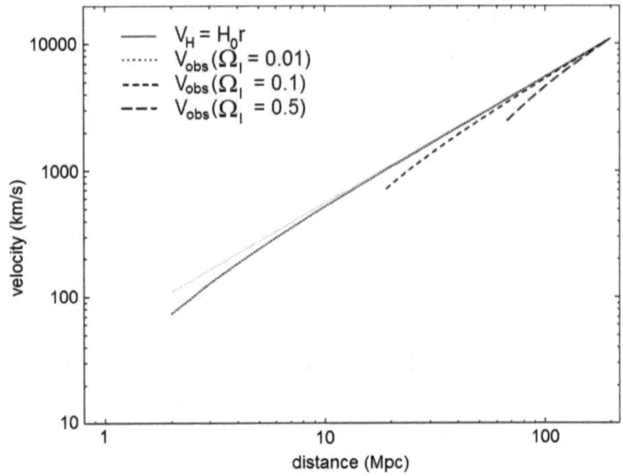

Fig. 9.1 The behaviour of the velocity field within an inhomogeneous (fractal) matter distribution for various values of Ω_{lum} when the total density has the critical value $(= 1)$. In this calculation the fractal dimension for the luminous matter is $D = 2$ up to the scale $r = 200$ Mpc. (From Baryshev et al. 1998)

If $\Lambda > 0$, then dynamic mass determinations (like the virial theorem) lead to an underestimate of the gravitating mass (the "lost gravity" effect of local dark energy; Chernin et al. 2009), also when one uses the zero-velocity method (Peirani and de Freitas Pacheco 2008).

Hierarchical Cosmologies Bonnor (1972) was the first to apply the LTB model to the hierarchical cosmology. He used de Vaucouleurs's density law $\rho \sim r^{-\gamma}$ with $\gamma = 1.7$. Ribeiro (1993) and Gromov et al. (2001) developed the approach to include a fractal matter distribution on small scales and a homogeneous distribution on large scales. The main result was that the fractal structure leads to a deflection from the Hubble law at small distances, unless the global density parameter of matter is low $(\Omega_m < 0.1)$ (Fig. 9.1). The low density was mentioned already by Sandage et al. (1972) as a possible cause of the smooth expansion within the lumpy local universe.

However, the LTB model leads to a conceptual problem, as the original formulation contained a central point of the universe. A fractal distribution has no unique centre, but every structure point may be treated as a local centre, surrounded by a spherically symmetric (on average) matter distribution. In a sense, applying the LTB model to fractals means that there is an infinity of centres with slightly different initial conditions (for any fixed scale the average density—a power law—is only approximately constant). A good thing is that this excludes geocentrism and points the way towards exact general relativistic models where space expansion becomes scale dependent: the expansion at distance r from a fixed point of the structure is determined by the average mass of the sphere around this point.[3]

[3]An analogous thing exists for Friedmann models where the rate of expansion within distance r from a galaxy is determined by the mass of the sphere around this galaxy.

9.3.3 Field Gravity and Fractality

In a more radical variant one takes a fractal matter distribution and changes the gravity theory, an idea proposed by Baryshev (1981). Though still a developing subject (Baryshev et al. 1994b), a qualitative picture has emerged, with some testable results. Here gravitation is described as a relativistic quantum field in Minkowski space (Chap. 6). The second hypothesis is fractal ($D = 2$) matter from the scales of galactic halos up to the Hubble radius.

Cosmological Solution in Field Gravity In Chap. 6 we treated field gravity in weak-field approximation using iterations. A specific feature of this theory is that there is the case of a weak force with $\nabla\varphi \to 0$ while $|\varphi| \to c^2/2$. This is what happens in the cosmological problem, and we can obtain some quantitative results even at the post-Newtonian level.

Let us inspect static homogeneous ($\rho = \text{const}$) dust-like cold matter ($p = 0$) in infinite space. Using expressions for the post-Newtonian energy-momentum tensors of matter (where the field and interaction EMTs are traceless), we get the equation for the $\psi^{00} = \varphi$ component:

$$\Delta\varphi = 4\pi G\left(\rho + 2\rho\frac{\varphi}{c^2} + \frac{2(\nabla\varphi)^2}{8\pi G c^2}\right). \tag{9.4}$$

The main terms in the right-hand side of Eq. (9.4) are the positive rest mass density ρ and the negative interaction mass density $(2\rho\varphi)/c^2$. The last term is negligible, because for $\varphi \to \text{const}$, $\nabla\varphi \to 0$. Hence the simple equation (left) which is equivalent to Einstein's equation (right):

$$\Delta\varphi - \frac{8\pi G\rho}{c^2}\varphi = 4\pi G\rho \leftrightarrow \Delta\varphi - \lambda\varphi = 4\pi G\rho. \tag{9.5}$$

A comparison shows that the Λ-term in field gravity theory is

$$\lambda = \frac{8\pi G\rho}{c^2} \tag{9.6}$$

and it comes from the contribution of the energy-momentum tensor of the interaction. This leads to the static cosmological solution

$$\varphi = -\frac{4\pi G\rho}{\lambda} = -\frac{c^2}{2}, \tag{9.7}$$

with the Λ-term twice the similar term in Einstein's static model (Sect. 7.2.2). Now $\varphi = \text{constant}$ is consistent with an initially homogeneous infinite gas distribution in the static Minkowski space, without the gravitational potential paradox. This solution can be also derived as a limiting case ($r \to \infty$) of the exact solution of (9.5) for a matter ball with radius r (Appendix C).

9.4 Physical Laws, Fundamental Constants and Large Numbers

A century ago Poincaré posed the question about the constancy of physical laws in the universe. He argued that unchanging physical laws give the basis of scientific knowledge; otherwise there were no repeatable experience and the physical theory would be devoid of predicting power. If some laws were changing under higher-level laws, then these should be fixed and so on.

Known physical laws contain also fundamental constants as parameters which may be determined from experiments. Though observations strongly limit possible variations of those constants, Poincaré's question is still with us. Another enigma comes from the coincidences between some combinations of microphysical constants and values of some cosmological quantities—emphasizing what modern cosmology has shown: both directions, "inwards" and "outwards", are important for understanding the universe.

9.4.1 Fundamental Constants in Cosmology

A physical law may be defined through a mathematical relation between physical quantities which can be measured empirically. Any abstract physical theory has a finite range of validity in reality. Experiments measuring physical quantities are needed to find the borders of validity for the theory.

Dimensional Quantities The measurement of a physical quantity gives a dimensional number. Founders of natural science from Newton to Maxwell were aware of the deep link between physical dimensions and the structure of fundamental physical laws (Bridgman 1920, 1936; Whitney 1968).

Corresponding to three parts of physical reality—space, time, and matter—there are three units: "cm", "sec", "gram". Buckingem's π-theorem, based on the scale invariance of the dimensional physical quantities, states that the dimension of any physical quantity may be presented as a product of power-law functions of those fundamental dimensions:

$$(\mathrm{cm})^n (\mathrm{sec})^m (\mathrm{gram})^l, \tag{9.8}$$

where n, m, l are rational numbers. For example, the physical dimension of the charge of an electron is $[e] = \mathrm{erg}^{1/2}\,\mathrm{cm}^{1/2} = \mathrm{cm}^{-1}\,\mathrm{sec}^{-1}\,\mathrm{gram}^{1/2}$.

In astronomy the Gaussian system of units $[cm]$, $[s]$, $[g]$ is generally used, while the International System is based on $[m]$, $[s]$, $[kg]$. Planck (1899) suggested three particular physical constants as a "natural system of units"—the velocity of light c, the gravitational constant G, and the constant of action h (now called the Planck constant; also $\hbar = h/2\pi$). Then the values of the Planck units of length, time and mass are:

$$l_{\mathrm{Pl}} = (G\hbar/c^3)^{1/2} \approx 1.6 \times 10^{-33}\ \mathrm{cm}, \tag{9.9}$$

$$t_{\mathrm{Pl}} = (G\hbar/c^5)^{1/2} \approx 5.3 \times 10^{-44}\ \mathrm{s}, \tag{9.10}$$

$$m_{\mathrm{Pl}} = (c\hbar/G)^{1/2} \approx 2.2 \times 10^{-5}\ \mathrm{g}. \tag{9.11}$$

In terms of the Planck units, the other constants of nature become dimensionless numbers and important equations in physics obtain simplified forms, e.g., Newton's law of gravitation ($F = Gm_1m_2/r^2$) becomes $F = m_1m_2/r^2$ where the force F is now relative to the unit Planck force ($= c^4/G$). As we mentioned in Chap. 5, the Planck system $c, G, \hbar$ has also the deep interpretation as a representation of basic physical theories.

Variability of Fundamental Constants The values of such constants as

$$G, c, h, e, k, m_{\mathrm{e}}, m_{\mathrm{p}}, \alpha_{\mathrm{e}}, \alpha_{\mathrm{w}}, \tag{9.12}$$

may be measured empirically. They link physical phenomena to mathematical models, i.e. the real world of physics to the ideal world of mathematics.[4]

The modern idea of possible variation of physical constants came from cosmological models describing the expanding universe. It was much discussed in 20th century physics starting from the works by Milne (1935, $G(t), e(t)$), Dirac (1937, $G(t)$), and Gamow (1967, $e(t), \alpha(t)$). The problem became more complex when it was realized that varying one constant implies the need to consider the changes in all other constants (Troitskii 1987; Barrow 1990; Okun 1991).

In modern physics there are theories where fundamental constants are changing as a consequence of the expansion of 3D space in higher dimensions. Kaluza-Klein, supergravity, and brane/string theories predict different types of time-dependencies (Uzan 2003).

The fine structure constant α, which plays a fundamental role in electromagnetic interaction, was suspected to change as $\Delta\alpha_{\mathrm{e}}/\alpha_{\mathrm{e}} \approx (-0.7 \pm 0.2) \times 10^{-5}$ over the redshift range 0.5–3.5 (Murphy et al. 2001). Such studies use the fine structure splitting in quasar spectra. The splitting ratio $r = (\lambda_2 - \lambda_1)/(\lambda_2 + \lambda_1)$ at different cosmic epochs (redshifts) gives the relative change of α between these epochs (Uzan 2003). In particular, for the redshifts z and 0 one may write:

$$\frac{\Delta\alpha}{\alpha} = \frac{1}{2}\left(\frac{r(z)}{r(0)} - 1\right). \tag{9.13}$$

Gutiérrez and López-Corredoira (2010) give a summary of the results from the different variants of the fine structure splitting method. The change of α claimed by Murphy et al. (2001) has not been confirmed, and there is no clear evidence for changes in α over the timescales considered (however, see Levshakov et al. 2007). Because $\alpha_{\mathrm{e}} = e^2/\hbar c$, the same restrictions as for α separately hold for e^2, $\hbar$, and c. Another possibility is that $e^2(t) \propto \hbar(t)c(t)$.

---

[4]Even "dimensionless" combinations of physical constants, say the fine structure constant $\alpha_{\mathrm{e}} = e^2/\hbar c$, actually depend on the dimensions of the constants that they contain. Thus $\alpha_{\mathrm{e}} \approx 1/137$ presupposes that c, h, e, are measured in the same system of units.

Along with α, the proton-to-electron mass ratio $\mu = m_p/m_e$ can be studied from the spectra of high-redshift quasars. Using the Keck telescope, Malec et al. (2010) derived from the molecular hydrogen absorption spectrum towards a $z = 2.059$ background quasar that $\Delta\mu/\mu = (+5.6 \pm 5.5_{stat} \pm 2.9_{sys}) \times 10^{-6}$, "indicating an insignificantly larger μ in the absorber". This result was consistent with the earlier study by King et al. (2008) (for $z_{abs} > 2.5$) using the VLT. There are also limits on other constants and their combinations (Uzan 2003). Observations with large next-generation telescopes will be needed for more accurate data.

Constants of Cosmological Physics Such quantities as the Hubble constant $H \approx 10^{-18}$ sec^{-1}, the Hubble radius $R_H \approx 10^{28}$ cm, the mass of the universe within R_H, $M_{univ} \approx 10^{56}$ g, the Hubble mass density $\rho \approx 10^{-28}$ g/cm^3, the background radiation temperature $T_0 \approx 2.726°$ K, and the fractal dimension of the galaxy distribution $D \approx 2$ are derived from observations and enter world models as free parameters. Within the Friedmann model these depend on time. Some other approaches to cosmology might view their values as linked to the phase of evolution of the matter in non-expanding space.

9.4.2 The Puzzle of Large Numbers in Cosmological Physics

Among the great puzzles of cosmology is the strange coincidence between the parameters of the universe and simple combinations of certain constants of microphysics. Known as the "Large Numbers Coincidences" (LNC), this still has no theoretical explanation within the standard cosmology, and the Anthropic Cosmological Principle is often used to obtain some kind of understanding.

How it Started Weyl (1919) noticed that the ratio of the electron radius to its gravitational radius is $\sim 10^{40}$ and the ratio of the world radius to the electron radius may also be of the same order. Eddington (1923b) gave Weyl's guess a more rigorous form, and attracted the attention of scientists to the problem of Large Numbers. Indeed, the ratio of electromagnetic to gravitational forces between a proton and electron is a big number B and the number of protons in the universe is the square of the same big number:

$$B = \frac{e^2}{Gm_p m_e} \approx 10^{40}, \qquad \frac{M_{univ}}{m_p} \approx 10^{80} = B^2. \tag{9.14}$$

Eddington never attempted to explain the LNC by invoking varying constants; he regarded it as a signal for a new foundation of physics. One can also write B as (Chandrasekhar 1937; Andreev and Komberg 2000)

$$B \approx \frac{e^2}{Gm_p m_e} \approx \frac{\hbar c}{Gm_p^2} \approx \left(\frac{m_{Pl}}{m_p}\right)^2, \tag{9.15}$$

where $e^2 \approx \hbar c/137$ and $m_{\mathrm{Pl}} = (\hbar c/G)^{1/2}$. Different quantities in micro- and cosmophysics are related as powers of B, so that $Q_{\mathrm{cos}} = B^n Q_{\mathrm{micro}}$.

Coincidences in Friedmann Cosmology Dirac (1937), impressed by the Weyl-Eddington findings, proposed "a new basis for cosmology" by considering, in addition to B, the dimensionless numbers B_1 and B_2 combined from constants of microphysics ($e, m_{\mathrm{e}}, m_{\mathrm{p}}, G$) and cosmological quantities (the size $R = c\tau$, the mass density ρ, and the age τ of the universe):

$$B_1 = \frac{c\tau}{(e^2/m_{\mathrm{e}}c^2)} \approx 10^{40} \quad \text{and} \quad B_2 = \frac{4\pi(c\tau)^3\rho}{3m_{\mathrm{p}}} \approx \frac{c^3\tau}{Gm_{\mathrm{p}}} \approx 10^{80}. \tag{9.16}$$

Here $\tau = 2 \times 10^9$ yrs and $G\rho\tau^2 \approx 1$. Dirac realized that within the expanding universe one gets a relation between fundamental constants of laboratory physics and continuously changing cosmological quantities $R(t), \rho(t), \tau$. The strange coincidence is that just at the present epoch we have

$$B \approx B_1 \approx \sqrt{B_2} \approx 10^{40}. \tag{9.17}$$

9.4.3 Possible Explanations of the Coincidencies

Dirac considered the possibility that $B = B_1 = \sqrt{B_2} \propto t$ and concluded that in this case the quantity $e^2/Gm_p \propto t$. He suggested that the gravitational "constant" is decreasing with time as $G \propto 1/t$. Then the number of particles in the universe will increase with time $B_2 \propto B^2 \propto t^2$, which he interpreted as a continous creation of protons. However, such a behaviour of $G(t)$ conflicts with geological data on the Earth's temperature. Gamow (1967) suggested that it is the charge which varies, $e^2 \propto t$, so the epoch of boiling oceans is far enough in the past. Modern limits on the changes of the fine-structure constant also exclude such a possibility.

Dicke (1961) proposed another explanation. He noticed that the stellar evolution, which produces the chemical elements necessary for carbon-based life, gives the time scale τ_* not far from the age of the universe:

$$\tau_* \approx \left(\frac{m_{\mathrm{Pl}}}{m_{\mathrm{p}}}\right)\frac{h}{m_{\mathrm{p}}c^2} \approx 10^{10} \text{ yrs}. \tag{9.18}$$

Hence $B_1 \approx 10^{40}$ would just follow from the anthropic principle, a selection effect of the presence of the very observer (e.g., Barrow and Tipler 1988). This explanation views the coincidence as having no fundamental theoretical reason.

Large Numbers and the Hubbloid The LNC can be formulated as an equilibrium condition for the galaxy universe, so it might not be a coincidence at all (Baryshev and Raikov 1988; Baryshev et al. 1994a, 1994b, 1996a, 1996b). Let us consider an extreme self-gravitating superobject ("Hubbloid") for which the Eddington and Planck luminosities are equal:

$$L_{\rm Edd} \approx L_{\rm Pl}. \tag{9.19}$$

The *Eddington luminosity* is the maximum luminosity of an object (having the mass M) for which there is a balance between the gravity force and radiation pressure:

$$L_{\rm Edd} \approx \frac{GMm_{\rm p}c}{\sigma_{\rm T}} \approx \frac{c^5 G m_{\rm p}^2 m_{\rm e}^2}{e^4} N. \tag{9.20}$$

Here N is the number of protons in the object, $\sigma_{\rm T} = (8\pi/3)R_{\rm e}^2$ is the Thompson scattering cross-section, and $R_{\rm e} = e^2/m_{\rm e}c^2$ is the classical radius of electron.

The *Planck luminosity* means the luminosity of an object which radiates its rest-mass energy Mc^2 during the time interval equal to the crossing-time of its gravitational radius by light $t = R_{\rm g}/c = 2GM/c^3$:

$$L_{\rm Pl} = \frac{Mc^2}{R_{\rm g}/c} = \frac{c^5}{G} \approx 3.6 \times 10^{59}\ \text{erg/s}. \tag{9.21}$$

Therefore Eq. (9.19) gives a rough estimate for the number of protons in the Hubbloid, directly explaining Eddington's large number B^2:

$$N_{\rm H} \approx \left(\frac{e^2}{Gm_{\rm p}m_{\rm e}}\right)^2 = B^2 \approx 10^{80}. \tag{9.22}$$

The mass $M_{\rm H}$ and the (gravitational) radius $R_{\rm H}$ of this object are

$$M_{\rm H} \approx m_p N_{\rm H} \approx 10^{56}\ \text{g} \approx 10^{22} M_{\odot}, \tag{9.23}$$

$$R_{\rm H} \approx \frac{GM_{\rm H}}{c^2} \approx \frac{e^4}{Gm_{\rm p}m_e^2c^2} \approx R_{\rm e}B \approx 10^{27}\ \text{cm}, \tag{9.24}$$

i.e. the Hubble radius. Hence, around this limit of the deep universe the Newtonian gravitational potential $GM(R)/R$ would reach the value c^2 (and where any non-relativistic Newtonian description of the expanding universe would break down; Sect. 12.4.3).

The density of the Hubbloid is

$$\rho_{\rm H} \approx \frac{3M_{\rm H}}{4\pi R_{\rm H}^3} \approx \frac{Gm_{\rm p}^2 m_e^4 c^6}{e^8} \approx \rho_{\rm p} B^{-1} \approx 10^{-27}\ \text{g/cm}^3. \tag{9.25}$$

Equating the Planck luminosity to the black body power $L_{\rm H} = 4\pi R_{\rm H}^2 \sigma T_{\rm H}^4$ gives the temperature of radiation from the Hubbloid's "surface"

$$T_{\rm H} \approx \left(\frac{L_{\rm H}}{4\pi R_{\rm H}^2}\right)^{1/4} \approx T_{\rm Pl} B^{-3/4} \approx 10\ \text{K}, \tag{9.26}$$

an equilibrium background temperature of this extreme object.

A Link to Fractality Finally, the density of a proton $\rho_p \approx m_p/\lambda_C^3$ (λ_C is its Compton length $h/m_p c$) and the product radius × density is the same for the proton and the Hubbloid:

$$f_H \approx \rho_H R_H \approx \rho_p R_e \approx (2\pi)^{-1}\ \text{g/cm}^2, \tag{9.27}$$

where $\rho_p \approx 10^{13}$ g/cm^3. The last equality uses the expression $H_g = (2\pi)^{-1}\rho_0 r_0 G/c$ for the gravitational Hubble constant (Sect. 9.5.1), so the fractal density (for $D = 2$) $f_H \approx (2\pi)^{-1}$ g/cm^2 may be considered as a fundamental cosmological constant, which can be expressed via microphysics as

$$f_H = \rho_0 r_0 \approx m_p \left(\frac{m_e c^2}{e^2}\right)^2 \approx m_p \left(\frac{hc}{m_e}\right)^2. \tag{9.28}$$

Hence, the Hubbloid could have, in principle, a fractal structure with $D = 2$ from the proton scale up to the Hubble radius. We find it interesting that within the idea of a fractal universe the constants of microphysics are naturally related to the observed global values of cosmological parameters.

Using the Planck scale, the quantities characterizing the Hubbloid become $R_H \approx 10^{60} l_{Pl}$, $M_H \approx 10^{60} m_{Pl}$, $t(\text{crossing})_H \approx 10^{60} t_{Pl}$, $T_H \approx 10^{-30} T_{Pl}$, and $\rho_H \approx 10^{120} \rho_{Pl}$. These relations, which also approximate the observable galaxy universe, may be seen as another formulation of the puzzle of large numbers.

9.4.4 Other Cosmological Coincidencies

The "classical coincidencies" discussed above were found when the composition of the cosmic matter was poorly known. New questions have arisen after the discovery and measurement of different matter components: baryonic, dark matter and the cosmic vacuum energy (dark energy). The physics of each of these involves their own riddles. Byrd et al. (2007) conclude that "the energy content of the Universe is well measured but poorly understood" and note that only the thermal background radiation, contributing less than 0.01% of the current cosmic energy, can be well treated on the basis of standard physics within the Friedmann model.

The relative importance of the cosmic components in terms of energy density also leads to a puzzle, because the energy fractions vary in time when the universe expands. Why is it so that currently especially the densities of dark matter and dark energy are so close to each other?

The Theoretical Problem with the Small Λ The energy-momentum tensor corresponding to the cosmological constant Λ can be written as:

$$T_{ik} = \Lambda \frac{c^4}{8\pi G} g_{ik} \tag{9.29}$$

so the cosmological constant as a kind of substance has

$$\rho_{\mathrm{vac}} = \frac{\Lambda c^2}{8\pi G}, \qquad \Lambda = \frac{8\pi G}{c^2}\rho_{\mathrm{vac}}. \tag{9.30}$$

From the observations of distant SNIa supernovae one has derived that $\rho_{\mathrm{vac}} \approx 0.7\rho_{\mathrm{crit}} \approx 6.4h_{70}^2 \times 10^{-30}$ g/cm^3 (from Eq. (7.33)) and so $\Lambda \approx 2 \times 10^{-56}$ cm^{-2}. From the theoretical point of view the energy density of "empty" space is the sum of the zero point energies of the normal modes of fields up to a wave number cutoff $k_{\mathrm{max}} \approx E_{\mathrm{Pl}}/\hbar c$, where $E_{\mathrm{Pl}} = m_{\mathrm{Pl}}c^2$ and $m_{\mathrm{Pl}} = \sqrt{\hbar c/G}$, so that the expected energy density of the vacuum is close to the Planck density ρ_{Pl}:

$$\rho_{\mathrm{vac}} \approx \frac{\hbar k_{\mathrm{max}}^4}{16\pi^2 c} \approx \rho_{\mathrm{Pl}} = \frac{c^5}{G^2\hbar} \approx 10^{94}\,\mathrm{g\,cm^{-3}}, \tag{9.31}$$

124 orders of magnitude larger than the observed value! Weinberg (1989) considers possible solutions of this problem, showing that the cosmological constant riddle has a great impact on other areas of physics and astronomy. However, "many theorists have tried to invent adjustment mechanisms to cancel the cosmological constant, but without any success so far".

One aspect discussed by Weinberg (1989) is the anthropic constraint, meaning that the value of Λ should be small enough to allow the formation of galaxies, stars, heavy elements and, finally, human beings. However, the constraint is not necessarily so tight as to exclude a range of values larger than what is now adopted in standard cosmology. On the other hand, Starkman and Trotta (2006) argue that one cannot use the anthropic principle to explain the value of Λ—much depends on how one weighs the different "universes" in the ensemble from which our universe is just one realization. In their specific example it would be very improbable to have Λ equal to or greater than what we observe.

Insight from Friedmann Integrals Chernin (2002) writes the first equation of Friedmann for the scale factor S in the form

$$\frac{\dot{S}^2}{c^2} = \left(\frac{S}{A_{\mathrm{V}}}\right)^2 + \left(\frac{S}{A_{\mathrm{D}}}\right)^{-1} + \left(\frac{S}{A_{\mathrm{B}}}\right)^{-1} + \left(\frac{S}{A_{\mathrm{R}}}\right)^{-2} - k, \tag{9.32}$$

where the A-terms with the unit of length are, mathematically, constants of integration which were physically determined by the initial conditions in the early universe. In fact, they can be written in terms of the present total density ρ and the equation of state parameters w as

$$A_i = (\kappa\rho S^{3(1+w_i)})^{\frac{1}{1+3w_i}}, \tag{9.33}$$

where $\kappa = 8\pi G/3c^2$. Especially, for the vacuum energy with $w = -1$

$$A_{\mathrm{V}} = (\kappa\rho_{\mathrm{V}})^{-1/2} = 1.6 10^{28}\mathrm{cm}. \tag{9.34}$$

This length is not far from the Hubble distance ($R_{\mathrm{H}} = c/H = 3000h^{-1}$ Mpc $= 0.925\,10^{28}h^{-1}$ cm) and the current particle horizon. Its expression does not depend on the scale factor S, while the other lengths (for the components with $w > -1$) depend on a combination of S and ρ in such a way as to remain constant. Thus one may also state the coincidence of the vacuum and the dark matter densities (as well as the baryonic matter and radiation, within two orders of magnitude) so that we are living during the period when the lengths (Friedmann integrals) are roughly equal to the Hubble distance. Chernin asks why the present Hubble distance should have such significance and links it to the possibility that the universe might actually have a topology making its volume finite (Luminet et al. 2003). This finite size would be nowadays near the Hubble (and the horizon) distance. Then one might say (Byrd et al. 2007) that "in terms of the standard cosmology, the Anthropic Principle requirements may perhaps all be reduced to only one condition: the present epoch is the epoch when the size of the Universe reaches the universal value of the Friedmann integral".[5]

9.5 The Nature of Cosmological Redshift

When the high redshifts were found, before the concept of expanding space redshift was known, high speeds plus the Doppler effect was a popular view. But one can also find reservations about the Doppler explanation, even before the de Sitter effect (Sect. 9.5.1). In letters to the discoverer, V.M. Slipher, William Campbell (Lick Observatory) underlined that the high velocities are very important, though it was not yet clear for what they are important. In 1914 Campbell wrote "As you have already stated, one naturally seeks for an explanation other than that of a Doppler effect. There is great interest in determining whether these high velocities are all of the same sign ... or whether you have cases opposite in sign. The former condition would encourage us to look further for explanations not based on radial velocities, whereas the latter condition would practically decode the question in favor of the Doppler effect" (Brémond 2008).

Attempts to find a mechanism producing a "plus-sign" spectral shift (other than the de Sitter effect and the expansion Lemaître redshift) started with the tired light effect (Zwicky 1929). This concept refers to some unknown process making photon energy decrease during its flight. So, one could say (Jaakkola et al. 1979) that "redshift is not an effect associated with space (its expansion) but one associated with the presence of matter". After Zwicky, there have been attempts to find a viable mechanism for the photon energy loss (e.g., LaViolette 1986; Brynjolfsson 2006). Also the gravitational cosmological redshift would belong to this general class where photon interacts with matter (gravitons).

There are a few tests in principle able to refute the tired light models, if expansion is real. We have discussed these in Chap. 8 (the surface brightness test, time dilation,

[5]Chernin (2002) suggests that the near-similarity of the four Friedmann integrals (for the four energies) may reflect a deep time-independent cosmic internal symmetry (COINS) characterizing our universe.

change of redshift). The first two tests have gone towards confirming the presence of expansion effects. The test suggested by Sandage (1962)—the slow change of redshift for individual objects in expanding space—remains an important challenge for the future.

9.5.1 Cosmological Gravitational Redshift

Space expansion ruled by uniform matter produces both the redshift and the linear Hubble law. It is interesting to ask what a global gravitational effect as a redshift mechanism would require from cosmological physics and the distribution of matter. In fact, there are theoretical hints that such an effect can exist and that for a smooth matter distribution it could lead to a quadratic r–z law for small redshifts.

De Sitter Effect of Gravitational Redshift In early history of relativistic cosmology de Sitter (1917) found a static solution of Einstein's equations for an empty universe with cosmological constant Λ:

$$ds^2 = \left(1 - \frac{r^2}{R_\Lambda^2}\right)c^2dt^2 - \frac{dr^2}{1 - r^2/R_\Lambda^2} - r^2(d\theta^2 + \sin^2\theta d\phi^2). \tag{9.35}$$

Here r is the distance from the source to the observer, and $R_\Lambda^2 = 3/\Lambda$ is the characteristic radius corresponding to the cosmological constant Λ.

The de Sitter effect is caused by the g_{00} component of the metric and from the definition $1 + z_g = 1/\sqrt{g_{00}}$ it is the cosmological gravitational redshift for a homogeneously distributed substance with positive mass density $\rho_\Lambda = \Lambda c^2/8\pi G$. Einstein (1917) used an extra condition on the metric $g_{00} = 1$, thus he lost the gravitational redshift in his static model.

Eddington (1923b) emphasized that in "De Sitter's theory ... there is the general displacement of spectral lines to the red in distant objects due to the slowing down of atomic vibrations which ... would be erroneously interpreted as a motion of recession". In his discovery study Hubble (1929) refers to the de Sitter effect as a possible explanation of the distance-redshift law.

We see that the cosmological gravitational redshift was considered as an explanation of the observed spectral shifts already before the expansion interpretation. In a sense this is a new effect in cosmological physics, due to the non-local observation. It appears only on cosmological scales and is not related to the Pound-Rebka redshift experiment in the local gravity field. The global effect is always redshift, while locally a photon has two possibilities—redshift or blueshift, depending on the direction of propagation.

Small Redshifts Within expanding space cosmology Bondi (1947) in a sense rediscovered the de Sitter effect. He considered a spherically symmetric distribution of dust matter and showed that at small distances the observed cosmological redshift, i.e. the Lemaitre-effect, can be expressed as a sum of two parts. The first one is the velocity effect due to the relative motion of the source and the observer. The second part is due to the global gravitation effect of the total mass inside the spherical ball with the light source at the centre of the ball and the observer at the surface of the ball. The infinitely distributed matter outside the ball does not affect the value of the redshift.

Bondi showed that for homogeneous matter and small redshifts ($z \ll 1$) the gravitational cosmological redshift is

$$z_{\text{grav}} = \frac{\delta\phi(r)}{c^2} = \frac{1}{2}\frac{GM(r)}{c^2 r} = \frac{1}{4}\Omega_0 \left(\frac{r}{R_{\text{H}}}\right)^2, \tag{9.36}$$

where $\delta\phi(r) = \phi(r) - \phi(0)$ is the gravitational potential difference between the surface and the centre of the ball and $R_{\text{H}} = c/H_0$ is the Hubble distance.

Bondi (1947) noted that "the sign of the velocity shift depends on the sign of v, but the Einstein shift is easily seen to be towards the red". In the case of a static universe the velocity term is zero, hence the only term is the global gravitational redshift (Eq. (9.36)), which is the "de Sitter" effect but now for the ordinary matter distributed in all infinite space.

Why is the cosmological gravitational effect a redshift? From causality it follows that the event of emission of a photon (or a spherical pulse) by the source which marks the centre of the ball, must precede the detection of the photon. The latter event marks the spherical surface where all potential observers are situated after the time $t = r/c$. Therefore to calculate the gravitational shift within the cosmologically distributed matter one should cut a matter ball with the centre at the *source*, with the radius equal to the source–observer distance. Then the shift is towards red.

It is interesting to mention that some discussions placed the *observer* to the centre of the ball and hence a blueshift was obtained instead of Bondi's and de Sitter's redshift (Zeldovich and Novikov 1984, p. 97; Peacock 1999, Problem 3.4). However, we surmise such a choice of the reference frame violates the causality in the process considered: the ball with the source on its surface has no causal relation to the emission of the photon.

Equation (9.36) shows that when $c \to \infty$, the redshift drops to zero. In Newtonian physics one may put the sphere either around the source or the observer, without causality problems, and thus infer that $z_{\text{grav}} = -z_{\text{grav}} = 0$.

Einstein's Local Gravitational Shift Consider a static spherically symmetric mass distribution inside a ball (radius r, mass M). The light emitted by a source at the ball's surface is observed to have at infinity the redshift

$$(1+z)_{\text{grav}} = \left(1 - \frac{2GM}{rc^2}\right)^{-1/2} \tag{9.37}$$

from $\nu_{\text{obs}} = \sqrt{g_{00}}\,\nu_{\text{em}}$ where Einstein's gravitation factor g_{00} is taken for the Schwarzschild coordinates. Note that $z_{\text{grav}} \to \infty$ when $r \to R_{\text{g}} = 2GM/c^2$.

Einstein's gravitational shift changes the sign when the light changes the direction of propagation. For an observer at a large distance from the body the spectral line shift is redshift, but for an observer at the surface of the body (and the source at a large distance from it), the observed shift is towards blue. So the Einstein effect will lead to extra noise in the global gravitational redshift due to local fluctuations of the gravitational potential.

Fractal Matter Distribution In order to have a linear redshift law, instead of the square law as expected within uniform matter, one may consider a fractal distribution (Baryshev 1981). Generalizing Eq. (9.36) to the distribution $M(r) \propto r^D$ one can derive for small redshifts a relation for the gravitational part of the redshift. For a fractal structure with $D = 2$ this cosmological gravitational redshift is a linear function of distance:

$$z_{\mathrm{grav}}(r) = \frac{4\pi G\rho_0 r_0^2}{c^2 D(D-1)}\left(\frac{r}{r_0}\right)^{D-1} = \frac{2\pi G\rho_0 r_0}{c^2} r = \frac{H_{\mathrm{g}}}{c} r \quad (D = 2). \tag{9.38}$$

"The gravitational Hubble constant H_{g}" may be expressed as

$$H_{\mathrm{g}} = 2\pi\rho_0 r_0 \frac{G}{c}. \tag{9.39}$$

For a structure with fractal dimension $D = 2$ the constant $\beta = \rho_0 r_0$ may be viewed as fundamental. Rost (2004) suggested the concept fractal density $\rho(r)r^{3-D}$ having a constant value for a general fractal. In this case $\beta = \rho_0 r_0$ is the fractal density for $D = 2$, determining the value of the gravitational Hubble constant. For instance, if the fractal density $\beta \approx 1/2\pi$ g/cm^2 (say, $\rho_0 = 5 \times 10^{-24}$ g/cm^3, $r_0 = 10$ kpc), then $H_{\mathrm{g}} = 2\pi\beta G/c \approx 69$ km s^{-1}/Mpc. A linear redshift law would seem possible, but it would require a huge amount of dark matter, organized fractal-like with $D = 2$ (see below).

For *large redshifts* there is still no exact field gravity theory and we consider only a hypothetical approximate formula. The PN approximation suggests that the strong gravity redshift could be given by the relation $1 + z = 1/\sqrt{1 + 2\varphi(r)/c^2}$. This describes a spectral shift for light radiated at point r, and detected by an observer at infinity. Hence for a source at the centre of a matter ball ($r = 0$) and an observer at its surface ($r = R$), the observed redshift will be

$$1 + z_{\mathrm{obs}} = \frac{1 + z(0)}{1 + z(R)} = \left(\frac{1 + 2\varphi(R)/c^2}{1 + 2\varphi(0)/c^2}\right)^{\frac{1}{2}}. \tag{9.40}$$

Inserting the expressions for the gravitational potential one can predict:

$$z_{\mathrm{obs}}(x) = \left(\frac{1}{2\sqrt{x}} I_1(4\sqrt{x})\right)^{\frac{1}{2}} - 1 \equiv W(x). \tag{9.41}$$

Here $x = r/R_{\mathrm{H}}$ and $R_{\mathrm{H}} = c/H_{\mathrm{g}}$. An approximation for calculating the modified Bessel function $I_1(y)$ is given by Abramowitz and Stegun (1964), Eqs. (9.8.3), (9.8.4).

The Total Mass-Radius Relation The equation (C.4) for the gravitating mass in Appendix C has two characteristic limiting cases. For small distances ($r \ll R_H$)

$$M(r) = 2\pi \rho_0 r_0 r^2 = 4.8 \times 10^{11} M_\odot (r/10\ \text{kpc})^2. \tag{9.42}$$

It is an interesting coincidence that this mass is close to a galaxy mass within $r \approx$ 10 kpc, and also to the mass within the Hubble radius R_H, if the density is close to the critical value. In order to produce the gravitational Hubble law on scales of about 10 Mpc the total mass within such a ball should be $M(10\ \text{Mpc}) \approx 5 \times 10^{16} M_\odot$, and one has to postulate large amounts of dark matter. The organization of galaxies on scales from 10 kpc up to 100 Mpc resembles a fractal with $D \approx 2$. If dark matter follows luminous matter, then the condition $D = 2$ for the total mass could be met.

Dark matter, both baryonic and non-baryonic, is an essential part of today's cosmological thinking, and its appearance also in this non-standard approach is interesting. However, there are two problems in this scheme, pending solution. First, to have small fluctuations in the Hubble law the fractal should be special: isotropic with small lacunarity. Second, and this is especially serious: the huge amount of dark matter on small scales (on large scales its average density is around the critical value, as noted above). This is best seen on the scale corresponding to the Local Group of galaxies, $M(1\ \text{Mpc}) \approx 5 \times 10^{14} M_\odot$, a hundred times more than even the highest current estimates (Chernin et al. 2009) for the mass of the Local Group.

9.5.2 Anomalous Redshifts

During the last decades some observations have challenged the view that the redshift of a distant extragalactic object always depends on its distance or that the redshift is a continuous function of the distance.

Anomalous Redshifts In 1966 Halton Arp, at the Palomar Observatory, noticed that radio sources, among them quasars, tended to be close to, or aligned across, some of the galaxies in his *Atlas of Peculiar Galaxies*. If true, such associations between high-redshift quasars and low-redshift galaxies would be violations of the Hubble law: the quasars would have a large extra component in their redshift, in addition to the cosmological one.

The current idea is that quasars indeed are associated with galaxies, though not as nearby companions, but as active galactic nuclei having essentially the same redshift as their host galaxies. This mainstream view regards Arp's configurations as chance associations. Indeed, when looking at the space populated by objects at different distances one necessarily finds "optical double stars". But are the associations more frequent than expected from chance alone? One can calculate the expected number of such cases, if one knows how many galaxies and quasars there are in the sky. Such estimates have been made and debated during the last decades.

Burbidge's (1996) Estimate The number of known cases when a quasar is close ($<3'$) to a bright galaxy was 46. So, draw a circle with a radius of $3'$ around every inspected galaxy. Then throw randomly all the "quasars" on the celestial globe with its total area of 41253 deg^2. Counting the scores, how many of the quasars would be found inside some 3 arcmin circle?

There are 18 000 bright galaxies ($m < 14.5$) over the whole sky and 12 500 quasars having $m < 18.5$. 18 000 circles with $3'$ radius cover 141 deg^2. Hence, one expects that $(141/41253) \times 12500 = 43$ quasars would be found by chance close to galaxies—not far from the mentioned 46. However, only a tiny fraction of the vicinities of all the 18 000 bright galaxies had been thus far inspected, especially by Arp about 200 galaxies. Hence, the expected number of coincidences will be only $(200/18000) \times 43 = 0.5$.

Goeffrey Burbidge concluded that in general quasars have an intrinsic redshift component plus a cosmological one. However, the last part of the above calculation illustrates the need of systematic studies of the galaxy-quasar distribution over large areas of the sky. In fact, Burbidge (2001) lists several statistical investigations which use well-defined samples of quasars and compare their positions with bright low-z galaxies. These suggested statistically significant, even strong correlations.

With large numbers of galaxies and quasars now available from the Sloan Digital Sky Survey and other surveys, studies have generally indicated correlations in the mentioned sense, usually interpreted in the framework of weak gravitational lensing by foreground large-scale structures (e.g., Gaztañaga 2003; Scranton et al. 2005).

From Statistics to Physics A small number of genuine cases of anomalous redshifts would have negligible influence on statistical analyses of the associations. Their message might be important, but it would go unnoticed. In order to be detected and appreciated, such cases would need some other physical characteristics than just the small angle between the objects.

It must be noted that for quasars there are no easy way to tell their distances independently of the redshift. To say for sure that a quasar is at the same distance as a galaxy, one should see a physical connection between them. Arp has pointed out galaxy–quasar pairs where such a link seems to exist in the form of a luminous "bridge" or "tail". A famous, but debated case is the spiral galaxy NGC 4319, at $z = 0.005$, with the quasar Mrk 205 ($z = 0.072$) almost "touching" it (see López-Corredoira and Gutiérrez 2006 for some other examples).

One may also ask whether the galaxies (or quasars) in the associations are somehow special. This is actually what Arp has claimed—he noticed the associations when inspecting peculiar galaxies. López-Corredoira and Gutiérrez (2006) have emphasized such a physical approach: it complements the statistical analysis which has often led to debates on the *a posteriori* nature of the probability calculations. In a series of case studies they re-observed several Arp's systems and nearly in half of them they found some new anomalies, perhaps something expected from the theories presented by Arp and Narlikar and by Hoyle and Burbidge and based on the idea that quasars are ejected from the nuclei of galaxies and are composed of freshly created matter, which clearly would require non-standard physics.

A detailed study of the companion objects in Arp's systems may also reveal peculiarities which could be a clue to the true nature of the apparent associations. As the associations have typical sizes of the order of 100 kpc, the higher-z companions are within the massive halos of galaxies or, if they are actually distant background objects, their line-of-sights intersect the halos. This leads to the possibility to explain galaxy-quasar associations using familiar physics of gravitational lensing, especially "mesolensing" (Baryshev and Ezova 1997), which amplifies the flux from background objects and thus makes them apparently more frequent in the vicinity of foreground galaxies.

We remind that within the paradigm of the Friedmann model striking exceptions to the Hubble law are anomalies that more generally in physics "may be either genuine or (more often) wrong observations or true observations having conventional but not yet known explanations" (Aurela 1973). In fact, even conventional explanations may have interesting new implications as illustrated by Arp's effect which possibly tells about the haloes of galaxies rather than about the redshift. This would show how rare phenomena, ignored by almost everybody, may be decisive keys in science.[6]

Gravitational Lensing and Galaxy–Quasar Associations The idea that halo objects could act as lenses was used by Barnothy (1974) with globular clusters, Canizares (1981) with dark condensed objects, and Baryshev et al. (1993) with globulars and other intermediate (*meso*) mass ojects, to explain the properties of galaxy-quasar pairs.

Baryshev and Ezova (1997) calculated probabilities of strong lensing of compact background objects by King lenses in galaxy halos (King 1966 model profiles provide good fits to globular cluster surface brightness).[7] The lenses were taken to be globular clusters, dwarf galaxies, and clusters of dark matter with masses of 10^3 to 10^9 $M_{\odot}$. They pointed out that the meso-lensing effect by such sub-halo objects will magnify background point sources by about five magnitudes and would make initially star-like images split or spread to 10–100 mas (milliarcsecond) sized forms.

Yonehara et al. (2003) arrived at similar conclusions in their work based on the predicted cold dark matter lumps. Namely, high-resolution cosmological simulations of hierarchic gravitational clustering (Klypin et al. 1999; Moore et al. 1999) lead to the picture where "galaxies are scaled versions of galaxy clusters", so that the massive haloes contain large numbers of CDM lumps as satellites of galaxies. These hypothetical lumps may have masses in the range from 10^6 to 10^9 $M_{\odot}$ and galaxies

[6]Hawkins et al. (2002) wrote aptly: "More cynical critics also point out that the results [on the redshift periods of quasars] tend to come from a relatively small group of astronomers who have a strong predujice in favour of detecting such unconventional phenomena. This small group of astronomers, not unreasonably, respond by pointing out that adherents to the conventional cosmological paradigm have at least as strong a prejudice towards denying such results."

[7]The importance of the King distribution is that it has a conical caustic which can explain the enhanced probability for the galaxies harbouring the lenses to be near either the observer or the quasar. The point and isothermal-sphere lens models which are often used give an enhanced probability for the galaxy to be located in a central position between the observer and the source, in contradiction with the observed associations.

like the Milky Way could have about 5000 dark satellites with masses larger than about $10^8\ M_\odot$.

The expected mesolensing may allow one to detect the substructures observationally. Baryshev and Ezova (1997) noted that VLBI radio observations might show the predicted image splitting on several mas scales. Surdej et al. (1993) suggested the use of the VLT optical interferometric facilities to probe the mass range 10^6–$10^9\ M_\odot$ and Bukhmastova and Baryshev (2008) have seen this as an important test of the CDM substructure prediction.

A typical expected angular separation between multiple quasar images is $\theta = (4GM/c^2)^{1/2} D_{\rm eff}^{-1/2}$, where $D_{\rm eff} = (D_{\rm ol} D_{\rm os}/D_{\rm ls})$ is expressed in terms of the angular size distances between the observer, the lens, and the background source (quasar). Normalized to a mass of $10^8 M_\odot$, with $D_{\rm eff}$ in Mpc, we obtain for θ (in mas)

$$\theta = 100(M/10^8 M_\odot)^{1/2}(D_{\rm eff}/100\ {\rm Mpc})^{-1/2}. \tag{9.43}$$

Thus lumps of $10^8 M_\odot$ can produce images of 0.1 arcsec, when the effective distance of the lens galaxy is 100 Mpc.

One may make a rough estimate of the probability of mesolensing for a quasar image located near the line-of-sight towards a foreground galaxy halo: $P = N[(4GM/c^2)/R_{\rm halo}^2] D_{\rm eff}(D_{\rm ls}/D_{\rm os})^2 {\rm Bias}(V)$, where Bias(V) represents the magnification bias correction as a function of the magnitude V of the background source (Surdej et al. 1993). If we simply assume that the halo is made out of N lumps of mass M, then this may be conveniently expressed in terms of the halo mass $M_{\rm ha} = NM$ and size $R_{\rm ha}$:

$$P = 0.1\left[\left(\frac{M_{\rm ha}}{10^{12} M_\odot}\right) \Big/ \left(\frac{R_{\rm ha}}{0.1\ {\rm Mpc}}\right)^2\right]\left(\frac{D_{\rm eff}}{100\ {\rm Mpc}}\right)\left(\frac{D_{\rm ls}}{D_{\rm os}}\right)^2 \frac{{\rm bias}(V)}{{\rm bias}(V=16)}. \tag{9.44}$$

A systematic mapping of the dark halo substructure using gravitational lensing has not yet been done. If the excess of high-z quasars close to low-z galaxies is real and if lensing and amplification significantly contribute to it, then it is clear that in the observed samples of galaxy-quasar pairs mesolensing must be common. Even if the original Arp effect could be only partly explained in this way, one should make efforts to detect the expected image splitting (Bukhmastova and Baryshev 2008). Modern and near future techniques will allow one to study the substructure of galaxy halos. Large interferometers like the VLTI and KeckI-II telescopes have achieved an optical resolution of about 5 milliarcsec, though for relatively bright objects.

Anomalous Redshifts: Galaxies The question of non-cosmological redshifts has not been restricted to quasars; it has also been asked if the redshifts of ordinary galaxies show any anomalies. In particular, this has been studied from the differential velocities (redshifts) of companion galaxies with respect to the main galaxy in a group, since Arp (1970) noted that primary galaxies appear to have mainly smaller redshifts than their companions (also Bottinelli and Gouguenheim 1973). Jaakkola (1971) and Jaakkola and Moles (1976) pointed out that the effect seems to depend on the morphological type of the companion, excess redshifts (of the order of 100 km/s) appearing especially for Sbc and Sc galaxies.

These results were sometimes interpreted as evidence for anomalous redshifts, though much smaller than in quasars. Jaakkola and Moles (1976) considered also a possibility of selection effect: the group data might contain non-members (projected field galaxies taking part in the Hubble flow). Because the background volume is larger than the foreground volume, one might thus ascribe to the group more galaxies from the background (larger redshifts) than from the foreground (smaller redshifts). Jaakkola and Moles did not investigate such a selection quantitatively and for various reasons they concluded that it does not work. However, Byrd and Valtonen (1985) made a quantitative study where they showed that if most of the group population is composed of unbound expanding members then the way the groups are selected will lead to an artificial redshift excess for the companions. This work was especially motivated by the paper by Sulentic (1984) who studied spiral-dominated galaxy groups in the catalogue published by Huchra and Geller two years earlier and found a significant excess of positive redshifts for the companions.

One should note that the explanation given by Gene Byrd and Mauri Valtonen was based not only on a selection effect, but also on an idea on the dynamics of the galaxy groups (a pure field galaxy population, around a small bound group, would not offer a sufficient explanation). About 25 years later, this explanation could be essentially verified by Niemi and Valtonen (2009) using synthetic galaxy groups extracted from the cosmological N-body simulation called the Millenium Simulation performed by the Virgo Consortium (Springel et al. 2005). Niemi and Valtonen found that about one half of all groups in their mock catalogues are actually gravitationally unbound, and a large and statistically highly significant redshift excess appears only for such unbound groups, when in addition their first-ranked galaxies have been misidentified. On the contrary, gravitationally bound groups did not show any significant redshift excess.

Essentially, Niemi and Valtonen showed from numerical simulations of Λ-dominated cosmological models that groups of galaxies should possess expanding populations which reveal redshift asymmetries when viewed from outside at a reasonably close distance. Thus the observed redshift asymmetries may be viewed as indirect evidence for non-zero local dark energy (as implied by the ΛCDM model). Independent evidence for expanding populations around the Local Group and some other nearby groups, behaving roughly as expected from the repulsive "antigravity" of dark energy, have been also found from direct distance-velocity observations (e.g., Teerikorpi et al. 2008; cf. Sect. 12.3). As Valtonen et al. (2008b) noted, "the observation of 'anomalous redshifts' in 1970 proved highly significant in the case of groups of galaxies, and it pointed to the existence of dark energy even though it took many decades to fully understand the implications".

The Hubble law might also reveal anomalous redshifts if such exist for a good standard candle class. As compared with a class having pure cosmological redshift, observations of a class with a significant non-cosmological redshift in its spectrum would indicate an apparently too large Hubble constant. This is what Arp (2002) has suggested especially for ScI galaxies (very luminous Sc galaxies in van den Bergh's morphological classification). In particular, Arp inspected the redshift versus distance diagram for spiral galaxies, based on the Cepheid distances derived from the

HST observations (Freedman et al. 2001). He notes that for small distances (below about 10 Mpc) there is a tight Hubble relation corresponding to $H_0 = 55$ km/s/Mpc. At larger distances the scatter is increased, especially towards higher velocities (redshifts), which Arp interprets as anomalous redshift.

In the same year Teerikorpi and Paturel (2002) independently studied the same HST data and suggested that the Cepheid method is affected by a selection effect which leads to too low distances for distant galaxies (as we discussed in Sect. 3.4.3). This implies "too large" redshifts for such biased distances. It was also noted that unbiased (generally nearby) galaxies follow a tight Hubble relation with $H_0 \approx 56$ km/s/Mpc (Paturel and Teerikorpi 2005)—in fact, the same thing was emphasized by Arp. It is interesting that also here a plausible selection effect was found, making the possibility of anomalous redshifts less likely. At the end, both Arp and Paturel and Teerikorpi concluded that the canonical HST value $H_0 = 72$ km/s/Mpc is too high locally, though for quite different reasons!

Periodic Redshifts and Quantized Physical Quantities According to Guthrie and Napier (1996) "The term 'quantized redshifts' encompasses a set of claims which are surely amongst the most bizarre to have been made in modern astrophysics". Twenty years earlier William Tifft had found that the redshifts of galaxies in the Coma cluster seemed to occur preferentially in steps of about 72 km/s; this result was later extended to galaxy groups and even to binary galaxies (Tifft 1980). Guthrie and Napier (1996) analysed the Local Supercluster using a hundred galaxies with accurately measured redshifts. After corrected for the rotation of our Milky Way, the redshifts of galaxies had a periodicity of 36 km/s, one half of Tifft's original period (see Napier 2003 for a brief review of this result on the "global" period among field galaxies, which was earlier suggested by Tifft and Cocke 1984).

The redshift periodicities for galaxies have been little studied in recent years (Godlowski et al. 2006 and Bajan et al. 2007 found weak effects of redshift periodisation in the Local Group and in the Hercules Supercluster). Indeed, the topic is not very appealing within the standard paradigm. Leaving aside the unknown origin of the quantization, there is the problem of "Doppler smoothing": large enough real motions of galaxies should sweep away narrow redshift peaks within clusters and binary galaxies.

In standard cosmology periodic redshifts, if real, could reflect regularly spaced structures (the Hubble law: redshift $\sim$ distance). Thus Valtonen and Byrd (1986) suggested that the 72 km/s period in binary galaxies could be due to contamination by optical pairs, together with a preferred distance between concentrations of galaxies. This explanation would require 1 Mpc spacings between the concentrations, and in any case would not work for clusters. Some much larger quasi-periods in galaxy redshifts are more natural to ascribe to large-scale structure, like Einasto's 120 Mpc cells. If interpreted in this way, some periodicities claimed in quasar redshifts[8] would correspond to spatial scales of about $270h^{-1}$ Mpc (Burbidge and Napier 2001).

---

[8]Here we speak about the formula of Karlsson (1977), $\log(1+z) = 0.089$, with peaks lying at $z = 0.061, 0.30, 0.60, 0.96, 1.41, 1.96, 2.63, 3.45, 4.47\ldots$

On the other hand, the $\Delta z \approx 0.6$ periodicity claimed for the Sloan Digital Sky Survey quasars appears to be due a redshift dependent selection effect (variation of the identifying characteristics of the colours) as demonstrated by Bell and Comeau (2010). In fact, a pioneer since 1973 in the study of selection effects influencing quasar samples, Dipak Basu argues that three effects can explain all earlier claimed periodicities in quasar redshifts: (1) the availability of search lines, (2) changes in the observed $(U-B)$ and $(B-V)$ colour indices of a quasar due to the effect of emission lines entering the U, B, V bands, such that the colour indices become similar to those of main-sequence stars, and (3) changes in the U, B, V magnitudes due to the emission lines so that brighter quasars are more easy to detect (Basu 2005). It is evident that one must be very careful when trying to draw conclusions on the spatial distribution of quasars (or on their possible non-cosmological redshifts) on the basis of available large quasar samples.

We leave open the reality of the 36–72 km/s periodicities for galaxy redshifts, and just note that it is natural to turn one's attention to all properties of the redshift including subtle phenomena which, if real, would easily elude detection without dedicated studies. At the same time, one should keep in mind that Nature and observing methods can collaborate in many ways to produce interesting but physically not real effects which may seem to require "new physics".

Finally, we wish to point out an interesting parallel development in studies of the "large" and the "small". In 1990 Ari Lehto, a Finnish physicist, had come up with a general formula for quantization of physical quantities, without knowing of Tifft's astronomy studies (as narrated by Lehto 1996 and Tifft 1996). He then found that his formula describes well the redshift periods 36 and 72 km/s that Tifft has derived. If the energy of any photon is quantized, as Lehto suggests, then also the frequencies and, hence, the redshifts of light would occur at preferred values. Lehto had searched for a common rule for the properties of the micro- and the macroworld and he found that the ratios of the quantities involving lengths or energies may be expressed as $2^{n/3}$ $(n = 1, 2, 3, \ldots)$. What he did was to take the ratios of observed values of various stationary discrete systems (like Planck energy/electron rest mass etc.) and he noted that the exponents of two thus obtained seemed to group near 0, 1/3 and 2/3 (Lehto 1990). In a more extensive study (Lehto 2009), involving over 40 quantities, it was shown that it is quite unlikely that the groupings could be due to chance. The pattern is best revealed by quantities which have been most accurately measured (laboratory physics), but it seems to be visible also for cosmic quantities.

Lehto's system has the Planck units as natural starting points for making up the physical world. The Planck scale is absolute, based on constants of nature. For example, the observed temporary periods take values $t = t_{\mathrm{Pl}} 2^{n/3}$, while the electron mass is obtained as $m_{\mathrm{e}} = m_{\mathrm{Pl}} 2^{227/3} = m_{\mathrm{Pl}} 2^{75+2/3}$. The redshift steps of about 72 and 36 km/s would come from $V = 2^{-n/3} c$ where $n/3$ is 12 and 13.

That the "natural" ratios seem to be built on the basis of the number 2, Lehto interprets as a fundamental (though unknown) phenomenon analogous to that called period doubling in chaotic systems (Lehto 2009). That the ratio $2^{n/3}$ is actually a cubic root of 2^n, Lehto interprets as due to 3-dimensional time; our perceived time is 1-dimensional, or the cubic root of the 3-D temporal volume, and it is this volume which evolves through period doubling. This he regards as a novel aspect of time, and not as any direct generalisation of our ordinary experience of the flow of 1-D time.

We have mentioned Lehto's approach as another reminder of how macroscopic (astrophysical) and microscopic quantities might be connected and how at a first sight strange results of observational cosmology might make sense only within a new fundamental theory.

Chapter 10
Large-Scale Structure: Methods of Analysis

A major discovery of 20th century astronomy was the complex filamentary spatial distribution of galaxies, after large surveys of galaxy redshifts permitted astronomers to move from the study of how galaxies are scattered in the sky to their real spatial arrangement. A rich variety of structures have been revealed, described as binaries, clusters, walls, superclusters, voids, filaments, cells, soap bubbles, sponges, great attractors ... They can be viewed as natural appearances of one global master entity—stochastic hierarchical structure. To describe this novel landscape, a new empirical law was found: the power-law behaviour of the galaxy correlations.

Analysis of galaxy samples requires careful consideration of the method, its basic assumptions and sensitivity to the fact that in practice one always extracts a small part from the universe. Finite samples unavoidably lead to systematic errors when one attempts to derive the statistical properties of the spatial organization of galaxies (the spatial density distribution function).

10.1 From Simple Hierarchies to Stochastic Fractals

The ancients recognized constellations, which are not real systems of stars, just nice projections on the sky. But there are also genuine stellar groups, and the same is valid for the nebulae, faint patches at different though unknown distances. With his telescope, the comet hunter Charles Messier found a swarm of them in the constellation Virgo. William Herschel found hundreds of nebulae in Coma Berenices and elsewhere. His son John saw hints of what is now called the Local Supercluster, around the Virgo concentration, with the Milky Way "placed somewhat beyond the borders of its densest portion". Photographs began to reveal thousands of nebulae, and later, in modern terms, galaxies.

In 1927 Lundmark's map of clusters of nebulae hinted that the clusters themselves may be clustered. It took decades of arguings before this issue of superclusters was resolved. Even earlier, one had speculated on hierarchies.

Y. Baryshev, P. Teerikorpi, *Fundamental Questions of Practical Cosmology*, Astrophysics and Space Science Library 383,
DOI 10.1007/978-94-007-2379-5_10,

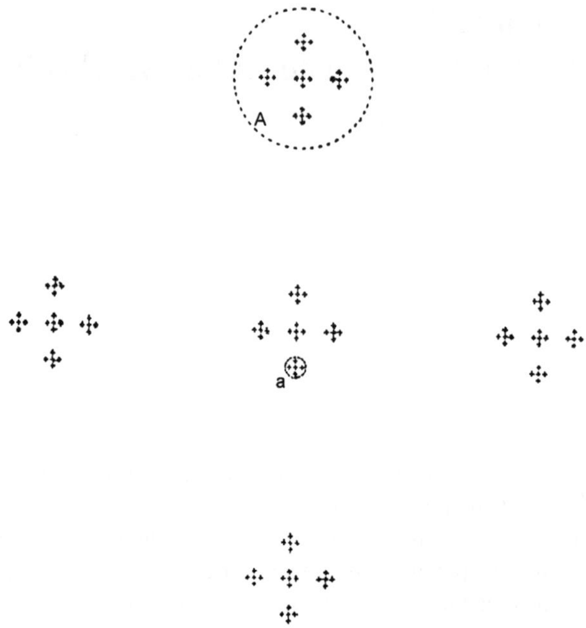

Fig. 10.1 Fournier's hierarchic world model as a protofractal: the elements are repeated on different levels in a self-similar way. The mean density decreases with increasing scale

10.1.1 Protofractal Worlds of Fournier d'Albe and Charlier

A self-similar structure of the cosmos was envisioned by Edmund Fournier d'Albe in his 1907 book *Two new worlds* which gave a mathematical description of a possible hierarchical distribution of stars. In his world (Fig. 10.1) the stars were distributed in a hierarchy of clusters, so that by construction the mass inside each sphere increases directly proportionally to its radius:

$$M(R) \propto R. \tag{10.1}$$

Dramatically different from the mass–radius behaviour in a homogeneous universe ($M \propto R^3$), this was Fournier d'Albe's idea of how to avoid paradoxes troubling Newton's universe. After inspection of Fournier's book, Swedish astronomer Carl Charlier (1908) considered other stellar distributions which also solve Olbers's paradox and the riddle of infinite gravitation (which appears in two forms, involving either forces or potentials).

The decisive factor is how fast the density decreases from one hierarchy level (i) to the next one ($i + 1$), and this depends on the ratio of the sizes of the successive elements and on the number N_{i+1} of the lower elements forming the upper element. Denoting the sizes (radii) with R_i and R_{i+1}, Charlier's first criterion may be written as

$$R_{i+1}/R_i \geq N_{i+1} \tag{10.2}$$

or the size of the upper level element divided by the size of the lower level element is larger or equal to the number of lower elements forming the upper elements.

Charlier (1922) derived, after a note by Selety (1922), a second criterion:

$$R_{i+1}/R_i \geq \sqrt{N_{i+1}}. \tag{10.3}$$

For identical particles with mass m the first criterion means $M(R) = mN(R) \propto R^1$. The second criterion, $M(R) \propto R^2$, is enough to cope with Olbers's paradox and the infinite gravity force in the Newtonian world. The original Fournier's (and the first Charlier's) criterion is a stronger condition, and allows also a finite gravitational potential and finite stellar velocities.

The Fractal Dimension of Fournier-Charlier Worlds Fournier's world was in fact an early attempt to make a protofractal. For regular, hierarchical clusters, the fractal dimension D is simply obtained from the number N of elements per cluster and the size ratio $q = R_{i+1}/R_i$. The fractal dimension is the ratio of their logarithms. Thus it is at once seen that Charlier's first criterion ($N \leq q$) means that the fractal dimension of such a universe is equal to or less than 1. The second criterion ($\sqrt{N} < q$) implies $D < 2$.

In Fournier's model mass grows proportional to radius. Hence the exponent D is equal to one. It is interesting to calculate D from Fournier's picture (Fig. 10.1). It gives $N = 5$ and $q = 7$, thus $D = 0.83$. And if we add two elements in the third dimension, then the resulting hierarchy has just the critical dimension 1!

10.1.2 Genuine Fractal Structures

Fournier d'Albe's and Charlier's hierarchies with a number of preferred scales, though overly simple for the real world, contained the seeds of the modern concept of fractal. Stochastic fractals can be used to model scale-invariant galaxy clustering without preferred scales.

Though fractal geometry emerged just a few decades ago, some elements of it can be found already in the works of Poincaré and Hausdorff about a century ago. Mandelbrot (1975) realized that fractal geometry is a powerful tool to characterize irregular systems. Nature is full of complex structures: trees, clouds, and lightnings are familiar objects, which have in common the property that if one magnifies a small portion of them, a complexity comparable to that of the entire structure is revealed. This is geometric *self-similarity*. Mandelbrot gave the following definition: *A fractal is a set for which the Hausdorff dimension strictly exceeds the topological dimension.*

The concept of fractal gives a convenient mathematical apparatus for treating complex structures with long-range power-law correlations. One can describe a non-uniform distribution by means of one number, the fractal dimension D. It appears in the (luminous) mass–radius behaviour $M \propto R^D$. A general approach which uses fractal as a model to describe large-scale structures at least on a limited range of spatial scales includes the theory of hierarchic formation of structures from galaxies to superclusters.

Fractals are simple but subtle, as Luciano Pietronero likes to say. Here we briefly describe their essential properties. We recommend Mandelbrot (1982) for an original presentation by the father of fractals and Falconer (1990) for a special mathematical treatment. We emphasize observational consequences of self-similarity so that if this important property is actually present in galaxy data, one will be able to detect it correctly.

Self-similarity and Power Law The difference between a self-similar distribution and one with an intrinsic characteristic scale was clearly discussed by Coleman and Pietronero (1992). Self-similarity implies that the rescaling of the length r by a factor b ($r \rightarrow r' = br$) leaves the considered property, presented by a function $f(r)$, unchanged apart from a renormalization that depends on b but not on the variable r. This leads to the functional relation

$$f(r') = f(b \cdot r) = A(b) \cdot f(r), \tag{10.4}$$

which is satisfied by a power law with any exponent. In fact for $f(r) = f_0 r^{\alpha}$:

$$f(r') = f_0 (br)^{\alpha} = (b)^{\alpha} f(r). \tag{10.5}$$

Here the exponent α defines the behaviour of the function everywhere. There is no preferred scale. It is true that the condition $f(r_0) = 1$ implies a certain length $r_0 = f_0^{-1/\alpha}$, and one might be tempted to call it a characteristic length. However, this is misleading for self-similar structures! The power law refers to a structure that was at the beginning constructed as self-similar and therefore cannot posses a preferred length. The value of r_0 from $f(r_0) = 1$ is just related to the amplitude of the power law, and the amplitude has nothing to do with the scaling property. Instead of $a = 1$, one could have used any other number in $f(r_0) = a$ to obtain other lengths. This is a subtle point of self-similarity; there is no reference value (like the mean density) with respect to which one can define what is big or small.

This behaviour of the power-law is in contrast with the exponential decay function $g(r) = g_0 e^{-r/r_0}$ where the intrinsic characteristic parameter r_0 fixes a preferred length scale for the function. In the power-law the dimensionless exponent α is not related to length scales at all.

Fractal Dimension The basic characteristic of a fractal structure is its dimension D. It is a measure of the "strength of singularity" around the structure points. If there is a zero-level of structure elements, as in physical fractals where there is no mathematical singularity, the rate of growth of density with a decreasing spatial scale still defines the fractal dimension.

Consider a simple regular fractal. Starting from a point occupied by an object we count the number of objects within a sphere of radius r. Suppose that in the structure of Fig. 10.1 we can find N_0 objects in a volume of size r_0. In a larger volume of size $r_1 = k_r \cdot r_0$ we will find $N_1 = k_N \cdot N_0$ objects. In a self-similar structure the parameters k_r and k_N will be the same also for other changes of scale. So, in general a structure of size $r_n = k_r^n \cdot r_0$ will have $N_n = k_N^n \cdot N_0$ objects. We can then write the number-size relation as

$$N(r) = B \cdot r^D, \tag{10.6}$$

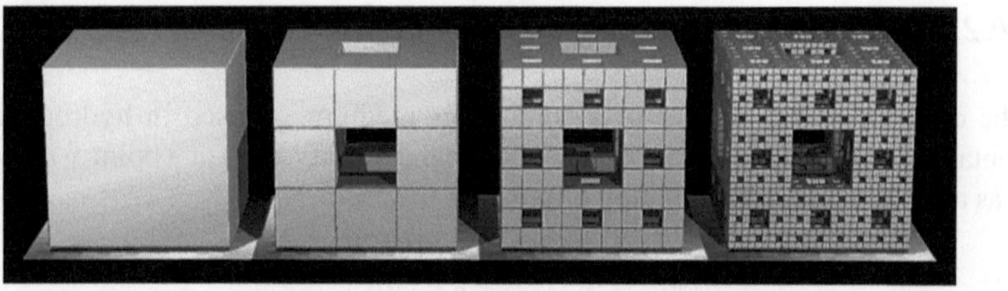

Fig. 10.2 The Menger sponge (from www.wikihow.com) illustrates how to make a regular (non-random) fractal: (1) an initial element, (2) a process of fragmentation or aggregation, following (3) an iteration rule. The true self-similar Menger fractal, the result of an infinite number of steps, is beyond imagination. However, if we apply this model to a physical object, Nature puts a size limit to the smallest elements and also the initial cube is finite. The fractal features can be seen within this range

where the fractal dimension is the exponent D of the power law, i.e.

$$D = \frac{\log k_N}{\log k_r}. \tag{10.7}$$

The prefactor B becomes fixed by the zero-level parameters N_0 and r_0, $B = N_0/r_0^D$. In terms of hierarchies, k_r is the size ratio of an upper and next lower hierarchy element and k_N is the number of elements.

Fractal structures, embedded in 3D space, no more "fill" the space in an ordinary manner. On smaller and smaller scales, where the eye would need a magnifying glass, the rising density (Eq. (10.6)) makes "new" points and rich structures appear there where uniformity would offer no surprises.

The Menger sponge. Take a cube of constant density (Fig. 10.2). Divide each face into nine equal squares. Then make three holes through the central squares of the cube. Repeat this procedure for all the remaining 20 smaller cubes. The new structure encompasses the same cubic volume, but the mass (and the density) has decreased by the factor 20/27. Now repeat again and again the same process, using the new smaller cubes. The structure resembles that of a bath sponge where all the holes are connected (this is what makes the sponge so useful: all the water is easily squeezed out). The fractal dimension $D \approx 2.7$ can be derived from (10.7) and the way the object was built: an element is always fragmented into $k_N = 20$ subelements so that the ratio of sizes is $k_R = 3$.

We note that the smooth average power-law (Eq. (10.6)) for a fractal is accompanied by large fluctuations and clustering on all scales (cf. Fig. 10.1).

10.2 The Concept of a Fractal Density Field

There are essential differences between ordinary and fractal density fields. The former kind of model is usual for the description of gas or fluid having short-range correlations, while the latter one emerges in physical systems with strong long range scale-invariant fluctuations.

10.2.1 Ordinary Fluid-Like Density Fields

The concept of the density of a continuous medium, as used in hydrodynamics, contains the assumption that one may define the density $\rho(\vec{x})$ at a point $\vec{x}$ and regard it as a usual continuous function of position in space:

$$\rho_{\text{fluid}} = \rho(\vec{x}) = \lim_{V\to 0} \frac{M(\vec{x}, V)}{V}, \qquad (10.8)$$

where M is the mass of the fluid inside the volume V around the point $\vec{x}$. For ordinary fluids the limit exists, because at sufficiently small scales homogeneity is reached. One can regard $\rho(\vec{x})$ as one realization of a stochastic process for which the usual statistical moments like average and dispersion are defined. Such ordinary *fluid-like correlated distributions* have a uniform background with correlated fluctuations superimposed.

An ordinary *stationary* stochastic density field $\rho(\vec{x})$ may be represented as a sum of density fluctuations $\delta\rho(\vec{x})$ and the mean density $\rho_0 = \langle\rho(\vec{x})\rangle$: $\rho(\vec{x}) = \rho_0 + \delta\rho(\vec{x})$, or in terms of the relative density fluctuation:

$$\delta(\vec{x}) = \frac{\rho - \rho_0}{\rho_0} = \frac{\delta\rho(\vec{x})}{\rho_0}. \qquad (10.9)$$

Note that the relative fluctuation $\delta(x)$ can have positive and negative values while the density field $\rho(x)$ is always positive for positive masses of particles.

One usually considers $\delta\rho(\vec{x})$ as a realization of a Gaussian stochastic process (with uncorrelated phases of fluctuations). Here the average density $\rho_0 > 0$ is fundamental. It should exist and be well defined and positive for each outcome of the process $\rho(x)$ (see Gabrielli et al. 2005).

Ordinary Stochastic Discrete Processes An ordinary density field may be also presented by a stochastic point-particle process. Here discreteness introduces new aspects, because particles are point-like singularities.

An important *homogeneous* stochastic discrete density field is the Poisson process. It creates a number density of particles $n(\vec{x}) = \sum_{i=1}^{N} \delta(\vec{x} - \vec{x}_i)$.

Poisson's law says that the probability P to find N particles in a volume is

$$P = \frac{\langle N\rangle^N \exp(-\langle N\rangle)}{N!}, \qquad (10.10)$$

where $\langle N\rangle = n_0 V$ is the average number of particles in the volume $V(r)$.

The only parameter of the Poisson distribution is the constant number density $n_0 = \langle n(\vec{x})\rangle$. It gives a characteristic scale λ_0 for this process

$$\lambda_0 \approx n_0^{-1/3} \approx R_{\text{sep}}, \qquad (10.11)$$

which is approximately the mean separation between the particles, R_{sep}.

The normalized number variance $\sigma^2(r)$ in a sphere with radius r

$$\sigma^2(r) = \frac{\langle N(r)^2\rangle - \langle N(r)\rangle^2}{\langle N(r)\rangle^2} \qquad (10.12)$$

characterizes a stochastic process on both small and large scales r.

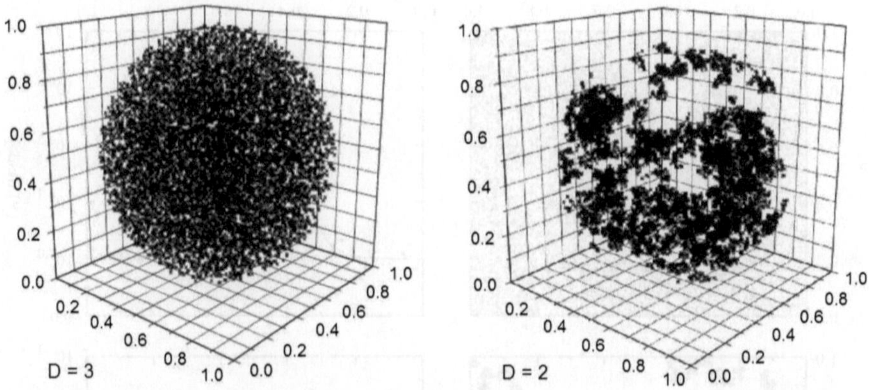

Fig. 10.3 Contrasting random homogeneity ($D = 3$) and fractality ($D = 2$) in spheres containing 10 000 points (constructed by N. Vasiliev)

A discrete stochastic process of point-particles is always accompanied by a *noise of discreteness* or *shot noise* which appear when the scale is less than the mean distance between particles $r < \lambda_0$. This fluctuation increases with decreasing distance and grows without limit on very small scales:

$$\sigma(r) \approx \frac{1}{\sqrt{N}} \approx \left(\frac{r}{\lambda_0}\right)^{-3/2} . \qquad (10.13)$$

On *large scales* r the normalized dispersion (10.13) approaches zero as $1/\sqrt{N}$. The homogenity scale $R_{\mathrm{hom}} \approx \lambda_0$ of the Poisson process is defined from the condition that the normalized number variance $\sigma^2(R_{\mathrm{hom}}) = 1$.

The Poisson process gives a homogeneous stochastic discrete density field *without correlations* (cf. Fig. 10.3). The eye may see apparent structures, but these are just random fluctuations in one outcome of the process.[1]

10.2.2 Fractal Density Fields

In the structure analysis one considers the density field in the form of a spatial distribution of N particles in positions $\vec{x}_a$ in a volume V:

$$\rho(\vec{x}) = \sum_{a=1}^{N} m_a \delta(\vec{x} - \vec{x}_a). \qquad (10.14)$$

For identical particles one may use the number density $n(\vec{x}) = \sum_1^N \delta(\vec{x} - \vec{x}_a)$.

---

[1] Another example is the *superhomogeneous discrete process* like particles in a lattice with small correlated shifts around the regular lattice knots. Such a process is used for making initial conditions for cosmological N-body simulations (Gabrielli et al. 2004).

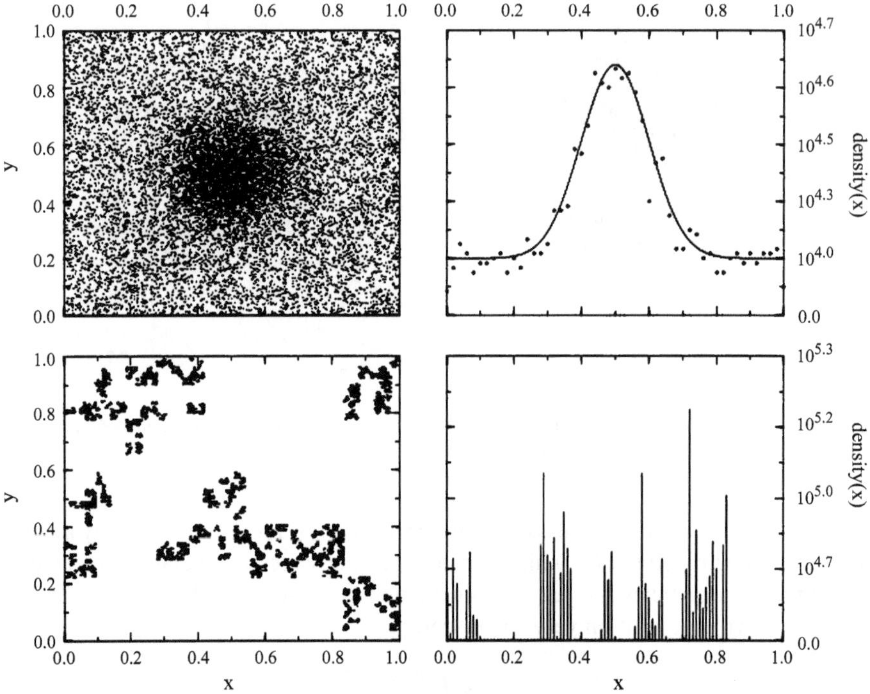

Fig. 10.4 *Top*: an ordinary (fluid-like) discrete density field: a fluctuation on a Poisson background. *Bottom*: a stochastic fractal density field. (Courtesy of F. Sylos Labini)

In order to describe the continuous hierarchy of clustering (a new characteristic of a fractal stochastic process), one may define the *mass density within a fractal* as a function of two variables, the position $\vec{x}_a$ of a particle and the radius r of the volume $V(r)$ in which the particles are counted:

$$\rho_M(\vec{x}_a, r) = \frac{M(\vec{x}_a, V)}{V}. \tag{10.15}$$

Here $M(\vec{x}_a, V)$ is the mass in the volume $V(r)$ around the particle at $\vec{x}_a$.

For a mathematical fractal the number of points of the structure in a finite volume is infinite. In physics there are some natural lower limits for sizes of elements, making the zero-level of the hierarchy. In this case the number of basic elements (point mass particles) within the volume $V(r)$ is finite, and Eq. (10.15) defines a measurable density. Though $\rho_M(\vec{x}_a, r)$ is highly fluctuating from one particle position to another, it is possible to consider a statistical average that is a more stable property of a fractal structure.

In Fig. 10.4 an ordinary density fluctuation is compared with a stochastic fractal fluctuation. For fractals the usual concept of the mass density of a continuous medium does not work. This is because the mass density can be defined only if both the position $\vec{x}$ and the volume V are considered. In every volume V containing a part of the structure there is a hierarchy of clusters and the mass density strongly depends on the size of the volume – quite different from the usual calculus. Now the mass-volume ratio or the density (Eq. (10.8)) increases indefinitely when the volume tends towards zero: $\{\rho_M \to \infty, \text{ for } V \to 0\}$. If the basic zero level elements exist then they determine the maximum value of the fractal density for the structure.

10.2.3 Exclusive Properties of Fractal Density Fields

A typical feature of the fractal density field is a *power-law density-radius relation* around structure points. We can illustrate this property for a regular fractal structure (Fig. 10.1), where the number of subelements within an element of the higher level is given by Eq. (10.6). In a continuous representation (Eq. (10.15)) the number density related to the radius r is

$$n_{V(r)} = \frac{N(r)}{V(r)} = \frac{3}{4\pi} B r^{-(3-D)}. \tag{10.16}$$

This power-law is valid around any structure point, apart from some fluctuation, depending on the actual position of the point within the structure.

The result is unexpected for our usual intuition—it seems like each point were the centre from which the density decreases outwards.

The Concept of Fractal Density Rost (2004) has introduced the *fractal density* ρ_D, the constant factor of proportionality k in the power-law $M(r) = kr^D$. It has the dimension g/cm^D and one may write

$$\rho_D = \frac{M(r)}{r^D} = \text{constant}. \tag{10.17}$$

So, for $D = 2$, the fractal density ρ_D has the dimension g/cm^2 corresponding to a surface or column density. For $D = 3$, this concept coincides with the usual average density defined for a homogeneous distribution. The fractal density ρ_D is directly related to the prefactor B in Eqs. (10.6) and (10.16).

Massive, Zero-Density Universes An important consequence of Eq. (10.16) is that in infinite space the fractal density field differs from an ordinary fluid-like density field at the limit of large volumes V where

$$\{\rho_{\text{fractal}} \to 0, \text{ for } V \to \infty\}. \tag{10.18}$$

This property is due to a growing dilution of the hierarchy with increasing scales, so that a fractal structure is asymptotically dominated by voids.

Hence an infinite fractal universe can contain an infinitely large number of objects (an infinite mass) simultaneously with the zero density of the whole universe. This unusual property of a hierarchical structure was exploited in old world models to avoid gravitational and photometric paradoxes of the infinite Newtonian universe (cf. $\varphi \propto M/R$ and $F \propto M/R^2$).

Lower and Upper Cutoffs In the realm of physics real structures usually have a lower scale $R_{\min}$ and an upper scale $R_{\max}$ between which the physical system follows self-similar behaviour. In particular, in studies of the large-scale galaxy distribution the *lower cutoff* $R_{\min}$ is assumed to be equal to the size of a galaxy (galaxies play a role of point-like particles).

The *upper cutoff* $R_{\max}$ presents a special problem. A good analysis method should allow one to determine from a galaxy survey the scale $R_{\max}$ where the galaxy distribution becomes uniform. However, such methods need a large survey volume with size several times the scale $R_{\max}$. We still do not know the upper cutoff scale for the galaxy distribution.

Lacunarity Two structures with the same fractal dimension may look very different. It is essential how large a relative volume is occupied by voids, on a given scale. This property was termed *lacunarity* by Mandelbrot (1982). It may be characterized by the factor F in the relation

$$N_v(\lambda > \Lambda) = F\Lambda^{-D}, \tag{10.19}$$

where N_v is the number of voids with size $\lambda > \Lambda$ within a fixed volume. For examples of structures of different lacunarities see Martinez and Saar (2002).

It was the high lacunarity of the Rayleigh–Lévy flight fractal that did not favour it as a model for the distribution of galaxies (Peebles 1980). Later Mandelbrot (1989) demonstrated that fractal structures with a small lacunarity resemble more closely the arrangement of galaxies.

Projections and Intersections The properties of orthogonal projections and intersections of a fractal object influence the analysis of galaxy samples with different geometries, both from angular 2D and spatial 3D catalogues.

Orthogonal Projection Let an object (structure) with a fractal dimension D, embedded in an Euclidean space of dimension $d = 3$, be orthogonally projected onto an Euclidean plane ($d' = 2$). Then according to a general theorem of fractal projections (see Mandelbrot 1982; Falconer 1990), the projection as a fractal object has the fractal dimension D_{pr} so that

$$D_{\text{pr}} = D \quad \text{if } D < 2 \quad \text{and} \quad D_{\text{pr}} = 2 \quad \text{if } D \geq 2. \tag{10.20}$$

This means that a cloud having the fractal dimension $D \approx 2.5$ gives rise to a homogeneous shadow ($D_{\text{pr}} = 2$) on the ground (Fig. 10.5). Consequently, the orthogonal projection hides from view fractal structures with $D > 2$. This may even affect the distribution of galaxies in the sky (Chap. 11).

Intersection of a Fractal If an object with a fractal dimension D, embedded in a $d = 3$ Euclidean space, intersects an object with the dimension D', then according to the law of co-dimension additivity (Mandelbrot 1982; Falconer 1990), the dimension of the intersection D_{int} becomes

$$D_{\text{int}} = D + D' - d. \tag{10.21}$$

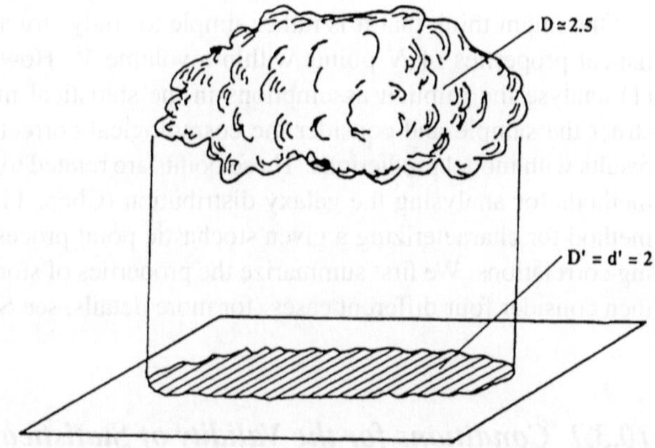

Fig. 10.5 The projection of a $D = 2.5$ cloud covers completely the ground

For example, if a fractal structure with $D = 2$ in 3-d space is intersected by a plane with $D' = 2$, then the fractal dimension of the thin intersection is $D_{\rm int} = 2+2-3 = 1$. This property of intersections explains why a fractal structure with $D \approx 2$ may look as a fractal with $D \approx 1$ when inspected on large scales from a sample coming from a thin slice-like galaxy survey.

The Fractal Dimension of a Subset Visible to the Observer Järvenpää et al. (2003) have proved an interesting theorem about visible parts and dimensions: Let F be a fractal set in R^3 with $D > 2$. The visible part of the set F from a point P is the subset F_V of those points lit by a spotlight at P. Then the part F_V that is visible to an observer cannot in general have a dimension more than 2.

Here one does not speak about the projection of F_V, but about its spatial distribution. Eckmann et al. (2004) have considered a related problem in the galaxy universe: in a deep sample a part of galaxies will remain "behind" more nearby galaxies and thus for this reason will drop away from the sample. The remaining sample can have D at most 2 even if as a whole the galaxy distribution has $2 \le D \le 3$. In practice, the galaxy samples are not yet so deep that this "shadowing" could essentially affect their constitution.

10.3 Methods to Detect Structures in Galaxy Distribution

From the maps constructed from redshifts one may see by eye various structures. However, a quantitative analysis of the spatial distribution of galaxies is a complex task, even leading to conflicting conclusions. Here we focus on critical questions arising in practical work with galaxy samples, related to idiosyncracies and limitations of the used mathematical tools. Before going to the standard ξ correlation function and the fractal-motivated Γ function we discuss general conditions for the structure analysis. A new scale-length method (Sylos Labini et al. 2009a) allows one to verify the reliability and statistical stability of the results from the standard analysis.

One might think that it is rather simple to study structures; just describe the statistical properties of N points within a volume V. However, one should carefully (1) analyse the implicit assumptions in the statistical method employed, (2) construct the sample and consider the cosmological corrections, and (3) compare the results with model predictions. These points are related to a debate about the suitable methods for analysing the galaxy distribution (Chap. 11). In fact, what is the best method for characterizing a given stochastic point process depends on the underlying correlations. We first summarize the properties of stochastic point processes and then consider four different cases (for more details, see Sylos Labini et al. 2009b).

10.3.1 Conditions for the Validity of Statistical Analysis

In statistical studies of galaxy samples, the galaxy distribution is viewed as a realization of a stationary stochastic point process (Gabrielli et al. 2005). A stochastic point process (SPP) is stationary (SSPP), if it possesses statistical translational and rotational invariance. It can be spatially uniform or non-uniform, and it is *ergodic* if the ensemble average of a statistical quantity characterizing its properties equals its infinite volume average. Ergodicity is needed when one wants to compare average quantities in finite volumes with theoretical predictions based on ensemble averages.

A SSPP is *uniform* if, in a large enough sample the density fluctuations are small. The scale λ_0 at which uniformity begins can be defined as the scale beyond which the fluctuations of the average density on that scale are of the same order as the average density itself, while smaller on larger scales. A uniform SSPP inside a given sample has a well-defined average density, representing the ensemble value within relatively small error. A SSPP has a *crossover to homogeneity*, if it is nonuniform on scales smaller than λ_0 and uniform on larger scales (the transition from the regime of large to small fluctuations). A SSPP can be uniform, but still have *long range correlations* (a non-zero two-point correlation function $\xi(r)$ on all scales). As a test of uniformity one can use conditional properties; these are defined also when the SSPP is not uniform. A SSPP is *non-uniform* inside a sample, if the conditional density does not converge to a constant value. If the distribution is self-averaging and nonuniform then the conditional density varies with the distance. *Self-averaging* in a finite volume and on a scale r means that the probability density function (PDF) of conditional fluctuations is statistically stable in different subvolumes of radius r.

Fluctuations and Correlations: 4 Cases For a uniform stochastic point process, the average density is well-defined inside a given sample—the density, measured in a sphere of radius r randomly placed inside the sample, has small relative fluctuations. The fluctuations may be correlated, and the correlation function can be (i) short-ranged (e.g., exponential decay) or (ii) long-ranged (e.g., power law).

If the distribution is not uniform, the density in a sphere probing the sample varies wildly in different regions. Such a point distribution can generally present long-range correlations of large amplitude. Then it may, case (iii), or may not, case (iv), present self-averaging properties, depending on whether the densities measured in different subregions show systematic differences depending on their locations. If so, there is no self-averaging in space: the PDF may differ in different subregions, because (1) the underlying distribution is not translationally or rotationally invariant, or (2) the volumes are not large enough for fluctuations to be self-averaging.

The Validity of Statistical Analysis The analysts of finite samples often use the assumption that the sample density provides a good estimate of the "true" space density, i.e. the situation corresponds to the cases (i) or (ii) above. This very strong assumption may lead one to belittle finite size effects in the statistical analysis. If the distribution inside the given sample is actually not uniform (cases (iii) and (iv)), then the results are biased by important finite-size effects, and all estimations of statistical quantities based on the uniformity assumption are affected, on all scales, by this incorrect assumption (Gabrielli et al. 2005).

Therefore, one should first study whether the galaxy distribution is self-averaging. If so, then one can study the possible transition to uniformity by using, e.g., the conditional density. If the distribution is uniform, or becomes uniform on a certain scale smaller than the sample size, one can characterize the correlations between density fluctuations by using the standard two-point correlation function which is appropriate only if one has proved that the distribution is self-averaging and uniform inside the given sample.

10.3.2 *The Scale-Length Analysis*

In a finite sample we should make a difference between statistical quantities that are normalized to the sample mean density and those that are not.

Strategy In order to determine whether a meaningful estimate of the average density is possible in the given sample, one should use statistical quantities that do not require the assumption of homogeneity inside the sample. An example is the conditional density $n_i(r)$ from the ith galaxy (the density in a sphere of radius r centered on the ith galaxy). Conditional quantities are well-defined both for homogeneous and inhomogeneous point distributions. If a distribution is self-averaging inside a given sample, or in a range of scales, then it is possible to consider the whole sample average of the conditional density: $\overline{n(r)} = (\sum n_i(r))/N$, the sum taken over all N galaxies of the sample. The amplitude of conditional fluctuations is quantified by the conditional variance $\overline{\delta(r)^2} = (\overline{n(r)^2} - \overline{n(r)}^2)/(\overline{n(r)}^2)$. When a distribution is inhomogeneous, then persistently $\overline{\delta(r)^2} \sim 1$ (Gabrielli and Sylos Labini 2001), while for homogeneous distributions, with any kind of small-amplitude correlations it is valid that $\overline{\delta(r)^2} \ll 1$.

To test whether a distribution is self-averaging inside a given sample one may measure the PDF of conditional fluctuations. If this is stable in different subregions, the statistical self-averaging permits one to determine whole-sample averages. Then if the conditional density is roughly constant inside the sample (i.e. the distribution is approximately uniform), one may safely determine fluctuations in amplitude and their correlations normalized to the sample density. For instance, the two-point correlation function (see below) measures the correlation amplitude of fluctuations relative to the sample average. For a non-uniform distribution one cannot estimate well the sample average, even if the distribution is self-averaging inside the sample (Gabrielli et al. 2005).

The SL Analysis We consider the conditional number of points in spheres and compute for each scale r the number of points $n_i(r)$ inside a sphere of radius r whose centre is on the ith galaxy. The number of centres N will depend on the radius r, i.e. $N = N(r)$. The random variable $n_i(r)$ depends on scale r and on the spatial position of the sphere's centre. We can express the location of the centre using polar coordinates, $n_i(r) = n_i(r; R_i, \alpha_i, \beta_i)$. When we integrate over the angular coordinates for a fixed radial distance R_i, we find that $n_i(r) = n(r; R_i)$ will depend on two variables: the length scale of the sphere r and the distance scale of the ith sphere centre R_i, hence the term "scalelength analysis" (Sylos Labini et al. 2009a).

The number of centres $N(r)$ will depend on the scale r. Why? This is because one should use only such spheres which are fully contained in the sample volume. If one uses incomplete spheres partially contained in the volume (and extrapolates to the outside), it may happen, especially for large spheres, that outside the sample there is a large-scale structure (or an under-density). This will introduce a bias in the measurements, affecting large-scale determinations. As the aim is usually precisely to study the properties of large-scale structures, one naturally should avoid a method that implicitly assumes that these are irrelevant (Gabrielli et al. 2005; Sylos Labini et al. 2009b).

One can study the statistical properties of the random variable $n_i(r)$ by determining its PDF, $P(n, r)$. This gives the probability distribution to find n points in a spherical volume of radius r centered on a distribution point. It can be estimated from the frequency distribution obtained from the studied sample. In practice, one inspects the form of the PDF (Gaussian or not) and its behaviour on different scales r in subsamples of different depth (volume limited samples) (for a case study, see Sylos Labini et al. 2009b).

For instance, if one finds that there is a non-Gaussian long tail towards large scales r in the PDF, this can be interpreted as an effect of large-scale structures (large fluctuations) within the volume in question.

The behaviour of the PDF in different VL samples, for large enough sphere radii, allows one to test for the self-averaging properties of the distribution, and whether, inside a given sample, it is meaningful to derive whole-sample average quantities and whether a certain estimator is expected to give a reliable and stable measurement of the ensemble properties of the distribution. One may divide the sample into two (or preferably more) non-overlapping subsamples of same size and compare the PDFs within them. If the PDFs differ systematically, one can conclude that there is no self-averaging in the sample (see Sect. 12.1.1).

10.3.3 Definitions for Correlation Functions

The theory of stochastic processes introduces and studies different functions intended for the correlation analysis.

Complete and Reduced Correlation Functions The *complete two-point correlation function* $R_{\mu\mu}$ of a stationary isotropic process $\mu(\vec{r})$ is defined as

$$R_{\mu\mu}(r) = \langle \mu(\vec{r}_1)\mu(\vec{r}_2)\rangle, \tag{10.22}$$

where $r = |\vec{r}| = |\vec{r}_1 - \vec{r}_2|$ is the mutual distance between considered points, and $\langle\cdot\rangle$ is the ensemble average over all realizations of the stochastic process.

Taking into account the truly constant mean value μ_0 of the process, $\mu_0 = \langle \mu(\vec{r})\rangle = \text{const}$, one may define the *reduced two-point correlation function* C_2 for the fluctuations around μ_0

$$C_2(r) = \langle (\mu(\vec{r}_1) - \mu_0)(\mu(\vec{r}_2) - \mu_0)\rangle = R_{\mu\mu}(r) - \mu_0^2. \tag{10.23}$$

For $r = 0$ it expresses the squared dispersion of the process as $\sigma_\mu^2 = C_2(0)$.

For example, as we shall see below, Peeble's two-point ξ correlation function actually is the reduced two-point correlation function $C_2(x)$, normalized to the squared mean value μ_0: $\xi(r) = C_2(r)/\mu_0^2$, while Pietronero's conditional density (Γ function) is the complete two-point correlation function $R_{\mu\mu}$, normalized to the mean value μ_0: $\Gamma(r) = R_{\mu\mu}/\mu_0$.

There is an important difference between the complete and reduced correlation functions. For a stochastic fractal (cf. Eq. (10.16)) the complete correlation function $R_{\mu\mu}(r)$ has a power-law form, while the reduced correlation function $C_2(r)$, according to its definition (10.23), cannot be a power law in this case. The same is also true for ξ and Γ functions: if one of these is power-law, then the other is not.

Mass Variation in Spheres and Characteristic Scales In the applications below, the stochastic process $\mu(\vec{r})$ will describe the density field $\rho(\vec{r})$.

One should distinguish between *conditional* and *unconditional* functions. When we inspect such statistics which are defined with the condition that there is a fixed point-particle relative to which other particles of a process are considered, then we have a conditional function. The major tools of structure analysis, Γ and ξ, are both conditional correlation functions.

As an example of unconditional statistics we consider mass (number) fluctuations inside a sphere of radius R using the stochastic variable $M(R)$:

$$M(R) = \int_{V(R)} \rho(\vec{r})d^3r. \tag{10.24}$$

For a given radius R fluctuations of this mass calculated in different positions in space can be characterized by the *normalized mass variance* $\sigma_M^2(R)$

$$\sigma_M^2(R) = \frac{\langle M(R)^2\rangle - \langle M(R)\rangle^2}{\langle M(R)\rangle^2}, \tag{10.25}$$

where

$$\langle M(R)\rangle = \frac{4\pi}{3}\rho_0 R^3, \qquad \langle M(R)^2\rangle = \int_V d^3r_1 \int_V d^3r_2 \langle \rho(\vec{r}_1)\rho(\vec{r}_2)\rangle. \qquad (10.26)$$

Here the location of the sphere is arbitrary. Its centre may be put anywhere within the sample, also "between" the particles.

The Scale of Homogeneity Our intuitive vision of uniformity may be formalized by means of the variable $M(R)$. Define the *homogeneity scale* R_{hom} as the scale at which $\sigma_M^2(R_{\text{hom}}) = 1$ (or some other threshold value). So a distribution of particles approaches homogeneity if the average mass fluctuation within spheres of radius R_{hom} is about the average mass $\langle M(R_{\text{hom}})\rangle$. This scale is well defined when $\sigma_M^2(R) \to 0$ for $R > R_{\text{hom}}$.

The Correlation Length Another scale is the *correlation length* R_{cor}. It does not depend on the amplitude of the correlation function and just characterizes the rate of decrease of the correlation function. The correlation length may be infinite, as it is for a power law correlation $R_{\mu\mu}(r) \propto r^{-\gamma}$, or finite, as for an exponential correlation function $R_{\mu\mu}(r) \propto e^{-r/R_{\text{cor}}}$.

The Zero-ξ Length The scale r_c where $\xi(r_c) = 0$ is from the theoretical point of view more fundamental than the above mentioned scales. Its true value is determined by the physics of the structure formation process.

10.3.4 The Method of the Reduced Correlation Function ξ

A classical approach to the analysis of the large-scale structure is the correlation function, first introduced to galaxies by Totsuji and Kihara (1969). They adopted this method from the statistical physics of density fluctuations of ordinary gas (e.g., Landau and Lifshitz 1958). It was further developed and extensively applied to galaxy data by Peebles (1980, 1993) and others.

Peebles' ξ-Correlation Function One may define the *two-point correlation function* $\xi(r)$ as the dimensionless *reduced* correlation function of the density fluctuations $\delta\rho(\vec{r}) = \rho(\vec{r}) - \rho_0$ around the average density ρ_0 (Peebles 1980).

$$\xi(r) = \frac{\langle \delta\rho(\vec{r}_1)\delta\rho(\vec{r}_2)\rangle}{\rho_0^2} = \frac{\langle \rho(\vec{r}_1)\rho(\vec{r}_2)\rangle - \rho_0^2}{\rho_0^2}. \qquad (10.27)$$

In fact, the ξ-function is simply the reduced correlation function (Eq. (10.23)) divided by the squared mean value of the process, i.e.

$$\xi(r) = C_2(r)/\rho_0^2. \qquad (10.28)$$

For a distribution of identical particles (with mass m_0) one uses a number density $n(\vec{r}) = \rho(\vec{r})/m_0$, whose average is $\langle n(r)\rangle = n_0$. Then

$$\xi(r) = \frac{\langle n(\vec{r}_1)n(\vec{r}_1 + \vec{r})\rangle}{n_0^2} - 1. \tag{10.29}$$

This dimensionless function measures correlations of fluctuations relative to a constant average number density n_0.

In the theory of stochastic processes one usually considers a *normalized correlation function* which is defined as $K_{\mu\mu}(r) = C_2(r)/\sigma_x^2 = (R_{\mu\mu}(r) - x_0^2)/\sigma_x^2$, and has the normalization condition $K_{\mu\mu}(0) = 1$. The definition of the ξ-function (Eq. (10.27)) implies the condition $\xi(0) = \sigma_\rho^2/\rho_0^2$.

Definition via Poisson Process The correlation function may also be defined as a measure of the deflection of a distribution of particles from a Poisson distribution (Peebles 1980). Consider two small spheres at points $\vec{r}_1$ and $\vec{r}_2$ with volumes dV_1 and dV_2 and with a mutual distance $\vec{r}_{12}$. Then the joint probability to find one particle in the volume dV_1 and another one in the volume dV_2 is proportional to the number of pairs dN_{12}

$$dN_{12} = n_0^2 dV_1 dV_2[1 + \xi(\vec{r}_{12})], \tag{10.30}$$

where n_0 is the average number density and $\xi(\vec{r}_{12})$ measures the deflection from the Poisson distribution (so, $\xi(\vec{r}_{12}) = 0$ for a Poisson process).

For a statistically isotropic distribution the function $\xi(\vec{r}_{12}) = \xi(r)$ depends on the separation r only. When an object is chosen at random from the sample, the expected number of neighbours dN at a distance r in dV is

$$dN = n_0 dV[1 + \xi(r)]. \tag{10.31}$$

Here $\xi(r)$ is considered as the same two-point correlation function as defined by Eq. (10.29) (see Sect. 33 of Peebles 1980). It is a measure of finding an excess number of particles relative to the Poisson distribution, at the distance $r > 0$ provided that there is a particle at $r = 0$. By definition (10.31), the correlation function $\xi = 0$ for a Poisson process.

ξ-Function Estimators Considering stochastic processes it is important to make a distinction between functions (e.g., $\xi(r)$) defined by ensemble averages and their estimators ($\hat{\xi}(r)$) applied to a finite galaxy sample.

To estimate the two-point correlation function from a sample of N_s objects within a volume V_s, one generally makes use of an artificial Poisson process, which fills the same volume V_s of the sample. Then the ξ-function for a given scale r is estimated as the ratio of the number of pairs with such mutual distance in the sample to the number of such pairs in the artificial Poisson distribution. Several different pairwise estimators exist (Kerscher et al. 2000; Martinez and Saar 2002; Gabrielli et al. 2005), and they differ mainly in their method of edge correction.[2]

[2]This means the way how to calculate the number of pairs when the points are so close to the border of the sample that in some pairs one point will lie outside the sample volume if we could repeat the counts in spheres outside the observable volume.

The Davis–Peebles estimator weights the points according to the part of the spherical shell volume contained in the sample volume. It is

$$\hat{\xi}(r) = \left(\frac{N_{rd}}{N_s - 1}\right) \frac{N_p(r)}{N_{p,rd}(r)} - 1, \tag{10.32}$$

where $N_p(r)$ is the number of data pairs of observed objects in the catalogue having their mutual distance in the interval $(r, r + dr)$. $N_{p,dr}(r)$ comes from the joint catalogue of data and artificial random distributions in the same volume V_s. It is the number of data-random pairs with the distance r in the joint catalogue. N_s and N_{rd} are the total numbers of objects in the real sample and the random distribution, respectively.

A central assumption of the correlation function method is homogeneity so that the true average n_0 is estimated from the observed sample as

$$\hat{n}_0 = \bar{n} = \frac{N_s}{V_s} \tag{10.33}$$

with a high formal accuracy $\sigma_{n_0} \approx 1/\sqrt{N_s}$. Here N_s is the total number of objects in the volume V_s of a "fair" sample that is assumed to be representative of the homogeneous distribution of galaxies in the whole universe.

The Normalization Condition for ξ Estimators A significant point related to ξ-function estimators was emphasized by Pietronero (1987) and Calzetti et al. (1988), and in more detail by Gabrielli et al. (2005). Namely, the definition of the correlation function as a deflection from the Poisson distribution (Eq. (10.31)) implies an integral condition for the ξ function estimated from a finite sample of galaxies.

For any sample with a finite number of galaxies N_s in a volume V_s one may define an average number density simply as $\bar{n} = N_s/V_s$. Then integrating the left side of Eq. (10.31) over the sample volume we get $\int_{V_s} dN = N_s - 1$, where $N_s - 1$ is the number of neighbours, i.e. the total number of particles in the volume V_s without the one whose neighbours are counted. Then the integration of the right side of (10.31) over the sample volume gives

$$N_s - 1 = \int_{V_S} \bar{n} dV + \bar{n} \int_{V_S} \hat{\xi}(r) dV. \tag{10.34}$$

The first term on the right side is $\int_{V_s} \bar{n} dV = N_s$, the total number of particles in the sample. Hence the second term will satisfy the condition $\int_{V_S} \hat{\xi}(r) dV = -1/\bar{n}$. For fluid-like correlated distributions the effective number density may be arbitrarily large and hence this condition becomes

$$\int_{V_S} \hat{\xi}(r) dV = 0. \tag{10.35}$$

These restrictions may lie behind some controversial results obtained by the ξ function method. In particular, we have in mind the inevitable non-power law behaviour of the ξ estimator. Equation (10.35) implies that there is a distance r_z where $\hat{\xi}(r_z) = 0$. Here the estimator changes its sign from positive to negative values, which is impossible for a power-law function.

A Distorted Power-Law Correlation Due to the $\hat{\xi}$-Estimator As was shown above, if the complete correlation function is a power-law then neither $\xi(r)$ nor $\hat{\xi}(r)$ can be a power-law function. Nevertheless, in practice $\hat{\xi}(r)$ is usually presented in the form valid for some range of scales $r_1 < r < r_2$:

$$\hat{\xi}(r) = \left(\frac{r}{r_0}\right)^{-\gamma}, \quad r_1 < r < r_2. \tag{10.36}$$

From such a power-law presentation one usually derives two numbers: the *unit scale* r_0 and the *correlation exponent* γ. We emphasize that due to the normalization condition (10.35) both numbers give systematically distorted values for the homogeneity scale and the power-law exponent of the true complete correlation function $R_{\mu\mu}(r)$ describing the density field.

The unit scale r_0 (often called, somewhat misleadingly, correlation length) is defined from the relation

$$\hat{\xi}(r_0) = 1, \tag{10.37}$$

which characterizes the amplitude of density fluctuations at the scale r_0. In fact, it is a distorted value of a true homogeneity scale of the distribution if the true value is larger than the maximal size of a sphere that may be completely embedded in the sample volume V_s.

The correlation exponent γ in the power-law representation of $\hat{\xi}(r)$ (10.36) describes correctly only a limited interval of scales $r_1 < r < r_2$. On scales $r > r_2$ this does not represent the true value of the exponent, because there the estimated value is distorted as the normalization condition (10.35) makes $\hat{\xi}(r)$ deflect from the inherent power-law and to cross zero level. For example, the estimated exponent γ_{est} at the unit scale r_0 is two times larger than the true correlation exponent (Sect. 10.3.6).

On scales $r < r_1$ the value of the exponent is affected by the noise of discreteness, behaving as $\sigma \propto r^{-3/2}$ (Eq. (10.13)). The error will essentially grow on scales smaller than the average distance between the particles.

10.3.5 The Method of the Conditional Density Γ

The conditional density has been used for analysis of fractal structures in modern statistical physics. Proposed for extragalactic astronomy by Pietronero (1987), it has been applied to 3D galaxy catalogues by many authors. On certain conditions, this method can give an undistorted estimation of the power law correlation and the fractal dimension. It may also be used for finding the true homogeneity scale of a galaxy sample.

Continuous Stochastic Processes The *conditional density* $\Gamma(r)$ is defined by means of the complete correlation function (Eq. (10.22)) as follows:

$$\Gamma(r) = \frac{R_{\mu\mu}(r)}{\mu_0} = \frac{\langle \rho(\vec{r}_1)\rho(\vec{r}_1 + \vec{r})\rangle}{\rho_0}. \tag{10.38}$$

Here $\rho(\vec{r})$ is the stochastic density field and ρ_0 is the ensemble average density. The Γ-function is a measure of correlation in the total density field without subtraction of the average density. The common interpretation of $\Gamma(r)$ as an average density law around each point of the structure makes its estimator a natural detector of fractality.

A statistical test of the nature of the galaxy distribution consists in measuring both correlation functions ξ and Γ for the same sample of galaxies. If the Γ function is power–law, this is evidence for a fractal distribution.

Discrete Stochastic Fractal Processes Let us consider a discrete stochastic process, one realization of which is a set of identical particles at randomly selected positions $\{\vec{x}_a\}$, $a = 1, \ldots, N$, so that the number density $n(\vec{x})$ is given by the expression

$$n(\vec{x}) = \sum_{a=1}^{N} \delta(\vec{x} - \vec{x}_a). \tag{10.39}$$

For a fractal, it is natural to define the number density as a function of two variables: $n = n(\vec{x}, r)$. The first variable describes the position $\vec{x}_a$ of a structure particle, and the second one gives the radius r within which one counts the number of particles. The variable r serves for constructing a statistics that measures the strength of the singularity around a particle of the fractal, where the number of particles grows as a power-law $N(r) \propto r^D$.

Denote by $N_V(\vec{x}_a, r)$ the number of particles in a sphere of radius r, centered at the particle a with coordinates $\vec{x}_a$, belonging to the structure:

$$N_V(\vec{x}_a, r) = \int_0^r n(\vec{x_a} + \vec{x}) 4\pi x^2 dx, \tag{10.40}$$

and $N_S(\vec{x}_a, r)$ is the number of particles in the spherical shell $r, r + \Delta r$, with the centre at $\vec{x}_a$:

$$N_S(\vec{x}_a, r) = \int_r^{r+\Delta r} n(\vec{x_a} + \vec{x}) 4\pi x^2 dx. \tag{10.41}$$

From one realization to another these quantities fluctuate, but after averaging over many realizations the stable power-law dependence on the scale r emerges. For ergodic processes averaging over many realizations may be replaced by many points in one realization. Following Pietronero (1987) we define the *conditional (number) density* of a stochastic fractal process as

$$\Gamma(r) = \left\langle \frac{N_S(\vec{x}_a, r)}{4\pi r^2 \Delta r} \right\rangle_{\vec{x}_a} = \frac{DB}{4\pi} r^{-(3-D)}, \tag{10.42}$$

and the *conditional volume density* as

$$\Gamma^*(r) = \left\langle \frac{N_V(\vec{x}_a, r)}{(4\pi/3) r^3} \right\rangle_{\vec{x}_a} = \frac{3B}{4\pi} r^{-(3-D)}, \tag{10.43}$$

where $\langle \cdot \rangle_{\vec{x}_a}$ means averaging over all points $\vec{x}_a$ in one realization with the condition that the centres of the spheres lie at the particles of a realization (this explains the word "conditional"). The last equalities in (10.42) and (10.43) relate to ideal fractal structures, for which $\Gamma^*(r) = \frac{3}{D}\Gamma(r)$.

Γ-Function Estimator Consider a stochastic fractal process where the number of particles $N_V(\vec{x}_a, r)$ in a sphere of radius r, centered at the point $\vec{x}_a$ and the number $N_S(\vec{x}_a, r)$ of particles in the shell $(r, r + \Delta r)$ are given by Eqs. (10.40) and (10.41). Taking into account the definitions of conditional densities (Eqs. (10.42) and (10.43)) one can use following two statistics for their estimation from one realization (a finite galaxy sample):

$$\hat{\Gamma}(r) = \frac{1}{N} \sum_{a=1}^{N} \frac{1}{4\pi r^2 \Delta r} \int_r^{r+\Delta r} n(\vec{x}_a + \vec{x}) 4\pi x^2 dx \tag{10.44}$$

for the shell conditional density Γ, and

$$\hat{\Gamma}^*(r) = \frac{1}{N} \sum_{a=1}^{N} \frac{3}{4\pi r^3} \int_0^r n(\vec{x}_a + \vec{x}) 4\pi x^2 dx \tag{10.45}$$

for the volume conditional density Γ^*.

So the conditional density method is quite simple in principle, just counting the number of particles inside a spherical volume $V(r)$ or shell Δr. This is done for each structure point and then the average is calculated. For the Γ-function estimation one need not generate artificial Poisson distributions, as is necessary for the ξ-function method.

Fractal Dimension and Co-dimension For a fractal structure both the Γ function (Eq. (10.42)) and the estimator (Eq. (10.44)) have a power-law form

$$\hat{\Gamma}(r) = \Gamma_0 r^{-\gamma}. \tag{10.46}$$

This very important property of the Γ-estimator allows one to obtain an undistorted value of the fractal dimension in a galaxy sample. The exponent that defines the decay of the conditional density $\gamma = D - 3$ is called the *co-dimension*, where D is the fractal dimension. The amplitude Γ_0 of the estimator $\hat{\Gamma}(r)$ does not change when the sample volume V_s is increased, only the range of available scales r increases. This corresponds to the meaning of $\Gamma(r)$ as characterizing the number density behaviour.

Homogeneity Scale For a fractal structure which has an upper cutoff at a homogeneity scale R_{hom}, beyond which the distribution becomes uniform, the estimator of the Γ-function is

$$\hat{\Gamma}(r) = \text{constant}, \quad \text{for } r > R_{\text{hom}}. \tag{10.47}$$

Thus the conditional density is a powerful tool when one searches for the crossover from the regime of fractal clustering to the realm of homogeneity.

Γ-Function for 2D Intersections If a 3D fractal structure is intersected by a plane then the fractal dimension for the intersection is given by Eq. (10.21):

$$D_{\text{int}} = D - 1. \tag{10.48}$$

To make the Γ-function analysis for the sample formed by the 2D intersection one may use the coordinate system $\vec{y} = (y_1, y_2)$ and calculate the Γ-function for the intersection, or $\Gamma_{\mathrm{int}}(y)$.

Such a situation may occur in a slice-like galaxy survey for scales r larger than the thickness of the slice. For instance, for the true fractal dimension $D = 2$ the fractal dimension of the intersection will be $D_{\mathrm{int}} = 1$, and one expects a power-law behaviour for the corresponding $\Gamma_{\mathrm{int}}(y) \propto y^{-1}$.

10.3.6 Comparison of the ξ and Γ Correlation Functions

When comparing the two methods, one should be aware that both are affected by finite volume effects (in fact, the ξ function more strongly than the Γ-function) and can be safely applied only in the range of scales where the required conditions (Sect. 10.3.1) are fulfilled for the galaxy sample.

The Relation Between Γ and ξ From the definitions of the conditional density Γ (Eq. (10.38)) and correlation function (Eq. (10.29)) we have the relation

$$\xi(r) = \frac{\Gamma(r)}{n_0} - 1 \tag{10.49}$$

if the mean number density n_0 of the stochastic process exists. This implies a similar relation between the *estimators* applied to a finite sample:

$$\hat{\xi}_{FS}(r) = \frac{\hat{\Gamma}_{FS}(r)}{\bar{n}} - 1. \tag{10.50}$$

Here $\hat{\xi}_{FS}$ is the "full shell" estimator as it is defined through the Γ estimator calculated using full shells completely embedded in the sample volume.

Both Γ and ξ functions are conditional characteristics, i.e. they are defined on the condition that the centres of counting spheres are set to structure particles. However, there is still a deep difference between them. $\Gamma(r)$ is a *complete* correlation function, while $\xi(r)$ is a *reduced* correlation function of the stochastic process. This makes the properties of their ξ and Γ estimators very different.

The estimator $\hat{\Gamma}$ (Eq. (10.44)) is always a positive function and has a power-law form for fractal structures. On the contrary, the estimator $\hat{\xi}$ (Eq. (10.32)) inevitably changes its sign and hence cannot be presented as a power law even for scale invariant structures. All estimators of the ξ-function, which are based on counting of pairs relative to an artificial Poisson distribution, have a common drawback. They give essentially distorted values for the correlation exponent of the complete correlation function of fractal distributions. But the Γ-function estimator is specially constructed in order to give true values of the correlation exponent and the fractal dimension.

The Γ-function estimator relates to *intrinsic properties* of the sample, while the ξ-function estimator depends on both intrinsic and external properties. In fact, $\hat{\Gamma}$

measures the behaviour of the total density inside spheres within a sample, while $\hat{\xi}$ measures fluctuations relative to the average density assumed to be valid for all space outside a finite sample. We illustrate this by considering counts around a fixed point. The expected number of points in a shell with radius r and volume dV is $\Gamma(r)dV$, as the conditional density $\Gamma(r)$ describes the density–radius law. The same expected number may be calculated with the correlation function $\xi(r)$ as $n_0(1+\xi(r))dV$. Hence,

$$\Gamma(r) = (1+\xi(r))n_0. \tag{10.51}$$

It is important that the right-hand side of (10.51) becomes defined only after the mean density n_0 is calculated for the whole sample, while $\Gamma(r)$ always exists locally. Remember that $\xi(r)$ is defined for fluctuations around n_0. For a study of how an incorrect value of n_0 can affect the measured large scale behaviour of the $\xi(r)$ correlation function, see Sylos Labini et al. (2009a).

Fractal Density Field For the case of a scale-invariant stochastic fractal density field, the complete correlation function $\Gamma(r)$ has the power-law form

$$\Gamma(r) = \frac{BD}{4\pi} r^{-\gamma}. \tag{10.52}$$

For the same fractal structure the reduced correlation function $\xi(r)$ will be

$$\xi(r) = \frac{BD}{4\pi n_0} r^{-\gamma} - 1 \tag{10.53}$$

which is not a power-law. This difference between complete and reduced correlation functions was pointed out by Pietronero and Kuper (1986).

Thus the ξ-function may be approximated by a power-law only when $\xi(r) \gg 1$, which corresponds to small scales $r \ll r_0$. However, on small scales the noise of discreteness deforms a true power-law. Hence a ξ-function estimation may give incorrect values of the correlation exponent not only on large scales (normalization), but also on small scales (discreteness).

The apparent slope for the correlation function ξ. It is instructive to calculate the exponent γ_ξ of the correlation function on scales close to the unit scale r_0 (Joyce et al. 1999). Take the logarithmic derivative of Eq. (10.53):

$$\gamma_\xi(r) = -\frac{d[\log \xi(r)]}{d \log r} = \frac{2\gamma (r/r_0)^{-\gamma}}{2(r/r_0)^{-\gamma} - 1}. \tag{10.54}$$

Therefore at the unit scale $r = r_0$ we get the remarkable result:

$$\gamma_\xi(r_0) = 2\gamma. \tag{10.55}$$

For a density power-law with $\gamma = 1$, one would infer an apparent slope $\gamma_\xi = 2$ for the ξ-function, if measured at scales close to the "correlation length" r_0! See Fig. 10.6.

The Sample Depth and r_0 For a fractal structure sampled inside a sphere (the average number density $\bar{n} = N_s/V_s = 3BR_s^{D-3}/4\pi$), Eq. (10.53) yields

$$\hat{\xi}(r) = \frac{D}{3}\left(\frac{r}{R_s}\right)^{-\gamma} - 1. \tag{10.56}$$

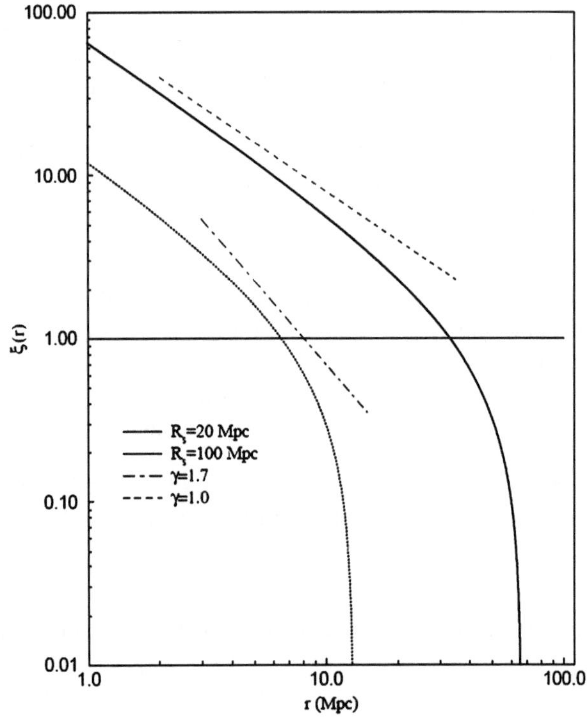

Fig. 10.6 The dependence of the derived slope for the ξ-function on the x range used for its determination

Inserting $r = r_0$ into (10.49) and noting that $\xi(r = r_0) = 1$ one gets $r_0 = ((DB)/(8\pi\bar{n}))^{1/\gamma}$. Hence $\Gamma(r_0) = \frac{DB}{4\pi} r_0^{D-3} = 2\bar{n} = \frac{6B}{4\pi} R_s^{D-3}$, from which follows the simple relation between the length r_0 and the sample depth R_s:

$$r_0 = \left(\frac{3-\gamma}{6}\right)^{1/\gamma} R_s. \tag{10.57}$$

Survey Geometry and Characteristic Scales In practice the Γ-function method is restricted by the requirement that there should be room for the whole sphere in the volume of the studied sample. For example, for galaxy surveys with slice-like geometry, this makes it impossible to measure the conditional density for scales larger than the thickness of the slice.

A galaxy sample is characterized by the following important scales (Fig. 10.7):

$$R_{sep}, R_{max}^{sph}, R_s. \tag{10.58}$$

The average distance R_{sep} between galaxies in the sample may be roughly estimated as $\bar{n}^{-1/3}$ or calculated from the nearest neighbour distribution. The radius of the maximal sphere R_{max}^{sph} refers to completely embedded spheres in the sample and has a crucial role. The depth R_s of a survey is related to the largest distances in the sample, and it essentially differs from the radius of the maximal sphere in case of slice-like surveys.

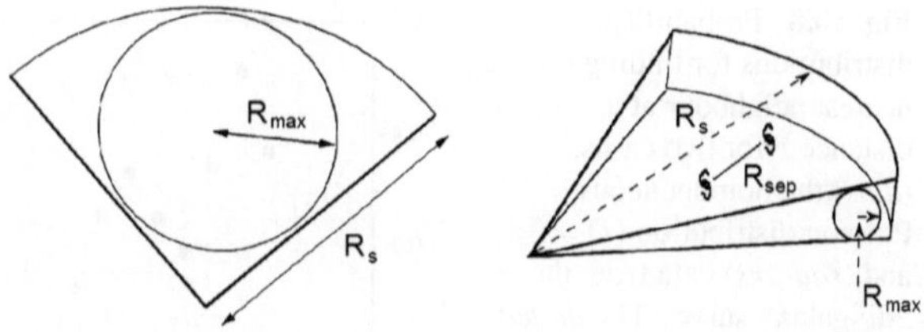

Fig. 10.7 Different geometries of the surveys and important parameters

It is a general rule that in a correlation analysis of slice-like samples one should consider separately two intervals of scales:

$$R_{\mathrm{sep}} < r < R_{\mathrm{max}}^{\mathrm{sph}}, \tag{10.59}$$

where one can use the Γ function method to estimate the value of D, and

$$R_{\mathrm{max}}^{\mathrm{sph}} < r < R_{\mathrm{s}}, \tag{10.60}$$

where the Γ function does not work. Here other possible methods are the Γ function for fractal intersections and the bi-conditional column density.

10.4 Other Methods

Other methods of structure analysis may complement the information from the ξ and Γ functions or can be used instead of them in some situations.

10.4.1 The Distribution of the Nearest Neighbour

The distribution of distances to the nearest neighbour point may be used to make a distinction between fractal and ordinary distributions.

Poisson Distribution For a Poisson process in 3D space the probability density $\omega(r)$ for finding the nearest neighbour at a distance r is

$$\omega(r) = 4\pi n_0 r^2 \mathrm{e}^{-\frac{4\pi n_0 r^3}{3}}. \tag{10.61}$$

Here the average distance between point-particles is $R_{\mathrm{sep}} \approx n_o^{-1/3}$.

Fractal Distribution Gabrielli et al. (2005) have derived a useful approximation for a fractal structure (here $C = DB/4\pi$):

$$\omega(r) = 4\pi C r^{D-1} \mathrm{e}^{-\frac{4\pi C r^D}{D}}. \tag{10.62}$$

Figure 10.8 presents $\omega(r)$ for two cases: a homogeneous Poisson distribution ($D = 3$) and the estimation from the 2dF galaxy survey ($D = 1.8$). The curves correspond to the predictions from (10.61) and (10.62). It is seen that the lower is the value of the fractal dimension, the closer are the neighbours within the structure. In fact, the nearest neighbour distribution may be used as an additional method for estimating the fractal dimension.

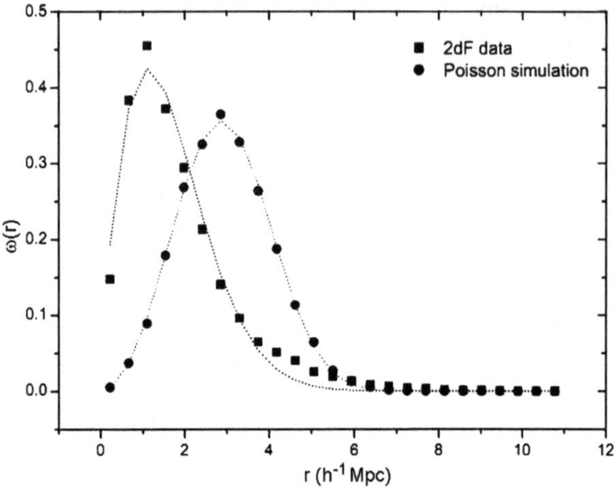

Fig. 10.8 Probability distributions for finding the nearest neighbour at a distance r for two cases: (*dots*) the homogeneous Poisson distribution ($D = 3$) and (*squares*) data from the 2dF galaxy survey The *dotted curves* correspond to predictions for the Poisson distribution (Eq. (10.61)) and for 2dF (Eq. (10.62)) with $D = 1.8$ (from Vasilyev et al. 2006)

10.4.2 The Bi-conditional Column Density

The conditional densities discussed above were one-point, as the counts are made around each one point of the sample. However, in some cosmological studies (e.g. gravitational lensing; Baryshev and Ezova 1997), it becomes necessary to use two-point conditional densities, whereby one simultaneously fixes two particles $\{a,\ b\}$ with the coordinates $\{\vec{x}_a,\ \vec{x}_b\}$ and counts galaxies within a thin cylinder between these points.

Definitions In order to define the distribution of particles along the cylinder whose axis connects two structure points $\{a,\ b\}$, Baryshev and Bukhmastova (2004) introduced the distribution between two occupied points (a, b), $\eta_{ab}(r)$ for a stochastic fractal process. If the particles a and b are independent, they are statistically equivalent and the probability to find a particle inside the cylinder will be equal to the sum of the two 1-point conditional densities (given by the expression (10.42)).

For another case, when the distance $r_{ab} = |\vec{x}_a - \vec{x}_b|$ between two structure points is fixed, Gabrielli (2005) derived the following formula for the bi-conditional (points a and b) one-point (one variable r) column density:

$$\eta_{ab}(r) = g\,[\Gamma_a(r) \times \Gamma_b(r_{ab} - r)]\,. \qquad (10.63)$$

Here $g = 1/(DB/4\pi)$ is a normalization factor and $\Gamma_a = \Gamma_b = (DB/4\pi)r^{-(3-D)}$ is the ordinary conditional density given by Eq. (10.42). The distance r is measured along the segment of the line connecting the particles a and b, and at the same time it defines the radius r of the sphere having its centre at the first point and the radius $r_{ab} - r$ for another sphere having its centre at the second point. In this method the volume elements are taken along the line connecting the two points—the coordinate r is a Cartesian coordinate labelling volume elements ("tablets") with thickness dr along this cylinder (Fig. 10.9).

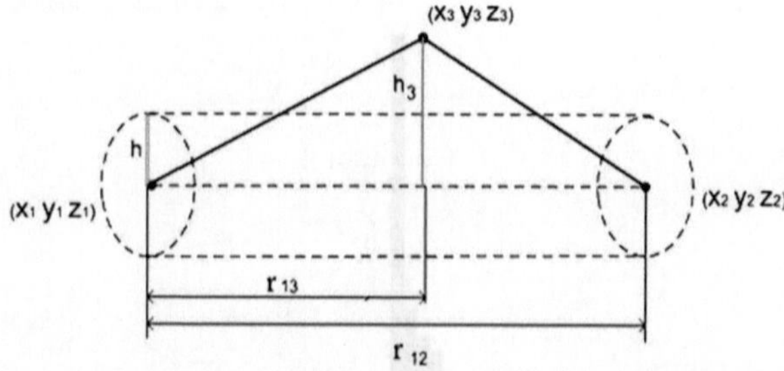

Fig. 10.9 Method of cylinder: the centres of two galaxies are fixed at the points (x_1, y_1, z_1) and (x_2, y_2, z_2). All the remaining galaxies with central coordinates (x_3, y_3, z_3) are checked to determine whether they belong to a given cylinder. In slice-like surveys this method allows one to study scales comparable to the depth of the survey

Estimation In order to estimate $\eta_{ab}(r)$ one may use the statistic

$$\hat{\eta}_{ab}(r) = \left\langle \frac{N_c(\vec{x}_a, \vec{x}_b, r, h, \Delta r)}{\pi h^2 \Delta r} \right\rangle_{\{a,b\}}$$

$$= \frac{1}{N_{ab}} \sum_{\{a,b\}}^{N_{ab}} \frac{1}{\pi h^2 \Delta r} \int_r^{r+\Delta r} \int_0^h n(\vec{x}) 2\pi h \, dh \, dr, \tag{10.64}$$

where N_c is the number of points in a volume element of the cylinder with a diameter h and height Δr and whose axis connects the structure points a and b. The volume element lies at the distance r from a (corresponding to the distance $r_{ab} - r$ from b). Averaging is performed for every pair of points with connecting cylinders having the length in the interval $(l, l + \Delta l)$. The parameters h and Δr are taken to be a fraction of the mean separation of the particles in the sample. Simulations show that the estimated fractal dimension is robust to reasonable variations of the tablet size.

If a fractal structure and a homogeneous background co-exist, then the fractal dimension D may be calculated by fitting on the observations the dimensionless probability distribution with free parameters A, γ, and B:

$$N(y)/N = R_1 \cdot [y^{-\gamma} \times (1 - y)^{-\gamma}] + R_2, \tag{10.65}$$

where $N(y)$ is the observed number of points in each tablet, i.e. within the intervals $(y, y + \Delta y)$ along the cylinder with a length l. The variable y is the relative distance measured along the line connecting the two points $(y = r/r_{ab} = r/l)$. N is the total number of points within cylinders of length l. For genuine self-similar structures it is possible to calculate these numbers for all cylinders of different length simultaneously (see Fig. 10.10). R_2 takes into account a possible Poisson background. One can measure the relative strength β of the fractal component as $\beta = (1 - R_2)/R_2$ (Vasilyev 2004). When $\beta = 1$, the contributions from the fractal structure and the Poisson background are equal $(R_2 = 0.5)$.

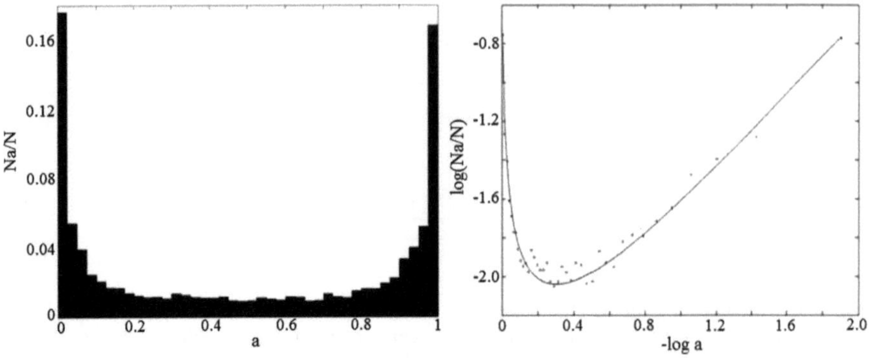

Fig. 10.10 *Left*: The histogram of the galaxy distribution $\hat{\eta}_{ab}(r)$ of the LEDA sample along cylinders with $l < 100$ Mpc. *Right*: The same distribution in logarithmic scale and the fitted theoretical curve $\eta_{ab}(r)$, with $D = 2.02$ (Baryshev and Bukhmastova 2004)

10.4.3 The Radial Distribution $N(z)$ and Its Fluctuations

The large number of photometric redshifts allows one to study structures of galaxies, which reach scales of thousands of Mpc, by analysing the distribution $N(z)$ in deep fields (Nabokov and Baryshev 2010a, 2010b). Fluctuations in the numbers of galaxies in the bins $\Delta z = 0.1$–0.3 are caused by Poisson noise, correlated structures, and systematic errors in estimated z. This method requires covering a sufficiently large region of the sky with a grid of deep multi-band surveys with a cell size $\sim 10 \times 10$ deg^2. The distribution of the photometric redshifts of the galaxies within each deep $\sim 10' \times 10'$ field at the nodes of the grid will yield information on the radial extent of superlarge structures; comparing the radial distributions in neighbouring fields will yield information on their tangential extent. This method must be accompanied by an analysis of the distortions in the $N(z)$ distributions arising from the technique for evaluating the photometric redshifts.

Modern multiband deep surveys of galaxies (Reshetnikov 2005), such as COMBO-17, COSMOS, FDF, HUDF, and ALHAMBRA, offer the possibility of measuring the sizes and contrast of superlarge inhomogeneities in the distribution of galaxies at redshifts of 0.5–5. These surveys contain 10^3–10^5 galaxies with measured magnitudes in several bands, so it is possible to use photometric redshifts. The accuracy of z_{phot} is usually $0.03(1+z)$, so that one may study scales greater than about 200 Mpc/h. By using large redshift bins (0.1–0.3) containing many galaxies (> 100), the Poisson noise ($\sim 1/\sqrt{N}$) will be small ($\sigma_P < 0.1$), and fluctuations corresponding to very large inhomogeneities in the distribution of galaxies can be observed with a contrast greater than the Poisson level.

Distribution of Redshifts For magnitude-limited surveys the redshift distribution is usually approximated (e.g., Massey et al. 2007a) as

$$N(z, \Delta z) = A z^{\alpha} e^{(-z/z_0)^{\beta}} \Delta z, \tag{10.66}$$

Table 10.1 Linear dimensions in the tangential direction for angle θ at $z = 1$ and in the radial direction for redshift bins Δz centered at $z = 1$, for the model ($h = 0.72$, $\Omega_V = 0.7$, $\Omega_m = 0.3$)

θ	$3'$	$10'$	$1°$	$10°$	$30°$
ΔL (Mpc)	2.8	9.3	55.7	557	1670
Δz	0.1	0.2	0.3	0.4	0.5
ΔL (Mpc)	235	470	706	943	1180

where the parameters α, β, and z_0 are found by least squares, and A normalizes the expression from the condition $\int N(z)dz = N_{\text{tot}}$. This formula has been tested on model samples of magnitude-limited distant galaxies, assuming a uniform spatial distribution of galaxies having luminosities according to the Shechter law (Lovyagin 2009).

For a uniform spatial distribution the observed redshift distribution may be determined by a model for the number and luminosity evolution of galaxies. One may also obtain a theoretical distribution of the type (10.66) from cosmological simulations of the distribution of galaxies. For example, from light cone simulations (Heinämäki et al. 2005; Kim et al. 2009), one can extract the radial distribution of dark halos ("galaxies") within a redshift range of 0–6. In this way one may obtain the form of the average distribution $N(z)$, as well as expected deviations from smooth behaviour due to structures.

The Scales Accessible for a Grid of Deep Fields In the flat Friedman model, the linear size L of a region in the sky at the present time, *which participates in expansion of space* (like a superlarge structure is expected to) and has now the observed angular size θ is given by $L(z) = \theta(z)r(z)$ where $r(z)$ is the metric distance of the region. Using the precipes for calculating the metric distance (Sect. 7.4), Table 10.1 shows a useful collection of calculated linear sizes L (Mpc) in the tangential and radial directions.

Selection Effects As the photometric redshift method is based on the continuum spectrum, it is independent of the visibility of spectral lines at different redshifts. However, there are some effects owing to the visibility of certain details of the continuum spectra and a degeneracy in the solutions for a fixed system of filters in a given survey (e.g., the visibility of the Balmer 3646 Å and Lyman 912 Å limits). The resulting systematic errors may show up as inhomogeneities in the radial distribution of galaxies. The errors in measuring magnitudes in different filters vary and depend on the individual features of galactic spectra, as well as on the accuracy with which the observations made at different angular resolutions are reduced (e.g., when combining optical and infrared data). In addition, the standards for the energy distribution in galactic spectra used to derive photometric redshifts may differ from the actual continuum spectra. Thus, when the systematic effects are included, the observed dispersion in the fluctuations will be

$$\sigma^2(z, \Delta z) = \sigma^2_{\text{corr}} + \sigma^2_{\text{P}} + \sigma^2_{\text{errors}}, \tag{10.67}$$

where the terms refer to correlated structures, to Poisson fluctuations and to the errors in the method for evaluating the redshift. In Chap. 12 we will present some results on the very large-scale structures, using this method.

10.4.4 Fourier Analysis of the Galaxy Distribution

If the spatial distribution of objects is given by a stochastic density field, then fluctuations of this field may be represented by the Fourier integral as a superposition of plane spatial waves with the wave number $k = |\vec{k}| = 2\pi/\lambda$:

$$\delta(\vec{x}) = \frac{\rho(\vec{x}) - \bar{\rho}}{\bar{\rho}} = \frac{1}{(2\pi)^3}\int \tilde{\delta}(\vec{k}) e^{-i\vec{k}\cdot\vec{x}} d^3k. \tag{10.68}$$

Here the Fourier transform $\tilde{\delta}(\vec{k})$ of the density fluctuations $\delta(\vec{x})$ is a complex quantity and thus may be given in the form

$$\tilde{\delta}(\vec{k}) = |\tilde{\delta}(\vec{k})| \exp(i\phi(\vec{k})). \tag{10.69}$$

For a complete description of the spatial distribution the analysis should include both the amplitude spectrum $|\tilde{\delta}(\vec{k})|$ and the phase spectrum $\phi(\vec{k})$.

For a Gaussian random process the phases of plane waves are distributed uniformly in the interval $[0, 2\pi]$ and to describe the density field it is sufficient to consider only the power spectrum $P(\vec{k}) = \langle|\tilde{\delta}(\vec{k})|^2\rangle$. When the distribution is isotropic, the power spectrum and the correlation function are connected by the relation

$$P(k) = 4\pi \int \xi(r) \frac{\sin(kr)}{kr} r^2 dr. \tag{10.70}$$

Thus for a power-law correlation function $\xi(r) \propto r^{-\gamma}$ the power spectrum has also a power law form $P(k) \propto k^{\gamma-3}$ for a restricted range of scales $k_1 < k < k_2$. For small k (large scales $\lambda > R_s$) there is a limit due to the size of the survey R_s, which causes $P(k) \to 0$ for $\lambda \to R_s$.

The main challenges of the power spectrum analysis are the same as for the correlation function: (1) the average density $\bar{\rho}$ should be well defined within the sample volume, and (2) one should be able to control the finite volume effects. However, a plus side is that the correlation exponent γ may be estimated without distortion from $P(k)$ for the scales $\lambda < R_{max}^{sph}$.

Fractal Density Fields A generalized power spectrum $\Pi(k)$ of a fractal process is defined by the Fourier transform of the conditional density $\Gamma(r)$, instead of $\xi(r)$ as in (10.70) (Sylos Labini and Amendola 1996; Sylos Labini et al. 1998). Due to the power-law form $\Gamma(r) \propto r^{-(3-D)}$ for a stochastic fractal, the power spectrum is also a power law $\Pi(k) \propto k^{-D}$. Hence the dimension D may be directly measured from the power spectrum.

The Role of Sample Geometry For a slice-like survey for scales larger than the maximum sphere completely embedded in the sample, the geometry of the studied structure approaches the case of an intersection of a 3D fractal by a plane. Hence one expects from the intersection theorem that for a structure with the fractal dimension D the measured dimension of the sample will be $D_{\rm int} = D - 1$. Therefore, there are three characteristic intervals of scales for the behaviour of the power spectra $P(k)$ or $\Pi(k)$:

- For scales $\lambda < R_{\rm max}^{\rm sph}$ where $P(k) \propto k^{-D}$ with the exponent equal to the true fractal dimension
- For scales $R_{\rm max}^{\rm sph} < \lambda < R_{\rm s}$ where $P(k) \propto k^{-(D-1)}$ with the exponent equal to the fractal dimension of the intersection
- For scales $\lambda > R_{\rm s}$ where $P(k) \to 0$ due to the finite size of the sample

10.4.5 Redshift-Space and the Peculiar-Velocity Field

From a redshift survey one obtains a redshift-space map, i.e. the position in the sky and $z_{\rm obs}$, the observed redshift of a galaxy. As was discussed in Sect. 4.1.1, the value of $z_{\rm obs}$ may contain contributions from several causes.

Distance Error Due to Peculiar Velocity For peculiar velocities $v \ll c$ the Doppler part of the observed redshift is determined by the radial component $v_{\rm r}$ of the velocity as $z_{\rm v} \approx v_{\rm r}/c$. So for small radial peculiar velocities $v_{\rm r}$

$$z_{\rm obs} = z_{\rm cos} + \frac{v_{\rm r}}{c}(1 + z_{\rm cos}). \tag{10.71}$$

Hence the derived distance $r_{\parallel {\rm obs}} = r_{\parallel} + \Delta r_{\rm v}$ will be distorted by a peculiar velocity field in the line-of-sight direction by the amount

$$\Delta r_{\rm v} = \frac{v_{\rm r}(1 + z_{\rm cos})}{H_0} = 5\ {\rm Mpc}\, {\rm h}_{100}^{-1} \frac{v_{\rm r}(1 + z_{\rm cos})}{500\ {\rm km/s}}. \tag{10.72}$$

Note that the factor $(1 + z_{\rm cos})$ leads to an increasing influence of $v_{\rm pec}$ on the distance error for deep redshift surveys.

Real-Space and Redshift-Space ξ Functions A ξ-function derived from redshift distances is termed the *redshift-space correlation function* $\xi_{\rm z}(s)$. In order to obtain the *real-space correlation function* $\xi_{\rm real} = \xi(r)$ one should extract all non-cosmological contributions to $z_{\rm obs}$, which is not easy.

The shape of the observed $\xi_z(s)$ is affected by the nature of the peculiar velocity field. In virialized clusters the velocity dispersion leads to the "finger-of-God" effect, i.e. an elongated shape along the line-of-sight direction $r_{\parallel}$. The mean tendency of galaxies on larger scales to approach each other, due to the gravity of large-scale structures, will appear as a compression of $\xi_z(s)$ in the direction $r_{\parallel}$. As these two effects operate on different spatial scales, they do not compensate each other.

Peebles (1980) and Davis and Peebles (1983) suggested a procedure for the restoration of both the true $\xi(r)$ and the relative peculiar velocity distribution $f(v)$ from the observed correlation function $\xi_z(r_\perp, r_{\|obs})$ where $r_\perp$ and $r_{\|obs}$ are the observed perpendicular and parallel to the line of sight components of the separation $s = (r_\perp^2 + r_{\|obs}^2)^{1/2}$.

The method is based on the calculation of the projected correlation function $w(r_\perp)$ which does not depend on the peculiar velocity field, if the distribution of the radial peculiar velocities is symmetrical around each galaxy of the sample. Integrating along the line of sight we obtain:

$$w(r_\perp) = 2\int_0^\infty \xi_z(r_\perp, r_{\|obs})dr_{\|obs}, \tag{10.73}$$

where in practice the interval of integration is restricted by chosen radial velocity limits. Then the required inverse is the Abel integral

$$\xi(r) = -\frac{1}{\pi}\int_r^\infty \frac{w' dr_\perp}{(r^2 - r_\perp^2)^{1/2}}, \tag{10.74}$$

where $w' = dw(r_\perp)/dr_\perp$. This is the restored real-space $\xi(r)$ function.

For the power law $\xi = (r_0/r)^\gamma$ the integral (10.73) gives

$$w(r_\perp) = A r_\perp^{1-\gamma}, \tag{10.75}$$

where $A = r_0^\gamma \Gamma_e(\frac{1}{2})\Gamma_e(\frac{\gamma-1}{2})/\Gamma_e(\frac{\gamma}{2})$ and $\Gamma_e(x)$ is the Euler gamma function.

Limitations of the Projection Method It is clear from (10.75) that such a solution for the real-space ξ function is valid only if the exponent $\gamma \geq 1$. For example, $\gamma = 1$ gives $w(r_\perp) =$ constant, which again shows that the projection of a real non-uniform distribution may be confused with a uniform background galaxy distribution. Such a method leads to elimination of information on structures with the fractal dimension $D \geq 2$. Therefore to take into account the peculiar velocity field within fractal structures with $D \geq 2$, one needs another method of restoration for the correlation function, which is free from the above limitation.

Estimating the velocity dispersion. When density and velocity fields are weakly coupled, the observed correlation function $\xi_z(s)$ can be modelled as a convolution of the real space correlation function $\xi(r)$ with the galaxy pairwise velocity distribution $f(v)$ (Peebles 1980). One presents this equation as

$$1 + \xi_z(r_\perp, r_{\|obs}) = H_0 \int \left[1 + \xi(\sqrt{r_\perp^2 + r_\|^2})\right] f(v) dr_\|. \tag{10.76}$$

Here $v = H_0 r_{\|obs} - H_0 r_\| + \bar{v}_{12}(r)$ and $\bar{v}_{12}(r)$ is the mean radial pairwise velocity of galaxies at separation r, represented by a model. Davis and Peebles (1983) used the model $\bar{v}_{12}(r) = H_0 r_\|/(1 + (r/r_0)^2)$ and an exponential form for $f(v)$:

$$f(v) = B\, \mathrm{e}^{-\sqrt{2}|v|/\sigma_{12}}. \tag{10.77}$$

As a result one obtains the *pairwise velocity dispersion* σ_{12}.

Redshift-Space $\Gamma_z(s)$ and $v_{\rm pec}$ Using the same notations as for the ξ function $(s, r_\perp, r_\|)$ we can write for the redshift-space conditional density

$$\Gamma_z(s) = \Gamma_z(r_\perp, r_{\|\rm obs}). \tag{10.78}$$

The relation between the real-space $\Gamma(r)$ and redshift-space $\Gamma_z(s)$ is

$$\Gamma_z(r_\perp, r_{\|\rm obs}) = \int g(\vec{r}, \vec{w})\Gamma(r) d^3 w, \tag{10.79}$$

where $\vec{w} = \vec{v}_1 - \vec{v}_2$ is the relative peculiar velocity of a galaxy pair at separation $\vec{r}$, and $g(\vec{r}, \vec{w})$ is the relative peculiar velocity distribution. Here the components of the relative distance $\vec{r}$ are given by the following formulae: $r_1 = r_\perp, r_3 = r_{\|\rm obs} - w_3/H_0$, and $r^2 = r_\perp^2 + (r_{\|\rm obs} - w_3/H_0)^2$.

In order to restore the real-space conditional density from the observed redshift-space Γ function one has to make computer simulations of artificial fractal structures with known peculiar velocities and compare the modelled redshift-space $\Gamma_{\rm mod}(s)$ with the observed $\Gamma_z(s)$.

10.5 A Summary: Requirements for Reliable Correlation Analysis

Let us collect together the main requirements for a reliable correlation analysis of the galaxy distribution. First of all, there is the basic assumption that the stochastic process behind the spatial distribution of galaxies is ergodic so that we can obtain information about the average properties of the infinite ensemble from its one realization in a finite volume. Furthermore, the finite sample should be self-averaging, so that we can really extract from it sufficiently accurate and stable whole-sample average quantities (for instance, the finite-volume effect can affect the mean number density and thus artificially shift the measured zero-ξ length). Fortunately, the property of self-averaging can be tested using the sample itself (Sect. 10.3; Sect. 12.2).

Besides these generalities, certain practical restrictions should be taken into account when one makes correlation analyses using the ξ or Γ functions. Borders of the finite volume restrict the sizes and numbers of the testing spheres, so that a narrow-angle survey contains only small spheres and the ξ or Γ functions can be derived only for small scales. Wide-angle surveys do contain larger spheres, allowing the analysis of larger scales; however, one should note that the largest spheres contain basically only the most luminous objects, hence a small number of galaxies. A particular problem is encapsulated in the integral constraint for ξ, which implies that one cannot model the ξ function as a power-law, except within finite intervals.

In the next chapter we will show, from studies of real galaxy data, how the correct interpretation of the results of correlation analysis depends on the understanding of the general requirements and idiosyncracies of the applied methods.

Chapter 11
The Inhomogeneous Galaxy Universe: Observational Results

The discovery of the surprisingly rich texture of galaxies, only faintly anticipated from photographies, is of deep cosmological significance. In fact, the homogeneity of the early universe is a fundamental hypothesis of the standard world model. One has to assume that the structures now seen in the galaxy universe were formed by gravitational clustering from initial small seeds. The predicted growth of the large-scale structure may be checked by current observational means thanks to the great progress in measuring distances by redshifts. At small redshifts ($z < 0.5$), huge redshift surveys have been completed. At redshifts up to 5, there are many deep multi-band galaxy surveys, some completed and others being planned.

11.1 From the 2D Sky to the 3D Map

The scarcity of available observations at any historical moment tends to lead to uncertain interpretations of the data. This also gave rise to a lengthy "fractal debate" around the nature of galaxy clustering and the scale where uniformity begins. It has involved Einstein, Hubble, Sandage, Peebles, Charlier, Lundmark, de Vaucouleurs, Mandelbrot, Pietronero and other students of the large-scale clustering. In fact, in the very decade when the galaxy universe was found and theoretical cosmology made its first steps, Selety's and Einstein's correspondence on cosmic hierarchies outlined the main directions for future studies of inhomogeneous world models. For résumés of the main events illuminating this process, see Tables 11.1 and 11.2.

New aspects and tools for tackling the clustering problem arrived with the concept of fractal (Mandelbrot 1975). Self-similar structures with long-range correlations had appeared in physics and geophysics and were then extended to astronomical scales, from the Solar System and interstellar clouds to galaxies and then to increasingly large cosmic scales.

The 2D and 3D Epochs Early counts of bright galaxies gave results close to the $N \propto 10^{0.6m}$-law, pointing to homogeneity. Started by Totsuji and Kihara (1969) and

Y. Baryshev, P. Teerikorpi, *Fundamental Questions of Practical Cosmology*, Astrophysics and Space Science Library 383,
DOI 10.1007/978-94-007-2379-5_11, © Springer Science+Business Media B.V. 2012

Table 11.1 Early milestones in the debate on large-scale fractality

Years	Authors	Subject
1900–1920s paradoxes	Fournier d'Albe	regular hierarchical models
	Charlier, Selety, Lundmark	criteria for infinite world
	Einstein, Selety	Mach, stability, middle point
1930s–1970s clusters and uniformity	Shapley, Zwicky, Abell	strong galaxy clustering
	Carpenter, Kiang, Karachentsev	superclusters up to 100 Mpc
	Neyman, Scott	2-level hierarchical model
	de Vaucouleurs	density-radius $\rho(r) \propto r^{-\gamma}$, $\gamma = 1.7$
	Hubble	uniformity from galaxy counts
	Ambartsumian, Holmberg, Fesenko	variable extinction in MW
	Sandage, Tammann, Hardy	linear Hubble law at < 30 Mpc
	Webster, Longair	isotropy of radio sources

then developed by Peebles, the angular correlation analysis gave a small homogeneity scale $R_{\mathrm{hom}} \leq 10$ Mpc within which the correlation exponent $\gamma \approx 1.8$ corresponds to a fractal dimension $D = 3 - \gamma \approx 1.2$. Beyond R_{hom} the galaxy distribution was thought to be homogeneous.

It was realized later that the correlation function estimator may result in distorted values for R_{hom} and γ (or D). Moreover, due to fractal projections, angular catalogues are not good for detecting a structure with $D \geq 2$. For this 3D maps are required and indeed on scales where the fractal analysis is possible in completely embedded spheres, redshift-based maps have revealed a power–law density field and have given evidence for the "hidden" fractal dimension of about 2.

The debate on the large-scale galaxy distribution has been revolving around two new major empirical numbers—the fractal dimension D and the crossover scale to homogeneity R_{hom}. The value of the correlation dimension $D = 1.2$ indirectly deduced from angular catalogues has been contrasted by $D = 2.2 \pm 0.2$ directly obtained from 3D maps and R_{hom} has expanded from 10 Mpc to scales approaching 100 Mpc. The narrow cones of the existing deep galaxy surveys and their finite depth have been the main limiting factors hampering precise estimates of D and R_{hom}.

11.2 Analysis of the Angular-Position Galaxy Catalogues

Already when galaxy catalogues gave just positions on the sky vault, and no distances, astronomers started to recognize superclusters of galaxies.

Table 11.2 Modern events in the debate on large-scale fractality

Years	Authors	Subject
1970s–1980s	Wertz, Bonnor, Wesson, Alfvén	physics of hierarchy
	Haggerty, Severne, Prigogine	N-body dynamics in hierarchies
	Totsuji, Kihara, Peebles	$w \propto \theta^{-0.8} \Rightarrow \xi \propto r^{-1.8}$
	Mandelbrot	fractals, multifractals
	Baryshev, Perdang	first evidence for $D = 2$:
	Lerner, Schulman, Seiden	$M(r)$, $z(r)$, stability, percolation
ξ and Γ function	Davis, Peebles	$\gamma = 1.8$, $r_0 = 5$ Mpc, 2000 galaxies
	Einasto, Klypin, Kopylov, Bahcall	r_0 depends on depth and type
	Pietronero	the method of Γ function
	Pietronero, Sanders, Coleman	the first fractal analysis of CfA
	Pietronero, Ruffini, Calzetti et al.	explanation of $r_0(R_s)$
	Jones, Martinez, Saar, Einasto	$D_2 = 1.2$, $D_0 = 2$, multifractal?
1990s–2000s	Sylos Labini, Montuori, Pietronero	$D \approx 2$ from all 3D catalogues
	Wu, Lahav, Rees	fractality at $r < 30$ Mpc
	Teerikorpi, Hanski, Theureau et al.	TF < 200 Mpc $\Rightarrow D = 2.2$, KLUN
	Paturel, Teerikorpi, Courtois	LEDA counts $< 15^m$: $0.44m$, $D = 2.2$
fractals in 3D maps	Baryshev, Bukhmastova	2-point column density: $D = 2.1$
	Zehavi (SDSS team)	$\xi(s) \propto s^{-1.2}$, 29 300 gal.
	Hawkins (2dFGRS team)	$\xi(s) \propto s^{-0.75}$, 200 000 gal., $D = 2.25$ distortion by peculiar velocities?
	Gott et al. (SDSS team)	500 Mpc Sloan Great Wall
	Hogg et al.	SDSS LRG $\Gamma^*(r) \propto r^{-1.0}$, $D = 2.0$ for scales $1 \div 30h_{100}^{-1}$ Mpc
	Vasilyev, Baryshev, Sylos Labini	2dF $\Gamma^*(r) \propto r^{-0.8}$, $D = 2.2$ SDSS DR4 $\Gamma^*(r) \propto r^{-1.0}$, $D = 2.0$ for scales $0.5 \div 30h_{100}^{-1}$ Mpc
	Sylos Labini, Vasilyev, Baryshev	SDSS DR6, 100 Mpc structures

11.2.1 Early Arguments for Galaxy Clustering

Inspecting his sky map for 11 475 nebulae from Dreyer's NGC and Index catalogues, Charlier (1922) concluded: *A remarkable property of the image is that the nebulae seem to be piled up in clouds.* Concerning the nature of nebulae, Charlier chose *to regard the spiral nebulae as foreign Galaxies similar to our own.* Thus he related the observed clustering to the global matter distribution.

After the discovery of the galaxy universe it soon became clear that in addition to field galaxies there are groups and clusters. In the 1930s clusters of galaxies were already routinely used for extending the redshift–distance law to larger distances by Hubble and Humason. In the following decades several observations were seen as evidence for large-scale clustering:

- Shapley's metagalactic clouds
- De Vaucouleurs's Local Supercluster and density-radius relation
- Abell's rich clusters and their superclusters
- Shane-Wirtanen clouds of galaxies in the Lick counts

Shapley's Wide Photographic Surveys of Galaxies Inspecting the distribution of galaxy clusters, Shapley came to conclude that there are "metagalactic clouds" (today's superclusters), e.g., in the constellations of Coma, Centaurus and Hercules. An interesting outcome of the 2D epoch, his book *The Inner Metagalaxy* (Shapley 1957) summarizes the work on the clustering of galaxies performed at the Harvard observatory. In a section about deep surveys he notes that though the distribution of galaxies on the surface of the sky is easily examined on photographs, a study of the distribution in the line of sight is much complicated "... a pair with equal apparent brightness may differ in distance by a factor of ten. ... In the study of the radial distribution of population it is necessary to use photometric methods for estimating distances, relative or absolute." The redshift is nowhere mentioned as a possible distance indicator—the local universe was not yet a subject of redshift surveys, the precious telescope time went to extending measurements into deeper space.

Gérard de Vaucouleurs (1953, 1958) presented evidence, from the Shapley and Ames (1932) catalogue of 1249 galaxies, for a local supercluster centered at the Virgo cluster, and having a diameter of 30 Mpc. It causes the well-known asymmetry in the numbers of bright galaxies in the two hemispheres.

A large increase in the number of known galaxy clusters came with George Abell's (1958) catalogue of 2712 *rich clusters* covering the sky north of $\delta = -27°$.[1] Together with southern rich clusters in Abell et al. (1989), these lists contain 4704 clusters. Abell's collection was one outcome of the photographic survey of the entire northern sky, made by the Schmidt telescope at Palomar Observatory. The nine hundred $60 \times 60\ \mathrm{cm}^2$ copies of the Palomar Sky Atlas photographs were a basic tool at observatories for decades. Rich clusters are rare, but can be seen from large distances. Now the question was: do Abell's rich clusters form superclusters?

Santa Barbara 1961 Conference In the meeting "On the stability of systems of galaxies" astronomers presented their works on galaxy clustering on different scales from binary galaxies to superclusters. Here de Vaucouleurs (1961) already discussed his density-radius relation for systems of different sizes (see below). He noted that there was no sign of a constant density level in the observed range, so the mean density might be reached only for scales 10 times larger than the Local Supercluster (i.e. 300 h_{100}^{-1} Mpc). Abell (1961) described his work on the rich clusters which form superclusters of typically 10 clusters. In the conference summary Neyman, Page and

[1] Abell's six richness classes were based on the number N of galaxies that are at most 2 mag fainter than the 3rd brightest galaxy: for class 0 $30 \leq N \leq 49$; for class 5 $N \geq 300$.

Scott wrote: *Both Abell and de Vaucouleurs feel that superclustering is established beyond doubt, and that the dimensions of second-order clusters (clusters of clusters of galaxies) are 30 to 60 Mpc. This means that clusters cannot be treated as isolated systems embedded in an isotropic, homogeneous medium of field galaxies...*

Here the statisticians Jerzey Neyman and Elizabeth Scott admit that their classical and simple 2-level hierarchy model of galaxy clustering (Neyman and Scott 1952) was not enough to explain the real situation.

Further evidence came from the Lick Observatory survey by Shane and Wirtanen (1967). Its results were summarized by Shane (1975), close to the end of the 2D epoch: *Clustering seems to be a general, if not a universal, property among the galaxies. We find larger aggregations comprising numbers of clusters that extend over linear distances up to 30 Mpc. There is suggestive evidence of still larger assemblages of galaxies on a scale of 100 Mpc or more.*

The Cosmological de Vaucouleurs Law Carpenter (1938) found an intriguing regularity among clusters of galaxies: larger clusters are less dense. In particular, the number of galaxies N in a cluster grows with the size r as $N(r) \propto r^{1.5}$. Carpenter viewed this relation as a cosmic restriction so that a cluster of a given extent may have no more than a limited number of members. The relation extends from pairs of galaxies to large clusters. This showed for him that small and large systems do not essentially differ.

After comparing the distribution of Abell's clusters with computer-made artificial maps, Kiang (1967) concluded that *simple clustering by uniform clusters fails to represent the world of Abell's objects, in the same way as it has failed in the world of galaxies*, referring to Neyman and Scott's model. He suggested that *clustering of galaxies occurs on all scales*, without clear-cut hierarchic levels, and visualized the arrangement so that the various clusters interpenetrate each other; this may guarantee that the mean density does not depend on the volume. He came close to the view of fractal clustering, but did not make the crucial step to the fractal concept, where the mean density depends on the volume. Later Kiang and Saslaw (1969) found that the clustering extends at least to scales of 100 Mpc.

Karachentsev (1966, 1968) added an important aspect to Carpenter's result. He estimated average properties of 143 systems from binary galaxies to superclusters and found evidence that both luminous and total (virial) mass densities are decreasing with increasing size of a system. This showed for the first time that the mass–radius behaviour of the dark mass is also a power law, but the exponent is different than for the luminous matter.

De Vaucouleurs made the decisive step in recognizing the cosmological significance of the clustering of galaxies. The results of his thinking he published in 1970 in "The Case for a Hierarchical Cosmology", where he suggested the existence of a universal density–size law in the galaxy universe:

$$\rho(r) = \rho_0 (r/r_0)^{-\gamma}, \tag{11.1}$$

where $\rho(r)$ is the mass density in the sphere of radius r and ρ_0 and r_0 are the density and radius at the lower cutoff of the structure. The available galaxy data led to $\gamma \approx 1.7$.

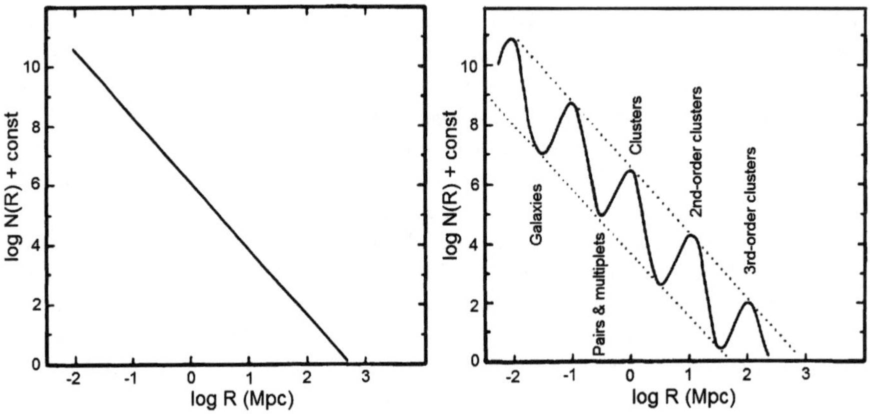

Fig. 11.1 De Vaucouleurs's concepts of smooth and preferred-scales density–radius relation (adapted from de Vaucouleurs 1971)

He considered two extreme cases for the possible behaviour of the density–radius relation (Fig. 11.1). First, the density may decrease smoothly. This would happen if there were no preferred sizes of systems (in modern terms, a stochastic fractal distribution). Second, the curve may have a series of peaks at several preferred scales, corresponding to different kinds of systems (from galaxies to superclusters). In fact, later the correlation analysis of observations pointed at the first case, with no preferred sizes.

11.2.2 Early Arguments for Homogeneity

Interestingly, many arguments also seemed to favour the uniformity of the galaxy distribution. In this view clusters were considered as exceptional objects. Evidence for homogeneity, as presented before the 1970s, included:

- Hubble's galaxy counts
- Fluctuations in numbers of galaxies due to variable dust extinction
- Sandage-Tammann-Hardy argument from the local linear Hubble law
- Isotropic distribution of distant objects

Hubble's Counts of Bright Galaxies Hubble (1926) concluded that his galaxy counts correspond to the expectation from homogeneity up to $m = 16$. This was widely regarded as evidence for a homogeneous world model.

The fundamental equation of stellar statistics (Eq. (3.2)), when applied to objects having a power-law radial number density law $n \propto r^{-\gamma}$, or a number–radius relation $N \propto r^{D}$, gives the number of objects brighter than m as $\log N(m) = 0.2Dm + \text{const}$, with $D = 3 - \gamma$. For a uniform distribution ($\gamma = 0$) one obtains the Seeliger law $\log N(m) = 0.6m + \text{const}$.

Hubble found that in the range 8^m–12^m and at the point 16.7 the "$0.6m$-law" was valid. For 10^m–13^m this was confirmed by Shapley and Ames (1932) in their catalogue, for the sum of the counts in the northern and southern skies. A shortcoming was the lack of data within 12^m–16^m.

Hubble thought that clusters of galaxies contain only a small fraction of all galaxies, and the true distribution is quite smooth. This view had a strong impact on theoretical cosmology, as expressed by Einstein in 1933: *Hubble's research has, furthermore, shown that these objects are distributed in space in a statistically uniform fashion, by which the schematic assumption of the theory of a uniform mean density receives experimental confirmation* (cited by Peebles 1980). From this time on the picture of a uniform galaxy field with clusters as rare fluctuations became a paradigm of the homogeneous universe.

The bright galaxy counts were used by Sandage et al. (1972) to show the conflict with de Vaucouleurs's hierarchical model. For its $\gamma = 1.7$, $\log N(m) = 0.26m +$ const, clearly different from the value $0.6m$ from the counts by Hubble and Zwicky, available at the time. The modern counts in the range 10^m–14^m differ from these old results (Sect. 12.1).

The Deep Counts Hubble (1934, 1937) extended the counts up to the magnitude 21 in his program on 1184 photographs (each covering 0.25 deg^2) of random positions in the sky. Now the counts followed a $0.5m$ law at faint magnitudes. Hubble did not abandon homogeneity, but tried to explain the deflection as due to a redshift effect on galaxy magnitudes. Considering a kind of K-correction he concluded that the $0.6m$ law may be obtained only in a static universe. In fact, to the end of his life, Hubble thought that the cosmological redshift might be caused by something else than expansion.

Sandage (2004) found three systematic errors that had affected Hubble's calculations: (1) systematic errors in photographic magnitudes made them increasingly too bright for fainter objects; (2) an error in the Hubble's K-term; (3) when calculating the spatial volumes in the Friedmann model for different redshifts Hubble used an incorrect kind of distance (the redshift distance $r \propto z$). Therefore, Hubble's old result on faint galaxies cannot any longer be used as evidence either for homogeneity or non-expansion.

Variable Extinction An argument against real superclusters referred to our "dusty window" to extragalactic space. As the early evidence came from the 2D sky, a natural source of worry was the variable light extinction in different directions, due to the cloudy dust foreground in the Milky Way.

Victor Ambartzumian (1940, 1951) developed the theory of a fluctuating Galactic extinction and applied it to the counts of galaxies. In their well-known article Neyman and Scott (1952) referred to Charlier and Ambartzumian and emphasized two views: (1) galaxies are really clustered in space; (2) the clustering in the sky is caused by variable extinction. Which factor dominates for clusters and superclusters?

Warwick (1950), Holmberg (1974), and Fesenko (1975) concluded that the observed clustering is much modified by dust. Zwicky (1955, 1957) proposed that also intergalactic dust within clusters of galaxies could shade the background. Karachentsev and Lipovetskii (1969) derived from counts of background clusters a mean value of 0.2 mag for the B-band extinction.

Mattila (1977) detected diffuse light in the Coma cluster and concluded that part of it could be scattered light from intergalactic dust in the cluster. Later together with Stickel et al. (1998), he measured the far-IR emission of the dust in Coma and calculated that the dust could cause only about 0.2 mag or less of extinction. Teerikorpi (2002) showed from the reddenings of quasars seen through galaxy halos (spectra of such quasars contain narrow absorption lines at lower redshifts) that there is about 0.2 mag of extinction per halo. Inside compact galaxy clusters and around the cores of rich clusters such as Coma the galaxy halos may almost overlap in projection; one might on this basis alone expect an extinction of the order of 0.1 mag.

The argument from extinction lost its power after redshift maps revealed the lumpy 3D galaxy distribution. However, Galactic extinction is still relevant for other extragalactic subjects like the distance scale and the fluctuations in the cosmic background radiation.

The "Linearity" Argument de Vaucouleurs (1970) expected density fluctuations to distort the Hubble law. In particular, in his hierarchical model the expansion rate should be reduced by gravitation inside the Local Supercluster. His PhD student Wertz developed a Newtonian expanding hierarchical model and predicted the deviation from linearity: the Hubble "constant" would grow with increasing distance within 20 Mpc from us.

When Sandage et al. (1972) confronted the observations with the predicted deviation from the Hubble law (Wertz 1971; Haggerty and Wertz 1972), the result was striking: the linear law was found on all tested scales, while the hierarchical model predicted so strong a deviation that one should not see any cosmological expansion at distances closer than 20 Mpc. Later studies by Sandage, Ekholm, Karachentsev and others strengthened the case: a linear flow starts close to the Local Group, already at 1.5 Mpc.

This test had a deeper meaning than being just a probe of the hierarchical distribution of matter. It brought into light a paradox: both empirical facts were true, i.e. the strongly inhomogeneous galaxy distribution and the linear Hubble law on the same local scales. Sandage et al. (1972) suggested two possible solutions: either the mass density parameter could be very small or there could be an invisible uniform medium of high density. In both cases, the perturbations of the Hubble law would be tiny. In fact, this cosmological test gave rise to a whole new approach, studying the properties of the very local Hubble flow (we return to this subject in Chap. 12).

de Vaucouleurs (1972) did present evidence for a curved Hubble law in the local universe. Teerikorpi (1975a, 1975b) argued that the phenomenon was caused by a selection effect due to the magnitude limited catalogue influencing distance estimates (Chap. 4).

Isotropy and Homogeneity Classically, from isotropy around one observer, together with the Copernican principle ("all points are alike"), one may infer the global homogeneity of the cosmic medium (Walker 1944; for a delightfully simple argument, see Weinberg 1977). Before the 1970s three major observations supported an isotropic matter distribution around us.

First, Hubble's counts of faint galaxies (up to $\approx 21^m$) did not show large differences in different directions, after corrected for Galactic extinction.

Second, the thousands of faint radio sources in the early catalogues (4C, Parkes, and others) were found to be uniformly distributed in the sky within statistical uncertainty (Holden 1966; Webster 1977).

Third, already the first measurements of the cosmic background radiation (Conklin and Bracewell 1967; Penzias et al. 1969) found a smooth isotropy at the level $\frac{\Delta T}{T} < 10^{-3}$ for all angular scales up to arc-minutes.

These isotropies have generally been interpreted as strong evidence for homogeneity on scales exceeding 1000 Mpc. However, a final verdict requires direct evidence from 3D-maps. It would also be useful to study which kind of a statistically isotropic fractal structure would be compatible with the observed level of isotropy as concerns galaxies and radio sources.

11.2.3 Results from Galaxy Angular Catalogues

The first deep wide area catalogues of galaxies and galaxy clusters appeared in the 1950s and 1960s, based on the Palomar Sky Atlas survey. Up to the 1970s, this was the golden age of 2D angular catalogues.[2]

Angular Correlations The main tool was the angular correlation function technique, described in detail by Peebles (1980). The angular two-point correlation function $w(\theta)$ is defined analogously to $\xi(r)$ as the excess probability relative to the Poisson law to find an object in the solid angle $\delta\Omega$ at the angular distance θ from randomly chosen objects in the sample:

$$\delta P = n\delta\Omega[1 + w(\theta)], \tag{11.2}$$

where n is the mean surface density of objects in the sky.

A practical estimate of $w(\theta)$ for a sample of N galaxies in the sky is obtained by counting objects in rings of radius θ and width $\delta\theta$. If Ω_s is the survey solid angle, the normalization condition is

$$N = n \int_{\Omega_s} [1 + w(\theta)] d\Omega. \tag{11.3}$$

An important point is that if the 2D projection of an inhomogeneous spatial distribution is close to homogeneous, then this method cannot detect the true inhomogeneity but regards it as a structureless Poisson distribution!

[2]The main sources were Abell's (1958) catalogue of rich clusters, the Reference Catalogue by de Vaucouleurs and de Vaucouleurs (1964), the Catalogue of Galaxies and Clusters of Galaxies by Zwicky et al. (1961–68), and the Lick counts by Shane and Wirtanen (1967).

Angular vs. Spatial Correlation Functions The aim of the analysis of angular catalogues was to infer the spatial correlation function $\xi(r)$ from the measured angular distribution. The relation between $w(\theta)$ and $\xi(r)$ was derived by Limber (1953), already in the context of galaxies (reviewed by Fall 1979).

Limber's integral equation may be inverted analytically in order to obtain $\xi(r)$. This procedure requires differentiation of observed data always containing some noise that amplifies errors. An important and useful case is the power-law solution. Then the angular and the spatial correlation functions $w(\theta)$ and $\xi(r)$ are simply linked:

$$w(\theta) = A\theta^{1-\gamma} = A\theta^{D-2} \Leftrightarrow \xi(r) = Br^{-\gamma} = Br^{-(3-D)}, \tag{11.4}$$

where the constant A comes from observations and B depends on γ and the selection function ϕ (Fall 1979).

Caution: The power-law solution exists only if $\gamma > 1$, or $D < 2$. This means that the method does not work for structures with the fractal dimension $D \geq 2$ (cf. the fractal projection theorem in Sect. 10.2.3). It is interesting to ask which side of $D = 2$ the real galaxy distribution prefers!

Hierarchical $D = 1.2$ Models for 2D Catalogues In their pioneering work Totsuji and Kihara (1969) derived from the angular data of the Lick galaxy counts a power law angular correlation function

$$w(\theta) = A\theta^{-0.8}, \tag{11.5}$$

from which they finally obtained

$$\xi_{\mathrm{TK}}(r) = \left(\frac{r}{4.7h^{-1}\,\mathrm{Mpc}}\right)^{-1.8} \tag{11.6}$$

In 1973 Peebles started an extensive programme analysing all angular galaxy and cluster catalogues, and concluded that these all are characterized by almost the same power law with $\gamma = 1.77$ and $r_0 = 5h_{100}^{-1}$ Mpc in the interval 0.1–10 Mpc (reviews may be found in Peebles 1980, 1993).

This was in fact a rediscovery of de Vaucouleurs's density–radius relation on small scales. Totsuji and Kihara (1969) and Peebles (1974) had for the first time found that the galaxy correlation function is a power law with no peaks, contrary to what would be expected from preferred scales (case 2 in de Vaucouleurs 1970). Thus it naturally reflects the self-similarity of fractal structures ($\gamma = 1.8$ corresponds to the fractal dimension $D = 1.2$).

The emerging picture of continuous hierarchy inspired the construction of protofractal models for the galaxy distribution. Soneira and Peebles (1977, 1978) studied a regular hierarchy of Fournier d'Albe's type with 12 levels of galaxy pairs, so that $k_N = 2$, $k_r = 1.76$, and (from Eq. (10.8)) $D = 1.23$ ($\gamma = 1.77$). Soneira and Peebles concluded that even this simple binary hierarchy model reproduced well the angular correlation function of the galaxy distribution in the Lick counts. They also inferred from the Zwicky et al. (1961–68) catalogue that there is no evidence for a spatially uniform population of field galaxies. This confirmed the general tendency for galaxies to appear only in clusters.

Tallinn 1977 Conference A landmark event in the study of clustering was the IAU Symposium 79 *The large scale structure of the Universe* held in Tallinn, Estonia, September 1977. This was also the first wide discussion of existing arguments for and against homogeneity on extragalactic scales.

Peebles (1978) gave a review of the angular correlation function analysis. He emphasized that *"a small systematic error in the angular distribution can be translated into a very large error in the estimate of the spatial clustering"* and that *"redshift data will allow us to avoid this problem"*. Peebles argued that at distances ≤ 10 Mpc the galaxy clustering was described by a power law, but on larger scales the galaxy distribution became homogeneous.

In a break-through work, Jõeveer and Einasto (1978) presented an analysis of existing redshift data and concluded that galaxies are arranged in filaments surrounding empty voids. They interpreted their surprising findings in terms of a cell structure, with the mean diameter of voids about $50h^{-1}$ Mpc. The presence of large structures and holes of various sizes was also demonstrated by de Vaucouleurs, Tully, Abell, and others. Indeed, in his concluding remarks Longair (1978) noted that *Everyone seemed to agree about the existence of superclusters ... systems on scales* $\sim$ 30–100 *Mpc*. Then, when referring to the angular correlation analysis leading to a homogeneity scale at 10 Mpc, he asserted *I am still a firm believer in the basic correctness of the covariance analysis*. Still today we are wondering if the two results are compatible!

Emerging Evidence for $D = 2$ Though the angular correlations suggested the fractal dimension $D \approx 1.2$ ($\gamma = 1.8$), other pieces of evidence hinted at $D \approx 2$. It is interesting that Lundmark (1927) made in effect the first estimate of D on the basis of Charlier's model, noting that the second criterion is fulfilled, which corresponds to $D = 2$. The scarce observations did not provide any convincing evidence, but Knut Lundmark liked the idea that "the world is constructed in such a way that it is not far from collapse on account of the total attraction being near one of the limits."

Baryshev (1981) discussed some arguments in favour of the fractal dimension $D \approx 2$. Using such data as: (1) the galaxy number counts $N(m)$ from 2^m up to 24^m; (2) the virial mass density—radius relation $\rho_{\mathrm{vir}}(R)$; (3) the peculiar-velocity dispersion—radius relation $\sigma_{\mathrm{v}}(R)$, he concluded that a hierarchical model of galaxy distribution with $\gamma = 1$ ($D = 2$) is consistent with the observations. A new theoretical idea on a special property of fractals with $D = 2$ was also presented, based on Bondi's (1947) consideration of the global gravitational part of the cosmological redshift (Sect. 9.5.1).

Why Did Angular Catalogues Miss the $D = 2$ Structure? An explanation of why it is difficult to study a fractal structure with dimension $D \geq 2$ from the angular distribution of galaxies was in essence given by Baryshev (1981). If we model a part of the fractal as a spherical cluster of particles inside the radius R, then one can derive the surface distribution $F(\sigma)$ of the particles, projected on the sky, using the

Abel equation

$$F(\sigma) = 2\int_{\sigma}^{R} \rho(r)\frac{r\,dr}{\sqrt{r^2-\sigma^2}}, \tag{11.7}$$

where σ is the projected distance from the centre of the sphere.

For a power-law representation of a spherically symmetric fractal structure $\rho(r) = \rho_0(r_0/r)^{3-D} \propto r^{-\gamma}$, analytical solutions exist in closed form for fractal dimensions 3, 2, and 1. For a homogeneous ball ($D = 3$, $\gamma = 0$)

$$F(\sigma) = 2\rho_0 r_0 \frac{R}{r_0}\sqrt{1-\frac{\sigma^2}{R^2}}. \tag{11.8}$$

For a structure with $D = 2$ ($\gamma = 1$) we get

$$F(\sigma) = 2\rho_0 r_0\left[\ln\left(1+\sqrt{1-\frac{\sigma^2}{R^2}}\right) + \ln\frac{R}{\sigma}\right]. \tag{11.9}$$

Finally, for $D = 1$ ($\gamma = 2$) the surface density will be

$$F(\sigma) = 2\rho_0 r_0 \frac{r_0}{\sigma}\arccos\frac{\sigma}{R}. \tag{11.10}$$

Hence for the values $\sigma \ll R$ the surface density behaves approximately as

$$F(\sigma) \sim \text{const} \quad (D=3), \qquad F(\sigma) \sim \ln\sigma \quad (D=2), \qquad F(\sigma) \sim \sigma^{-1} \quad (D=1). \tag{11.11}$$

For $D = 3$ and $= 2$ the surface density varies slightly. This means that the projected distribution appears in the sky as a smooth surface. Thus the angular correlation analysis becomes inefficient for structures with the fractal dimension close to or larger than the critical value 2, because the information on the spatial structures with $D \geq 2$ is lost. As we already emphasized this result reflects the general theorem on fractal projections (Sect. 10.2.3).

11.3 Redshift and Photometric Distance Surveys

For years, astronomers could make only indirect conclusions about the distribution of galaxies on the basis of their projected locations in the sky. The situation was changed when it became possible to make 3D maps of galaxies using data from massive surveys of redshifts.

Table 11.3 Some galaxy surveys. The columns: name of survey, solid angle Ω covered by survey, magnitude limit, number of galaxies N, distance indicator

Catalogue	Ω (sr)	m_{lim}	N	distance indicator	reference
CfA1	1.83	14.5	1845	z	Huchra et al. (1983)
CfA2 (North)	1.23	15.5	6478	z	De Lapparent et al. (1986)
PP	0.9	15.5	3301	z	Haynes and Giovanelli (1988)
SSRS1	1.75	14.5	1773	z	Da Costa et al. (1991)
SSRS2	1.13	15.5	3600	z	Da Costa et al. (1994)
LCRS	0.12	17.5	26000	z	Shectman et al. (1996)
IRAS 2Jy	4π	2. Jy	2652	z	Strauss et al. (1992)
IRAS 1.2Jy	4π	1.2 Jy	5313	z	Fischer et al. (1995)
ESP	0.006	19.4	4000	z	Vettolani et al. (1997)
KLUN	4π	15	6500	TF	Theureau et al. (1997b)
KLUN+	4π	16	(20000)	TF	Theureau et al. (2005)
Local Volume	4π	< 500 km/s	500	RGS	Karachentsev et al. (2003)
2dF	0.27	19.5	250 000	z	Colless et al. (2003)
SDSS	π	19	10^6	z	www.sdss.org
COSMOS	$77' \times 77$	25	500 000	photoZ	cosmos.astro.caltech.edu
HUDF	$3' \times 3'$	30	10 000	photoZ	Beckwith et al. (2006)

11.3.1 Large Redshift Surveys

Nature has given the astronomer, in the form of the redshift–distance law, a way to measure extragalactic distances, which is usually more precise than photometric methods for regions where peculiar velocities are low enough.

Over 2700 galaxies had their redshift listed in the *Second Reference Catalogue* by De Vaucouleurs et al. (1976). This was the breakthrough which made it possible to use redshifts for mapping the structures made by galaxies. During the last decades, technological advances and systematic surveys have spurred a tremendous explosion in the number of measured redshifts. Currently more than one million redshifts have been measured.

Large redshift surveys are known by a number of abbreviations (from CfA to SDSS; Table 11.3 shows the parameters for some of these). Based on these, several 3D maps have become available: both wide- and narrow-angle. The last decade saw the appearance of especially large surveys: the two-degree field survey 2dF and the Sloan Digital Sky Survey SDSS. The depth of these galaxy catalogues allows one to detect and analyse structures with sizes up to 100 Mpc (see Fig. 11.2). Even much larger scales have been probed using extremely deep surveys such as COSMOS and HUDF. In principle, these surveys are able to detect structures with sizes of about 1000 Mpc, at lookback-time of around 10^9 years (Chap. 12).

Magnitude limit of a survey. The number of galaxies increases steeply with magnitude (there is $\sim$ one galaxy/deg^2 up to 15.5^m and 85 up to 19^m). Fortunately, multiple object spectrographs allow a simultaneous detection of hundreds of spectra. Even if the sample is complete up to a fixed limiting magnitude m_l, its completeness in space depends on the absolute magnitude M: the limit is at the distance modulus $\mu_\mathrm{l} = m_\mathrm{l} - M$. So, with $m_\mathrm{l} = 15.5$ and $M = -20$, $\mu_\mathrm{l} = 35.5$ means the distance 125 Mpc, while galaxies with $M = -18$ are sampled only up to 50 Mpc. Thus a survey probes the distribution of only the most luminous galaxies at large distances, perhaps causing a biased picture. In order to cope with this problem, one has to push the surveys to fainter limits, as one cannot tell beforehand whether a galaxy is intrinsically bright or faint (distant or near).

11.3.2 Galaxy Catalogues with Photometric Distances

For the study of large-scale structure, redshift catalogues are generally superior to those based on photometric distances: redshifts are (1) much less time-consuming to measure, (2) usually more accurate than photometric distance indicators, and (3) measurable for all Hubble types of galaxies.

However, some questions (such as the Hubble constant and galaxy streams) require large samples of galaxies for which both redshifts and photometric distances are known. Even photometric distances as such are valuable. They can be used for mapping the environment of the Local Group and for deriving the average number density law around us.

Igor Karachentsev's program of distance measurements to nearby galaxies within about 10 Mpc has used especially the tip of the red giant branch (TRGB) method (Karachentsev et al. 2003). Over these years many nearby galaxies have been resolved into stars for the first time in this program requiring large ground-based telescopes as well as the Hubble Space Telescope.

The KLUN sample and its growing version KLUN+ (http://klun.obs-nancay.fr/KLUN+) contain spiral galaxies for which photometric magnitudes and HI 21 cm line widths have been measured. Using the TF relation, these quantities lead to an estimate of the distance (up to $\sim$ 100 Mpc). Originally planned for measuring the Hubble constant (e.g., Theureau et al. 1997b), the KLUN with its 5500 galaxies was also used to study the density law of galaxies around us (Teerikorpi et al. 1998). In the KLUN+ project an HI survey at the Nançay radio telescope is building a wide-angle sample of 20 000 spiral galaxies (Theureau et al. 2005). The photometry, from the DENIS (Near Infrared Survey) and 2MASS (2 Micron All Sky Survey), will be complete to magnitude limits in five bands (B, I, J, H, K).

11.4 Analysis of the 3D Distribution of Galaxies

During the 1980s several galaxy redshift catalogues (Table 11.3) became available for 3D analysis. The first survey, planned to produce a complete sample of redshifts on a large region of the sky, was carried out at the Harvard-Smithsonian Center

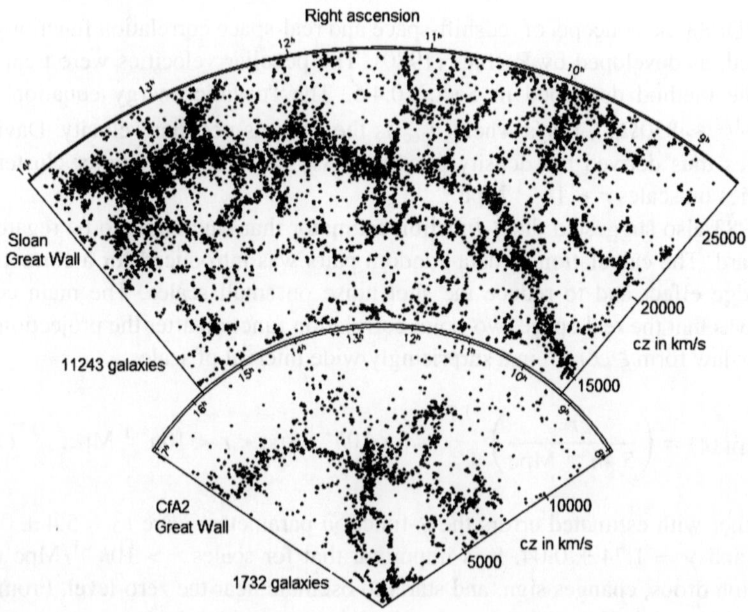

Fig. 11.2 A comparison of the CfA map (*below*) and the slice from the larger SDSS survey (*above*). The Great Wall of CfA and the 500 Mpc long Sloan Great Wall extend across these maps. Filamentary structures with sizes of 100 Mpc are evident. (Courtesy of J. Richard Gott III and Mario Juric)

for Astrophysics. The redshifts for all galaxies brighter than 14.5 mag in Zwicky's catalogue were measured. A CfA survey map (see Fig. 11.2), published in a paper titled *A Slice of the Universe*, became a symbol of the complexity of the galaxy universe (de Lapparent et al. 1986).

11.4.1 The Fractal Breakthrough in the 1980s

After the concept of fractal entered the scene of large-scale structure research, one part of the cosmological community continued to use the ξ-function method and explained its peculiarities as due to a luminosity dependence of clustering. The second, smaller, group started to apply the Γ-function, and this "fractal approach" has also given rise to new fruitful directions for observational and theoretical studies.

ξ-Function Analysis of the CfA Data The paper by Davis and Peebles (1983; DP83) was in many ways classical. It was the first systematic study of the CfA data by the method of reduced correlation function. Complete to $m_B = 14.5$ in the regions ($\delta > 0, b > 40°$) and ($\delta \geq -2.5°, b < -30°$)), the catalogue contained 2400 galaxies with redshifts.

In DP83 the concepts of redshift-space and real-space correlation functions were utilized, as developed by Peebles (1980). The peculiar velocities were treated using the method described in Sect. 10.4.5. The "cosmic energy equation" gave $\langle v_{\rm pec}\rangle^{1/2} \approx 850\Omega_0^{1/2}$ km/s where $v_{\rm pec}$ is the 3D rms peculiar velocity. Davis and Peebles thus derived the density parameter $\Omega_0 \approx 0.2$ for the matter clustered as galaxies on scales $r < 1h^{-1}$ Mpc.

DP83 also suggested the ξ-function estimator that later came to be regarded as standard. The calculation of data–random pairs was introduced for a correction of the edge effect and to reduce the shot noise on small scales. The main conclusion was that the real-space two-point correlation function after the projection has a power-law form $\xi \propto r^{-\gamma}$ in a surprisingly wide interval of scales:

$$\xi_{\rm DP}(r) = \left(\frac{r}{5.4h^{-1}\ {\rm Mpc}}\right)^{-1.74} \quad \text{for } 10h^{-1}\ \text{kpc} < r < 10h^{-1}\ \text{Mpc}. \tag{11.12}$$

Together with estimated errors the ξ-function parameters were $r_0 = 5.4 \pm 0.3h^{-1}$ Mpc and $\gamma = 1.74 \pm 0.04$. It is important that for scales $r > 10h^{-1}$ Mpc the ξ-function drops, changes sign, and starts to oscillate near the zero-level. From what we considered in Chap. 10, this is exactly as is expected for the ξ-function estimator.

After this pioneering work, the values of the unit scale $r_0 \approx 5h^{-1}$ Mpc (defined as $\xi(r_0) = 1$) and the correlation exponent $\gamma \approx 1.8$ have been generally considered as standard cosmological numbers.

The Puzzling Behaviour of the ξ-Function In the 1980s it seemed that the first 3D map gave results consistent with the analysis of angular catalogues and that the galaxy distribution becomes homogeneous on scales larger than $20h^{-1}$ Mpc. Actually, an unexpected problem appeared. The characteristic length r_0 was found to depend on certain properties of the samples, such as the depth of the survey, the type and luminosity of galaxies and clusters, and the mean separation between the objects in the sample.

Unexpectedly, it was found that the length r_0 (and hence the amplitude of the ξ-function) is quite large for clusters of galaxies. Bahcall and Soneira (1983) calculated the redshift-space ξ-function for a complete sample of 104 Abell clusters: $r_0^{\rm cl} \approx 25h^{-1}$ Mpc, $\gamma^{\rm cl} \approx 1.8$. Klypin and Kopylov (1983) obtained almost the same parameters, when they studied another Abell sample, containing 158 rich clusters of galaxies: $r_0^{\rm cl} \approx 25h^{-1}$ Mpc, $\gamma^{\rm cl} \approx 1.6$. These results revealed quite a discrepancy between the unit scales r_0 for galaxies (5 Mpc) and for clusters (25 Mpc). Moreover, when the ξ-function was calculated for superclusters of galaxies (Bahcall and Burgett 1986; Lebedev and Lebedeva 1988), an even larger scale was found: $r_0^{\rm cl} \approx 60h^{-1}$ Mpc, $\gamma^{\rm cl} \approx 1.8$. According to these data, the correlation length r_0 increases from 5 to 60 h^{-1} Mpc when one considers increasingly massive objects in the universe.

An important property of this effect was also found by Einasto et al. (1986). From galaxy and cluster samples having different volumes they found a rough relation between the derived scale r_0 and the depth R_s of the sample:

$$r_0 \propto R_s. \tag{11.13}$$

In the framework of Gaussian density fluctuations on a homogeneous background this behaviour of the ξ-function is enigmatic. The most popular explanation is Kaiser's (1984) idea of *biased galaxy formation*, based on a possible relation between the correlation functions for galaxy clusters and for the underlying mass density field. Here clusters are viewed as rare high density spots in the density field, so that one might expect

$$\xi_{\text{clust}} = b\xi_{\text{density}}(r). \tag{11.14}$$

The bias factor b is about 10 if $\xi_{\text{density}} = \xi_{\text{gal}}$. This means that galaxies in clusters are formed from rare peaks above some global threshold in the primordial density field. However, the validity of this explanation was criticized by Gabrielli et al. (2004). They showed that the increasing sparseness of peaks over the threshold in Gaussian random fields does not explain the observed increase of the amplitude of the correlation function $\xi(r)$.

Other possibilities, such as luminosity segregation (Davis et al. 1988), demand careful studies with larger galaxy samples.

Pietronero and the Mystery of r_0 A radically new interpretation of the behaviour of the ξ-function was found by Pietronero (1987). He introduced the Γ-function method for the 3D galaxy map analysis and derived the relation between the Γ and ξ functions (Chap. 10). For a spherical sample with the depth R_s and a fixed galaxy luminosity Pietronero derived that for a fractal structure one expects a linear dependence of r_0 on R_s. Hence the increasing amplitude of the ξ-function and the corresponding increase of r_0 for deeper samples may not be due to a larger correlation length, but may be an artifact caused by the definition of the reduced correlation function.

In the first Γ-function analysis of the CfA catalogue, Coleman et al. (1988) found a power-law $\Gamma(r)$ with $\gamma = 1.5 \pm 0.2$. The fractal behaviour was detected within the interval from 1 up to 20 h^{-1} Mpc without any characteristic scale, contrary to $r_0 = 5\,h^{-1}$ Mpc from the ξ-function.

Cellular Fractal Structure A few months after Pietronero's 1987 article, the fractal interpretation of the correlation length versus depth was applied to redshift data by Calzetti et al. (1987). They confirmed the linear relation between r_0 and R_s, when the depth changed from 5 up to $50h^{-1}$ Mpc.

Then Ruffini and collaborators developed a cellular model of the Friedmann universe: within cells with sizes of about 100 Mpc the distribution of galaxies has the fractal dimension $D \approx 1.2$, and on larger scales the universe becomes homogeneous (Ruffini et al. 1988; Calzetti et al. 1988, 1989). The main conclusion was that de Vaucouleurs's density law may be reconciled with the uniform Friedmann model if fractality has a maximum scale.

The Ruffini at al. model uses hypothetical massive dark matter particles "ino", which obey Fermi statistics and are responsible for the initial density fluctuations of the cellular structure formation. The "ino" rest-mass-energy is 0.4–10 eV, and the density parameter would be $\Omega_{ino} = 0.4$–1. The size of the fractal cell is about 100 Mpc, determined by the Jeans length at the epoch when "inos" decoupled from matter. The angular scale of the corresponding CBR fluctuations is about 1°. The low value of $D \approx 1.2$ from the ξ-function was used. It would be interesting to reconsider this model for $D \approx 2$.

Multifractals and the ξ-Function Method Jones et al. (1988) and Martinez and Jones (1990) noted that when they applied the methods of box-counting and minimal spanning tree for determining the Hausdorf dimension D_H of the CfA redshift survey, they obtained the value $D_H = 2.1 \pm 0.1$. However, from the ξ-function for the same galaxy catalogue they concluded that the correlation dimension differs from this value, $D_2 = 1.2$, hence *"the Universe is not a simple fractal. It is a more complex structure, a multifractal"*. An analogous conclusion was made by Klypin et al. (1989) and Balian and Schaeffer (1989).

Here we may have an example of how the method of data analysis can affect theoretical conclusions. Indeed, as discussed in Chap. 10, the estimation of the fractal dimension from the ξ function as $D_2 = 3 - \gamma_\xi$ on scales close to r_0 gives an erroneous result, because there $\gamma_\xi(r = r_0) \approx 2\gamma$ (here γ is the true value of the correlation dimension). In order to take into account this deflection in the observed slope $\gamma_\xi(r = r_0) = 1.8$, we calculate $D_2 = 3 - \gamma = 3 - (\gamma_\xi(r = r_0)/2) = 2.1$.

Therefore one may conclude that after all $D_2 \approx D_H$, i.e. the correlation dimension is consistent with the Hausdorf dimension for the CfA data and there is no need for multifractality based on unequal dimensions. Modern results of the Γ-function analysis for CfA, 2dF, SDSS and other galaxy redshift surveys confirm the value for the correlation dimension $D_2 \approx 2$, and hence eliminate the need for a multifractality of this kind.[3]

11.4.2 Further Steps in the Debate

In the 1980s, new evidence appeared for structures with sizes much larger than the scale r_0 derived from the ξ-function analysis.

Balatonfured 1987: Very Large Structures The results of a decade of research after the Tallinn 1977 meeting were discussed at the 130th IAU Symposium in Hungary.

[3]The multifractal picture, characterized by a continuous set of fractal dimensions, is a generalization of fractal properties (Paladin and Vulpiani 1987; Benzi et al. 1984). Multifractals may be viewed in different ways because of the complexity of the problem (even for fractals there is no unique definition). They are in contrast with homogeneity exactly like fractals are. Such an approach was first suggested for galaxy distributions by Pietronero (1987) (see Gabrielli et al. 2005; Martinez and Saar 2002; Jones et al. 2004).

In Balatonfured, new observations on the reality of structures with sizes of about 100 Mpc were presented. An extension of the CfA redshift survey (Huchra et al. 1988) showed that empty regions (voids) and filaments are common in the galaxy distribution. The sizes of voids achieve 50 h^{-1} Mpc.

Karachentsev and Kopylov (1988) reported the results of a survey of 245 galaxies in a narrow strip which passed through the Coma cluster. They detected a bubble-like structure within the Coma supercluster and found that the average size of 14 voids was about $25h^{-1}$ Mpc.

The biggest structures were revealed by studies of rich galaxy clusters. Tully (1986, 1987) analysed the distribution of 47 Abell clusters within a region up to $z = 0.1$ and found a flat structure having a size of about $300h^{-1}$ Mpc (the Pisces–Cetus Supercluster Complex). Similar results from an even deeper survey were obtained by Kopylov et al. (1988)—the program "The Northern Cone of Metagalaxy" with redshifts up to $z = 0.28$ for 58 rich compact clusters with $b^{II} > 60°$ suggested that there are structures on huge scales up to $500h^{-1}$ Mpc.

In the conference summary by Peebles (1988) one may read: *There is considerable evidence of structure on scales* $\geq 50\ h^{-1}$ *Mpc, but I think it is fair to repeat the old questions: could this be an artifact of errors in the catalogues? Could the eye be picking patterns out of noise? If the answers were definitely "no" it would be very damaging for scale invariant cold dark matter. We all will be following the debate with great interest.*

Princeton "Dialogues'96" In 1996 the conference *Critical Dialogues in Cosmology* in Princeton was opened by the dialogue between Marc Davis and Luciano Pietronero on the homogeneity of the galaxy distribution.

Davis (1997) discussed "overwhelming evidence for large scale homogeneity on scales in excess of approximately $50h^{-1}$ Mpc, with a fractal distribution of matter on smaller scales". He emphasized that the observed correlation function $\xi(r)$ is well characterized by a power law, $\xi(r) \approx (r/r_0)^{-\gamma}$, with $r_0 \approx 5h^{-1}$ Mpc and $\gamma = 1.8$. Hence the fractal dimension on scales $r < r_0$ is $D = 1.2$. His arguments for homogeneity included, e.g., the isotropy of X-ray and radio source counts. He argued that $\xi(r)$ can be reliably recovered from the observed angular correlation function $w(\theta)$ and he concluded: *The measured two-point galaxy correlation function* $\xi(r)$ *is a power law over three decades of scale and approximates fractal behavior on scales of* $0.01h^{-1}$ *Mpc* $< r < 10h^{-1}$ *Mpc, but on scales larger than* $\approx 20h^{-1}$ *Mpc, the fractal structure terminates, the rms fluctuation amplitude falls below unity, and the Universe approaches homogeneity, as necessary to make sense of a FRW universe.*

Pietronero defended such fractality which extends deeper. He described the Γ-function method appropriate for the study of fractal structures and compared it with the ξ-function. His main points were (Pietronero et al. 1997): (1) The method of ξ function gives artificially distorted values both for the fractal dimension D and the homogeneity scale r_{hom}, (2) Γ-function is appropriate when one wants to estimate the true value of D and to detect the crossover to homogeneity. (3) The Γ-function analysis of the available 3D catalogues gave the fractal dimension $D = 2.0 \pm 0.2$ on scales up to the radius $R_{\mathrm{max}}^{\mathrm{sph}}$ of the largest sphere contained in the sample. Pietronero also noted that the homogeneity scale was not yet reached in existing galaxy catalogues.

Both sides agreed that the galaxy distribution is fractal at least within scales 0.1–$10h^{-1}$ Mpc. But they disagreed about the values of the fractal dimension (Davis for $D \approx 1.2$, Pietronero for $D \approx 2$) and the homogeneity scale (Davis for $R_{\mathrm{hom}} \approx 20h^{-1}$ Mpc, Pietronero for $R_{\mathrm{hom}} \geq 150h^{-1}$ Mpc).

11.4.3 Sky Projection of Fractals: The Angular Γ-Function

The fractal dimension $D = 1.2$ and the homogeneity on scales larger than $r_0 = 5h^{-1}$ Mpc were supported by the analyses of angular catalogues. Indeed, starting with Totsuji and Kihara (1969) the analysts found in all angular galaxy catalogues the universal behaviour of the angular correlation $\xi_{\mathrm{ang}}(\theta) \propto \theta^{-\alpha}$ with $\alpha \approx 0.8$. For scales $r < r_0$ in 3D space this would correspond to the fractal dimension $D = 2 - \alpha \approx 1.2$. As this is less than 2, then the fractal projection theorem discussed in Chap. 10 allows one to estimate the true fractal dimension from the galaxy distribution projected on the sky.

However, this logic is not quite watertight. If the spatial galaxy distribution is actually a fractal structure with $D \geq 2$ it is not certain if we could detect it from the angular distribution of galaxies, as the projection would imitate a homogeneous surface density. Consequently, 3D maps are required for detecting and reliably measuring structures with $D \geq 2$.

A decade after the first hint of a possible "conspiracy" of structures with $D \geq 2$ (Baryshev 1981), detailed studies started to appear on the complex problem of angular projections (Dogterom and Pietronero 1991; Coleman and Pietronero 1992; Durrer et al. 1997 and others).

The Method of Angular Γ-Function The best demonstration of the consistency of the observed angular and spatial fractal structures with $D \approx 2$ was given by Montuori and Sylos Labini (1997). They studied 3D maps together with the corresponding angular distributions from several redshift catalogues. For an undistorted estimation of the correlation exponent in angular data they used the conditional surface density Γ_{ang}

$$\Gamma_{\mathrm{ang}}(\theta) = \frac{1}{S(\theta)}\frac{dN(\theta)}{d\theta} = \frac{B_{\mathrm{ang}}D}{2\pi}\theta^{-\alpha}, \tag{11.15}$$

where $S(\theta)d\theta$ is the solid angle element ($S(\theta) \approx 2\pi\theta$ for small angles $\theta \ll 1$), $N(\theta) = B_{\mathrm{ang}}\theta^D$ is the number of galaxies in the "polar cap" with the radius θ and D is the fractal dimension of the 3D structure. If $0 \leq D < 2$, then the fractal dimension D_{pr} of the projected structure coincides with D.

This analysis of angular and spatial Γ-functions demonstrated clearly that the fractal dimension of the observed structure is

$$D = 1.9 \pm 0.1. \tag{11.16}$$

The angular correlation exponent was $\alpha = 0.1 \pm 0.1$ and the spatial correlation exponent was $\gamma = 1.1 \pm 0.1$ for all studied catalogues.[4]

The most interesting fact shown by this study was that the previously derived "universal" value of the angular correlation exponent $\alpha_w = 0.8$ could be an artefact caused by the $w(\theta)$ method itself. This reduced correlation function gives distorted values of the true correlation exponent. Similarly as with the ξ-function, the underlying reason is the normalization condition.

Here again the story of how to find the true correlation exponent revolves around the difference between the power-law behaviour of the complete correlation function and the non-power-law reduced correlation function. This happens both for angular and spatial distributions.

11.5 Results from the 2dF and Sloan Digital Sky Surveys

The 2dF redshift survey (Colless et al. 2003) opened new prospects for different kinds of statistical analyses. This survey contains about 220 000 galaxies in two narrow slices of about 90° × 15° (SGP) and 75° × 10° (NGP) complete up to $b_j = 19.5$, with a typical redshift $z_s \approx 0.15$, and a typical absolute magnitude $M_s \approx -20.0 + 5 \log h$ (Norberg et al. 2002).

The Sloan digital sky survey (SDSS) with its million redshifts and wide sky coverage (about π steradians) is presently the best sample for the study of fundamental questions of the galaxy distribution (the web site of SDSS: http://www.sdss.org/).

In order obtain reliable information on the correlation exponent and the homogeneity scale from these large samples, one should be aware of the methodological limitations, as explained above. We will see that within the common range of applicability ξ and Γ functions give compatible results.

11.5.1 The 2dF Galaxy Survey

Redshift Space ξ- and Γ-Functions Hawkins et al. (2003) analysed the 2dF data using the reduced correlation function ξ. The redshift-space correlation function was approximated by two different power-law forms:

$$\xi_z(s) = \left(\frac{13h^{-1}\,\mathrm{Mpc}}{s}\right)^{0.75} \quad \text{for } 0.1 < s < 3h^{-1}\,\mathrm{Mpc}, \tag{11.17}$$

$$\xi_z(s) = \left(\frac{6.82h^{-1}\,\mathrm{Mpc}}{s}\right)^{1.57} \quad \text{for } 3 < s < 20h^{-1}\,\mathrm{Mpc}. \tag{11.18}$$

[4]The angular correlation exponent α is related to D as $\alpha = 2 - D = \gamma - 1$. The last equality follows from the relation between the spatial and angular Γ-functions (the 3D $\Gamma(r) \propto r^{-\gamma}$ with $\gamma = 3 - D$ and the angular $\Gamma_{\mathrm{ang}}(\theta) \propto \theta^{-\alpha}$ with $\alpha = 2 - D$.)

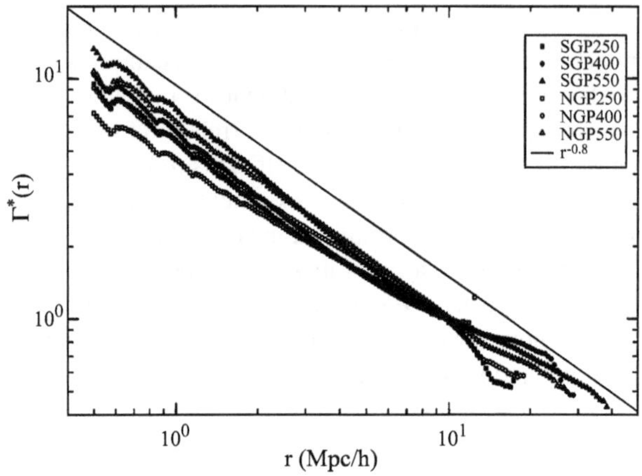

Fig. 11.3 Conditional density in spheres in six volume-limited samples of the 2dF GRS catalogue (Fig. 4 from Vasilyev et al. 2006; reproduced with permission © ESO). The reference line is a power-law with slope $\gamma = 0.8$, corresponding to the fractal dimension $D = 2.2$

On larger scales $30 \div 60h^{-1}$ Mpc, $\xi_z(s)$ becomes negative.

Such a behaviour of ξ agrees with our discussion in Chap. 10. It means that the distribution within the scales $0.1 < s < 3h^{-1}$ Mpc may be considered as a fractal structure with the dimension $D = 3 - \gamma = 2.25$. For scales $s > 3$ Mpc the ξ-function continuously changes its slope and at $r_0 \approx 5h^{-1}$ Mpc the exponent becomes $\gamma_{r_o} = 2\gamma = 1.5$ as predicted by Eq. (10.57).

Using the Γ-function in volume-limited samples, Vasilyev et al. (2006) derived that the conditional density for the 2dF data is a power-law with the exponent $\gamma = 0.8 \pm 0.2$ on scales $0.5 < s < 40h^{-1}$ Mpc (Fig. 11.3).

This value $D = 2.2 \pm 0.2$ is consistent with results obtained by Sylos Labini et al. (1998) for all redshift catalogues available at the time. In their Γ-function analysis the probed scales were limited by the radius of the maximum sphere $r_{\mathrm{sph}}^{\mathrm{max}}$ within the sample volume, i.e. about $20h^{-1}$ Mpc for slice-like surveys and $100h^{-1}$ Mpc for the LEDA sample.

11.5.2 Results from the SDSS Survey

Zehavi et al. (2002) performed the ξ-function analysis of a sample of 29 300 SDSS galaxies (redshifts $5700 < cz < 39000$ km/s, absolute magnitudes $-22 < M_r < -19 + 5\log h$). The derived redshift-space ξ function has a non-power law form (their Fig. 5). In the interval $2 < s < 8h^{-1}$ Mpc the ξ-function was approximated by the power-law $\xi_z(s) = (s/8.0h^{-1}\,\mathrm{Mpc})^{-1.2}$. Actually $\xi_z(s)$ has three characteristic intervals of scales: (1) the interval 0.1–0.5h^{-1} Mpc where $\gamma \approx 1.8$, (2) 0.5–5h^{-1} Mpc where $\gamma \approx 1$, and (3) 5–30h^{-1} Mpc where $\gamma \approx 1.8$. As we discussed above, such a behaviour of the ξ function is just as expected for a fractal structure if one takes into account the characteristic scales R_{sep}, r_0, and $R_{\mathrm{max}}^{\mathrm{sph}}$.

The SDSS Luminous Red Galaxy sample is so deep ($\langle z \rangle \approx 0.3$) that the radius of the maximal sphere $R_{\mathrm{max}}^{\mathrm{sph}}$ is about $100h^{-1}$ Mpc. Hogg et al. (2005) found for 3658

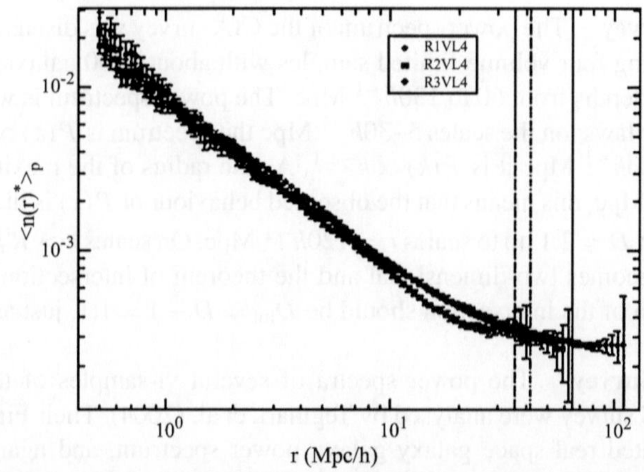

Fig. 11.4 Conditional density in spheres for the three largest volume-limited samples of SDSS DR4 main galaxies catalogue (from Sylos Labini et al. 2007; reproduced with permission © ESO). There is the power-law correlation with $\gamma = 1.0 \pm 0.1$ in the interval of scales [0.5, 30] Mpc/h and large fluctuations on scales [30, 100] Mpc/h. The apparent tendency to homogeneity after 30 Mpc/h may be related to the strongly decreasing number of large spheres inside the slice-like sample volumes

LRG galaxies that $\Gamma^*(r)$ is a power-law corresponding to $D \approx 2$ in the interval 1–25h^{-1} Mpc. On scales 25–70h^{-1} Mpc there is a deflection from the power law, and within 70–100h^{-1} Mpc Γ^* achieves a constant value. This was interpreted as a detection of the homogeneity scale (but see Chap. 12).

A Γ-analysis of the SDSS DR4 main galaxy sample (Sylos Labini et al. 2007) found power-law correlation with $\gamma = 1.0 \pm 0.1$ in the interval 0.5–30h^{-1} Mpc and large fluctuations in slope on larger scales (Fig. 11.4).

Other Results Tikhonov et al. (2000) and Tikhonov and Makarov (2003) presented results from the Γ-function method for several galaxy samples. They found the fractal dimension $D \approx 2$ on scales from 1 to 30 Mpc.

The method of the two-point conditional column density (Sect. 10.4.2) allows one to extend the fractal analysis up to the depth R_s of a survey with slice-like geometry. The studies of LEDA, SDSS, and 2dF galaxies (Baryshev and Bukhmastova 2004; Vasilyev 2004; Vasilyev et al. 2006) extended the maximum scale from 20h^{-1} Mpc (the Γ-function method) up to $\sim 100h^{-1}$ Mpc. The result was that $D \approx 2.2$ for the whole range of scales.

11.5.3 Power Spectrum and Intersection of Fractals

We illustrate the method of power spectrum analysis by considering its application to the galaxy data from the CfA and SDSS galaxy samples.

The CfA Survey The power spectrum of the CfA survey was discussed by Park et al. (1994) using four volume-limited samples with about 1000 galaxies per sample and with the depths from 60 to $130h^{-1}$ Mpc. The power spectrum is well described by two power laws: on the scales 5–$30h^{-1}$ Mpc the spectrum is $P(k) \propto k^{-2.1}$, while within 30–$120h^{-1}$ Mpc it is $P(k) \propto k^{-1.1}$. As the radius of the maximal sphere is about $30h^{-1}$ Mpc, this means that the observed behaviour of $P(k)$ is like for a fractal structure with $D = 2.1$ up to scales $r = 120h^{-1}$ Mpc. On scales $r > R_{\max}^{\mathrm{sph}}$ the survey effectively becomes two-dimensional and the theorem of intersection predicts that the dimension of the intersection should be $D_{\mathrm{int}} = D - 1 = 1.1$, just as observed.

The SDSS Survey The power spectra of several vl-samples of the early data release SDSS survey were analysed by Tegmark et al. (2004). Their Fig. 22 presents the decorrelated real-space galaxy-galaxy power spectrum, and again two power laws can be seen: on the scales 10–$60h^{-1}$ Mpc the spectrum is $P(k) \propto k^{-2}$, while within 60–$200h^{-1}$ Mpc the spectrum is $P(k) \propto k^{-1}$. Similarly as for the CfA, this is consistent with a fractal structure with $D \approx 2$ on all considered scales. The same behaviour was found also for the 2dFGRS sample (Tegmark et al. 2002).

11.6 Concluding Remarks

The history of the study of the large-scale structure demonstrates that the statistical parameters of the observed galaxy distribution should be derived using adequate mathematical techniques. This means that the samples of galaxies should first be tested for the necessary conditions required by the intended correlation analysis (Sylos Labini et al. 2009b). Earlier, when this was not done, discoveries of larger and larger structures from more extensive data were unexpected and surprising.

In a Nature review, Wu et al. (1999) gave an apt summary of the situation as follows: "The Universe is inhomogeneous—and essentially fractal—on the scale of galaxies and clusters of galaxies, but most cosmologists believe that on larger scales it becomes isotropic and homogeneous.". It is also true that we are still not quite certain about the large scale beyond which the lumpy galaxy universe looks really smooth. In Chap. 12 we will discuss further steps in the analysis of the 3D galaxy maps. One question is if one can see truly superlarge structures, on giga-parsec scales, in deep fields.

Chapter 12
Some Outstanding Problems of Cosmological Physics

We conclude with short descriptions of a few important problems of cosmological physics. These also serve to tie together several topics from previous chapters.

We first consider two questions about the large-scale structure: Has the crossover scale to homogeneity already been found? How certain is the detection of the Baryonic Acoustic Oscillations (BAO) on 100 Mpc scales in the distribution of galaxies? This feature is an important prediction of the standard cosmology.

Second, we look at more nearby space, and discuss whether dark energy, initially detected on 1000 Mpc scales, can also be observed on small scales, in the local galaxy universe. This would form a local test of the standard cosmology with its everywhere smooth dark energy (Einstein's Λ-term).

Third, we describe interesting conceptual problems in the ideal Friedmann model, related to the physics of space expansion. These may suggest new theoretical developments and cosmological tests.

12.1 Homogeneity Scale and Superlarge Structures

The Great Debate on the nature and distances of spiral nebulae, around 1920, heralded the change of our Milky Way into an ordinary galaxy, whereby the centre of the world finally disappeared into the expanse of galaxies. It soon went over to a discussion of the structure of the galaxy universe. Where does one find the smoothness that Newton and Einstein pondered about? How large is the scale where the universe becomes homogeneous? This question has always been at the heart of cosmology, linked as it is to cosmological principles and to the problem of structure formation on different scales.

Y. Baryshev, P. Teerikorpi, *Fundamental Questions of Practical Cosmology*, Astrophysics and Space Science Library 383,
DOI 10.1007/978-94-007-2379-5_12, © Springer Science+Business Media B.V. 2012

12.1.1 How to Establish the Homogeneity Scale?

Early counts of bright galaxies were close to the $10^{0.6m}$ law, pointing to smoothness at quite short distances of several megaparsecs. As one might now guess from large-scale 3D maps, showing much larger structures, those results were spurious. Modern LEDA galaxy counts in the B magnitude range 10 to 16 give shallower slopes.

All-Sky Counts of LEDA Galaxies Especially in the epoch of angular galaxy catalogues, the counts were regarded as evidence for a homogeneous galaxy distribution. Also, the local counts served as a reference value for deeper galaxy counts.

The Lyon extragalactic data base (LEDA), created by Georges Paturel in 1983, is a continuation of de Vaucouleurs's Reference Catalogue and its later editions. In fact, *The Third Reference Catalogue* (De Vaucouleurs et al. 1991) was already based on the LEDA data. The LEDA database offers currently a catalogue of homogeneous parameters of galaxies for the largest available whole sky sample. Among its more than million galaxies, about 50 000 galaxies have a measured B magnitude brighter than 16 mag.

Made from the amalgamation of all available catalogues and continually being revised with the flow of new data, the completeness of the LEDA sample has been studied over the years by inspecting the counts (Paturel et al. 1994, 1997b; Courtois et al. 2004). Along with the completeness, the counts give information about the slope of the bright end of the galaxy counts, hence on the spatial distribution law.

Teerikorpi (2004), in connection with a study of the influence of the Eddington bias (Sect. 3.1.2) on galaxy counts, investigated the LEDA galaxy counts in the B magnitude range 10–16. The analysis of the counts indicated a slope of 0.44 in the B range 10–14, which corresponds to $D = 2.2$ up to scales of about $100h^{-1}$ Mpc. Rather similar results were obtained by Courtois et al. (2004), who calculated the regression line up to $B = 16$ for a somewhat different LEDA sample, and derived the slope of about 0.5.

Radial Number Counts from the KLUN Sample Usually the large-scale structure is studied with redshift as a distance indicator. However, it is possible to use other distance measures. Teerikorpi et al. (1998) derived the radial spatial distribution of galaxies around our Galaxy, using over 5000 Tully–Fisher distance moduli from the KLUN program. The results gave clear evidence for a decrease in the average density consistent with the fractal dimension $D = 2.2 \pm 0.2$ in the distance range 10–$100h^{-1}$ Mpc.

The method of distance moduli differs from usual galaxy counts in the sense that the TF moduli probe with a better spatial resolution the distribution of galaxies. A magnitude measurement provides a very poor distance estimate, while a TF distance modulus has an error of about 0.5 mag. Furthermore, the method takes into account the incompleteness of the sample.

The results from the counts and the TF distance moduli have important consequencies: First, there is no slope of 0.6 in the B magnitude interval brighter than 14, i.e. no uniformity in the galaxy distribution up to $100h^{-1}$ Mpc. Second, the counts

$\log N(m) = 0.44m + \text{const}$ in LEDA, the radial distribution $N(r) \propto r^{2.2}$ for the KLUN sample, and the conditional density for main redshift catalogues $\Gamma(r) \propto r^{-1}$ (Chap. 11) are consistent with a fractal structure having $D = 2.2 \pm 0.2$ within the scales 10–100h^{-1} Mpc. Third, the inhomogeneity on small scales influences the estimation of average number and luminosity densities (Joyce and Sylos Labini 2001). For a fractal structure these densities depend on the radius of the volume in which they are calculated. This means, e.g., that the usual normalization of deep galaxy counts based on the local homogeneity should be revised to take into account the local radial inhomogeneity.

The Problematic Homogeneity Scale In the 1930s Hubble argued that galaxy clusters were the largest units in the distribution of matter, whereas Shapley presented evidence for still larger structures and he turned out to be right. In our times the superclusters of galaxies have put the border of uniformity to a scale of at least 100 Mpc. On the other hand, the isotropy of the cosmic background radiation and other arguments make us expect that on scales large enough the universe is homogeneous.

There are reports that the homogeneity scale has been already found in the available data, like the luminous red galaxy (LRG) sample of the SDSS. According to Hogg et al. (2005), a power law with $\gamma = 1$ ($D = 2$) gives a good fit to the LRG data up to at least 20h^{-1} Mpc after which the decrease of the density gets slower, and there seems to be a flattening around 70 Mpc up to 100 Mpc, the largest scale probed by the sample.

As we described in Chap. 10, a galaxy sample is characterized by several parameters, which are important for the correlation analysis of the spatial galaxy distribution, among these the radius $R_{\max}^{\mathrm{sph}}$ of the maximum sphere completely contained in the sample. Joyce et al. (2005) have discussed the difficulties in establishing the transition to a well-defined mean density. In a slice-like survey, such as the earlier data releases of SDSS, an artificial homogenization may happen starting from scales of about $0.25 R_{\max}^{\mathrm{sph}}$, when independent spheres in transversal direction cannot be completely embedded in the sample volume. Hence the findings of a homogeneity should in future be checked using larger spherical volumes.

Self-averaging Properties of Galaxy Fluctuations Analyses of finite sample distributions usually assume that fluctuations are self-averaging, which means that they are statistically similar in different regions of the given sample volume (Chap. 10). By using the scale-length method, Sylos Labini et al. (2009b) tested whether this assumption is satisfied in several samples of the SDSS Data Release Six. They found that for scales shorter than 30h^{-1} Mpc, the probability density function (PDF) was indeed statistically stable. However, on large spatial scales ($r > 30h^{-1}$ Mpc), the PDF showed systematic variations in different subvolumes of the survey. Thus while up to 30h^{-1} Mpc galaxy structures have well-defined power-law correlations, on larger scales it is not possible to consider whole sample averages as useful statistical descriptors. This means that the density fluctuations are too large in amplitude and too extended in space to be self-averaging on such large scales inside the sample volumes. Sylos Labini et al. (2009b) concluded that the galaxy distribution is inhomogeneous up to the largest scales ($r \approx 100h^{-1}$ Mpc) probed by the SDSS samples.

Sylos Labini et al. (2009a) studied also the statistical properties of large-scale galaxy structures in samples from the 2dF Galaxy Redshift Survey. They measured conditional fluctuations by means of the scale-length method and determined their probability distribution. It was found that the galaxy distribution in these samples is characterized by large-amplitude fluctuations with a large spatial extension, whose size is only limited by the sample's boundaries. These fluctuations were detected in two independent regions in the northern and southern galactic caps (see also Sylos Labini 2011).

12.1.2 *Fluctuations in Very Deep Fields—Gigastructures?*

Though the available data may not yet be sufficient to define the crossover scale to homogeneity, one can find evidence for occasional very large structures which may suggest some further surprises in this subject.

Spatial Tomography and Fluctuations in Deep Fields A perspective method of the structure analysis on the very largest scales is offered by photometric redshifts with an accuracy of about 0.003%. For example, in this way Padmanabhan et al. (2007) found for the large LRG sample of the SDSS that the power spectrum $P(k)$ has an exponential dependence extending to the gigaparsec scales $\lambda = 2\pi/k \approx 1200h^{-1}$ Mpc.

Nabokov and Baryshev (2010a, 2010b) used the COSMOS, FDF, HUDF, and HDF-N galaxy catalogs[1] to study the prospects for the method of very deep narrow fields (the 3-D space tomography as discussed in Sect. 10.4.3) for searching for superlarge structures in the spatial distribution of galaxies. Their analysis of the distribution $N(z)$ of photometric redshifts in a grid of deep fields pointed to the possible existence of superlarge structures with a contrast $dN/N \sim 50\%$ and having sizes of about 1000 Mpc (Fig. 12.1). This work demonstrated the usefulness of the space tomography method as a tool for studying gigaparsec scale structures. However, the reality of the detected candidate gigastructures must be verified by further observations with a finer grid of deep fields. Nabokov and Baryshev note that the influence of systematic errors can be reduced by observing the same deep fields with several 3–10 metre telescopes and using different methods for determining the photometric redshifts.

[1] These deep galaxy surveys are COSMOS (Cosmic Evolution Survey), FDF (FORS Deep Field of the ESO VLT), HUDF (Hubble Ultra Deep Field), and HDF-N (Hubble Deep Field North). The COSMOS survey contains about half a million photometric redshifts evaluated using 30 filters for over 600 000 galaxies in a field of $77' \times 77'$, with the limiting B magnitude of 26. The other samples are smaller in galaxy number and field size, but extend deeper, with the limiting magnitudes 27–30.

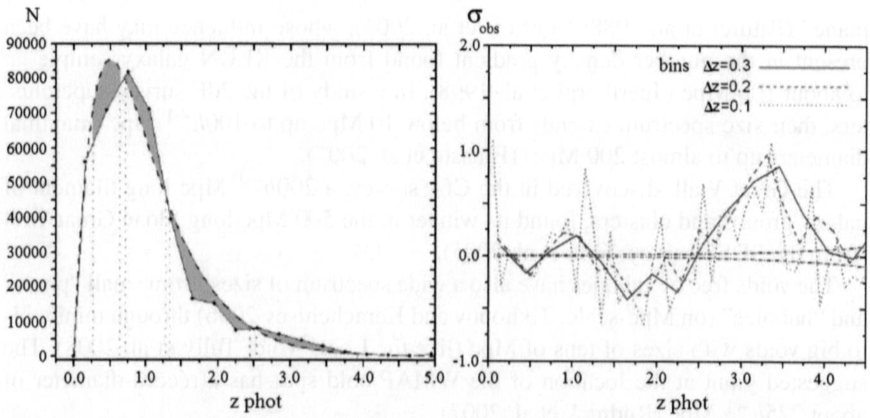

Fig. 12.1 *Left*: Radial distribution of galaxies from the COSMOS photometric redshift survey (382143 galaxies). The *solid curve* indicates superlarge regions with densities higher or lower than the expected average one (the *dashed curve*). Right: Relative deflections of galaxy numbers in redshift bins for three resolutions in redshift. The expected Poisson level is also shown. From Nabokov and Baryshev (2010b)

It is well known that the distribution of the redshifts of high-z quasars show non-uniformities due to selection effects (Sect. 9.5.2). In order to study systematic effects in galaxy photometric redshifts, artificial catalogs of uniformly distributed galaxies with parameters corresponding to those of the surveys should be compiled to accompany the deep field observations, so that selection effects and systematic distortions in the observed photometric red shift distributions can be evaluated quantitatively.[2]

Anyway, it is interesting that the relative density fluctuations in the COSMOS-HUDF-FDF fields behave similarly (Nabokov and Baryshev 2010b). Since the data for these fields were obtained with different instruments and processed in different ways, it is possible that real superlarge structures are making a significant contribution to the observed fluctuations (Fig. 12.1). For example, the HUDF and FDF fields are separated from one another in the sky by 36°, so that the transverse size of a superlarge structure at a distance of $z = 1$ would be about $1700h^{-1}$ Mpc.

Observations of the redshifts of gamma bursts can also serve as a grid that covers the entire sky. Available redshift data for more than 100 SWIFT gamma bursts are compatible with the existence of superlarge structures in the distribution of their parent galaxies (Nabokov and Baryshev 2008b). Raikov et al. (2010) concluded that gamma bursts with $z < 3$ have an inhomogeneous spatial distribution with signs of fractality ($D = 2.2$–2.5).

Known Large Structures The existence of superlarge structures is also consistent with the already known large structures in the universe discovered by different observational techniques. In the local universe, there is Paturel's "hypergalactic

---

[2]The ALHAMBRA project (Moles et al. 2008), with observations of 8 deep fields in 20 filters and a total of 6.6×10^5 galaxies, is currently in the process of completion.

plane" (Paturel et al. 1988; Courtois et al. 2004), whose influence may have been present in the number density gradient found from the KLUN galaxy sample up to about 100 Mpc (Teerikorpi et al. 1998). In a study of the 2dF survey superclusters, their size spectrum extends from below 10 Mpc up to $100h^{-1}$ Mpc (maximal diameters up to almost 200 Mpc) (Einasto et al. 2007).

The Great Wall, discovered in the CfA survey, a $200h^{-1}$ Mpc long filament of galaxy groups and clusters, found its winner in the 500 Mpc long Sloan Great Wall (from the SDSS survey; Gott et al. 2005).

The voids free of galaxies have also a wide spectrum of sizes, from small "pores" and "bubbles" (on Mpc-scale; Tikhonov and Karachentsev 2006) through minivoids to big voids with sizes of tens of Mpc (like the Local Void; Tully et al. 2008). The suggested giant at the location of the WMAP cold spot has a record diameter of about $225h^{-1}$ Mpc (Rudnick et al. 2007).

We note that the unexpected discovery of a large scale "dark flow", based on the observed Sunyaev-Zeldovich effect for X-ray galactic clusters (Kashlinsky et al. 2008) and on the peculiar velocities of the galaxies (Watkins et al. 2009), means that the entire local volume, of size $300h^{-1}$ Mpc, is undergoing a large-scale motion, a fact consistent with the existence of superlarge structures on scales of 1000 Mpc, but problematic for the standard ΛCDM cosmology.

The trend in observational cosmology toward the discovery of structures of ever larger size reminds one of the words by de Vaucouleurs (1970, 1971) who summarized, forty years ago, the properties of galaxy clustering concluding: *In the 1930s astronomers stated, and cosmologists believed, that, except perhaps for a few clusters, galaxies were randomly distributed throughout space; in the 1950s the same property was assigned to cluster centres; now the hope is that, if superclusters are here to stay (and apparently they are), at least they represent the last scale of clustering we need to worry about…*

The 3D Dark Matter Galaxies are the visible tracers of the large-scale structure, but in modern cosmology dark matter and dark energy are the greatly dominating substances. There are various indications of dark matter within galaxy systems of different scales. However, in order to determine the overall distribution of gravitating dark matter, the gravitational lensing method is the most promising one, being based on the analysis of rays of light influenced by dark matter along trajectories passing through voids as well as superclusters. First important steps towards revealing the 3D organization of dark matter were made via an analysis of the weak gravitational lensing in the HST survey COSMOS (Massey et al. 2007a). The shapes, orientations, and photometric redshifts of half a million distant galaxies in a $1.3° \times 1.3°$ field were measured. The main result was that the dark matter generally repeats the distribution of the luminous matter both on small and large scales (galaxies, hot gas in galaxy clusters). If confirmed by more extensive future studies in future this will underline the relevance of the large-scale structure as measured using the luminous tracers. Remarkable filamentary superlarge clusters of dark matter with sizes up to 1000 Mpc were found. Another intriguing result was that sometimes a cluster of dark matter can appear without luminous matter and what is more challenging, in some cases clusters of luminous matter are devoid of dark matter. If true, one might see here a sign of the same nature of both dark and luminous matter.

12.2 Detecting Baryonic Acoustic Oscillations

According to the standard cosmological model, the early universe contained a hot plasma of electrons and baryons (protons and neutrons), in addition to dark matter and photons. Some 400 000 years after the big bang the temperature had fallen low enough for formation of neutral hydrogen atoms. At this recombination epoch the sound speed dropped and the acoustic oscillations which were propagating in the cosmic fluid became frozen.

12.2.1 Baryonic Oscillations as a Crucial Test

The length scale l_{BAO} of the baryonic acoustic oscillations is basically obtained from the integral

$$l_{\mathrm{BAO}} = \int_0^{t_{\mathrm{rec}}} c_{\mathrm{s}}(t)(1+z)dt \approx 100h^{-1}\mathrm{Mpc}, \qquad (12.1)$$

where t_{rec} refers to the time of recombination and $c_{\mathrm{s}}(t)$ is the sound speed at different cosmic times t.

The BAO imprints in the matter distribution should be present in the galaxy correlation function. Because the prediction is rather straightforward and gives a well-defined spatial scale (about $100h^{-1}$ Mpc), the search for the BAO is an important test of the standard model. In addition, this length scale could serve as a standard meter stick for cosmological studies (Beutler et al. 2011 derived $H_0 = 67 \pm 3$ km s^{-1}/Mpc).

$\xi(r)$ and $P(k)$ for the ΛCDM model Standard models of galaxy formation predict the two-point correlation function $\xi(r)$ of density fluctuations in the early universe, and they also can make a prediction for $\xi(r)$ at the present time, in the regime of weak perturbations, where fluctuations have been linearly amplified by gravitational clustering in the expanding universe (Peebles 1980). The various models differ in the predicted value of the characteristic length scales and in the scale-behaviour of $\xi(r)$. One may point out three length scales and three different regimes (Sylos Labini and Vasilyev 2008; see Fig. 12.2):

(1) On scales smaller than r_0 (where $\xi(r_0) = 1$), the matter distribution is characterized by strong clustering, about which little is known analytically and which is generally constrained by N-body simulations where typically for $r < r_0$, $\xi(r) \sim r^{-\gamma}$ with $\gamma \approx 1.5$ (Springel et al. 2005).

(2) The second length scale at $r_c \gg r_0$ is such that $\xi(r_c) = 0$ (Peebles 1993; Gabrielli et al. 2002). In the range $r_0 < r < r_c$, $\xi(r)$ is characterized by positive

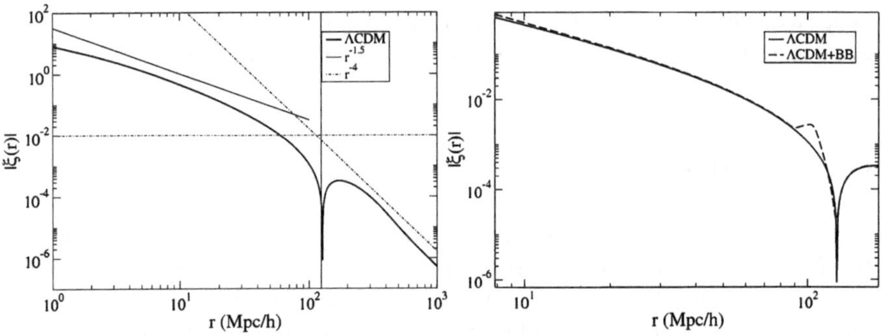

Fig. 12.2 Expected *module* of the two-point correlation function ξ for a ΛCDM model, (*left*) when there is no baryonic component, and (*right*) when there is a baryonic component (in this case, the matter density is $\Omega_m = 0.12h^{-2}$ and the baryonic density is $\Omega_b = 0.024h^{-2}$). Note that beyond the point r_c, where $\xi(r_c) = 0$ (the *vertical line*), the correlation function is negative (from Sylos Labini and Vasilyev 2008; reproduced with permission © ESO)

correlations, which rapidly decay to zero when $r \to r_c$. The scale r_c is an imprint of the early universe physics. It corresponds to the size of the horizon at the time when matter and radiation were equal. The third length scale r_{BAO} is located on scales somewhat smaller than r_c. This corresponds to the baryon acoustic oscillations at the recombination epoch. Its precise location depends on the matter density parameters, the baryon abundance and the Hubble constant (Eisenstein and Hu 1998).

(3) In the third range of scales, for $r > r_c$, $\xi(r)$ has typically a negative power-law behaviour, i.e. $\xi(r) \sim r^{-4}$ (Gabrielli et al. 2002, 2005).

The importance of the scales r_{BAO} and r_c as a crucial test for the ΛCDM model is emphasized by the fact that these are rather insensitive to biasing, i.e. they are the same both for the underlying dark matter and the visible luminous matter (see, e.g., Sylos Labini et al. 2009c).

From the samples provided by the Sloan Digital Sky Survey it has become possible to try to estimate the galaxy 2-point correlation function on scales large enough for measuring the important lengths r_{BAO} and r_c. Thus Eisenstein et al. (2005) determined the Landy and Szalay estimator of $\xi(r)$ using the luminous red galaxy (LRG) sample of the SDSS. The result appeared to reveal the baryon bump at around $110h^{-1}$ Mpc. Martinez et al. (2009) used the LRG-DR7 sample and also found the bump, though the shape of the correlation function around r_{BAO} is slightly different from the one measured by Eisenstein et al.

Sylos Labini et al. (2009c) discussed the intriguing fact that in the ξ function of Martinez et al. (as well as of Cabré and Gaztañaga 2009) the correlation is positive up to $200h^{-1}$ Mpc, while the ΛCDM prediction makes it negative beyond $r_c \approx 120h^{-1}$ Mpc (cf. Fig. 12.3). In fact, Sylos Labini et al. argued that even in the newest SDSS samples, the large-scale behaviour of the correlation function cannot be confidently derived because of intrinsic errors and volume-dependent systematic effects. They claim that even in the LRG sample, deeper than the MG sample, fluctuations in the density field are so large that one cannot detect the scale corresponding

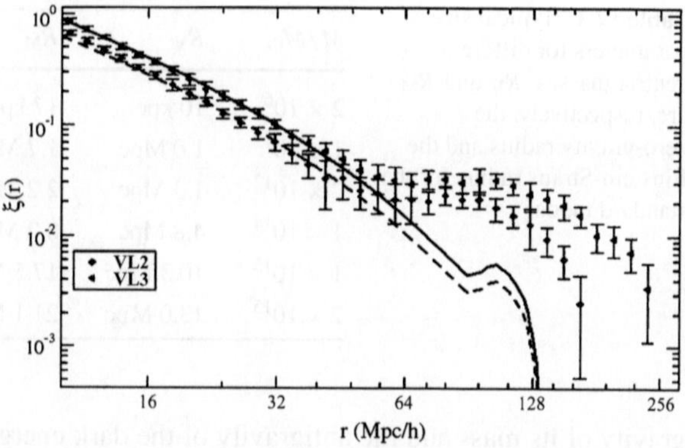

Fig. 12.3 Two-point correlation function ξ from the SDSS DR7 sample keeps positive beyond $r \approx 120h^{-1}$ Mpc, where it is expected to be negative. (Courtesy of F. Sylos Labini)

to r_{BAO}. In view of these big differences in the conclusions, one should not think that the scale of the baryon acoustic oscillations has already been reliably detected.

12.3 Dark Energy in the Neighbourhood of the Local Group

The ΛCDM cosmology views dark energy as having constant density everywhere so that it fills the voids as well as the interiors of galaxy systems, consistent with the conception of uniform and constant vacuum density (Einstein's Λ-term, the cosmological constant). Thus it may represent, together with other relativistic components of the universe, the naturally homogeneous substance on which the Friedmann model is based. If this is true, its local density should be identical to that inferred from the cosmological recession of distant galaxies using Ia supernovae as distance indicators, i.e. $\rho_V \approx 7 \times 10^{-30}$ g cm^{-1}.

The possible alternative explanations of the SNIa Hubble diagram (Sect. 8.4.4), requiring no dark energy, typically refer to phenomena on large scales. If any of them is valid and there is no dark energy (or $\Lambda = 0$), then one does not expect any true dark energy effect on much smaller scales, neither. But, if dark energy exists like suggested by the large-scale observations and is smooth and constant like the vacuum energy (with the equation of state $\rho_V = -p_V$), then it should exist also locally. According to general relativity, gravity depends on pressure as well as density: the effective gravitating density $\rho_{\mathrm{eff}} = \rho + 3p$ is negative for a vacuum ($= -2\rho_V$), and this leads to repulsion ("antigravity"). Hence the study of the cosmic antigravity in our neighbourhood is an important local test of the standard cosmological model, somewhat analogous to the attempts to detect non-baryonic dark matter particles in laboratories.

12.3.1 Towards a Local Measurement of Dark Energy

As a first step towards the local study of dark energy, Artur Chernin (2001) asked a simple but highly interesting question: at what distance from the Local Group do the

Table 12.1 Typical size parameters for different central masses. R_V and R_M are, respectively, the zero-gravity radius and the Einstein-Straus radius for the standard model

$M/M_\odot$	R_V	R_M	example
2×10^6	10 kpc	17 kpc	a globular cluster
1×10^{12}	1.0 Mpc	1.7 Mpc	a compact binary galaxy
2×10^{12}	1.3 Mpc	2.2 Mpc	the Local Group
1×10^{14}	4.8 Mpc	8.2 Mpc	the Fornax cluster
1×10^{15}	10.3 Mpc	17.5 Mpc	the Virgo cluster
2×10^{15}	13.0 Mpc	21.1 Mpc	the Coma cluster

gravity of its mass and the antigravity of the dark energy (if its local density equals its globally measured value) balance each other?

Zero-Gravity Radius Treating the Local Group as a point mass M on the background of the antigraviting dark energy, its gravity produces the radial force $-GM/r^2$, where r is the distance from the group barycentre. The antigravity of the vacuum produces the radial force $G2\rho_V(\frac{4\pi}{3}r^3)/r^2$; this Newtonian expression can be justified from general relativity, inspecting a static space-time for a mass embedded in the uniform dark energy (e.g., Chernin et al. 2006).

The gravity force ($\propto 1/r^2$) dominates the antigravity force ($\propto r$) at small distances, and the acceleration is negative there. Taking the mass of the Local Group as $M \approx 2 \times 10^{12} M_\odot$, then at the distance

$$R_V = \left(\frac{3}{8\pi} M/\rho_V\right)^{1/3} \simeq 1.3 \text{ Mpc} \tag{12.2}$$

the gravity and the antigravity balance each other (the "zero-gravity distance"). At larger distances the acceleration is positive.

A Gravitating System Within Dark Energy Calculations (Table 12.1) show that for star clusters and individual galaxies, even for tight binary galaxies the zero-gravity distance R_V is much larger than the size of the system which is located deep in the gravity-dominated region. At the same time, the region of the Huble flow is well beyond R_V. For galaxy groups and clusters, the zero-gravity radius is near or within the region where the outflow of galaxies begins to be observed (Fig. 12.4). It is on such scales where the system and its neighbourhood might shed light on the dark energy. In fact, Chernin et al. (2010) have suggested that different systems (groups and larger) may have a universal two-part design: a quasi-stationary bound central component with an expanding outflow around it on the scales up to the "Einstein-Straus" radius R_M (will be defined below; Table 12.1). The dynamical structure of such systems could reflect the gravity-antigravity interplay.

What happens to the test particles (dwarf galaxies) that have left the central region of the system? The particles move radially practically as predicted by the Newtonian equation of motion, where the forces are the gravity of the central mass and

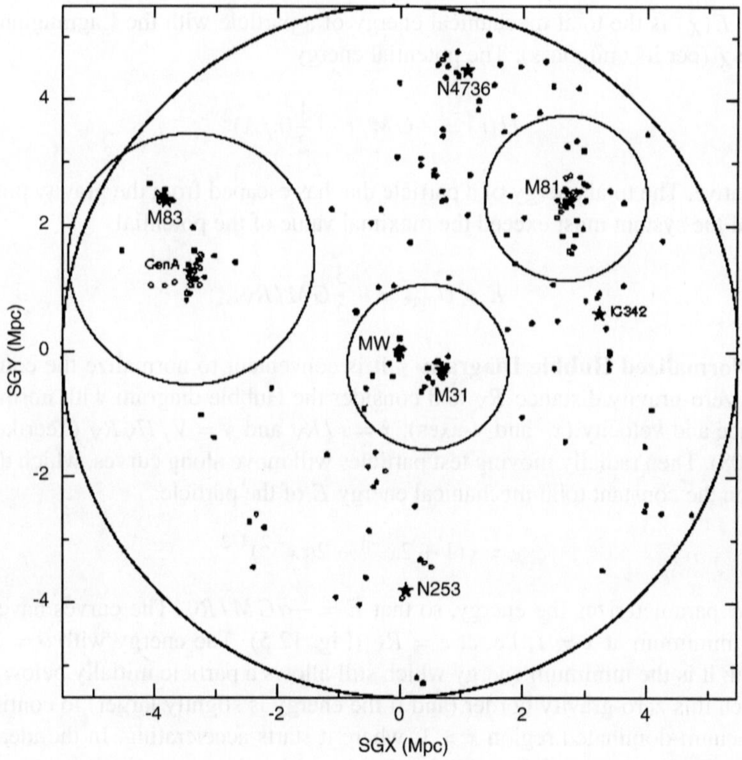

Fig. 12.4 The zero-gravity spheres around the Local Group (*at the centre*) and two nearby galaxy groups, as calculated using the masses $2 \times 10^{12} M_\odot$ for the LG and the M81/M82 Group and $7 \times 10^{12} M_\odot$ for the CenA/M83 Group (the map presenting the local environment up to about 5 Mpc is adapted from Karachentsev et al. 2003)

the antigravity of the dark energy:

$$\ddot{r}(t,\chi) = -GM/r^2 + \left(\frac{8\pi G}{3}\rho_V\right) r, \tag{12.3}$$

where $r(t,\chi)$ is the distance of a particle to the barycentre of the Local group and χ is the Lagrangian coordinate of the particle.[3]

The first integral of (12.3) expresses the mechanical energy conservation:

$$\frac{1}{2}\dot{r}^2 = GM/r + \frac{1}{2}(r/A)^2 + \bar{E}, \tag{12.4}$$

[3] $A = 1/H_V = (\frac{8\pi G}{3}\rho_V)^{-1/2} \simeq 5 \times 10^{17} s \simeq 1.5 \times 10^{28}$ cm is a characteristic vacuum time/length, the inverse value of the "vacuum Hubble constant" H_V.

where $\bar{E}(\chi)$ is the total mechanical energy of a particle with the Lagrangian coordinate χ (per its unit mass). The potential energy

$$U(r) = -GM/r - \frac{1}{2}(r/A)^2 \tag{12.5}$$

is negative. The total energy of a particle that has escaped from the gravity potential well of the system must exceed the maximal value of the potential:

$$E > U_{\rm max} = -\frac{3}{2}GM/R_{\rm V}. \tag{12.6}$$

The Normalized Hubble Diagram It is convenient to normalize the equations to the zero-gravity distance $R_{\rm V}$ and consider the Hubble diagram with normalized distance and velocity (x- and y-axes): $x = r/R_{\rm V}$ and $y = V/H_{\rm V}R_{\rm V}$ (Teerikorpi et al. 2008). Then radially moving test particles will move along curves, which depend only on the constant total mechanical energy E of the particle:

$$y = x(1 + 2x^{-3} - 2\alpha x^{-2})^{1/2}. \tag{12.7}$$

Here α parameterizes the energy, so that $E = -\alpha GM/R_{\rm V}$. The curves have a velocity minimum at $x = 1$, i.e. at $r = R_{\rm V}$ (Fig. 12.5). The energy with $\alpha = 3/2$ is special: it is the minimum energy which still allows a particle initially below $x = 1$ to reach this zero-gravity border (and if the energy is slightly larger) to continue to the vacuum-dominated region $x > 1$, where it starts accelerating. In the ideal case one does not expect particles with $x > 1$ below this minimum velocity curve. If one changes the values of the parameters M and $\rho_{\rm V}$, the $y - x$ curves do not change, but the normalized positions of test particles do change as $R_{\rm V}$ and $H_{\rm V}$ change.

The physical sense of the minimum energy curve is that it corresponds to ejections an infinitely long time ago. Hence, any upper limit for the age provides a still stricter lower-limit curve. The present position of the particle (in practice, a dwarf galaxy) on its energy curve depends on how long ago it was ejected from the centre. This, together with the fact that the masses of the galaxy groups (including our Local Group) are not yet very accurately known, complicates this way of deriving the local amount of antigravitating dark energy.

12.3.2 The Hubble Law and Dark Energy

After Lahav et al. (1991) derived a formula for the growth rate at the present epoch $f(z=0) \approx \Omega_{\rm m}^{0.6} + \frac{1}{70}\Omega_\Lambda(1 + \frac{1}{2}\Omega_{\rm m})$, it has been often thought that dark energy has insignificant local effects. Namely, the formula shows that for a fixed matter density parameter $\Omega_{\rm m}$, adding the cosmological vacuum into the model has practically no effect for the present growth rate, which also determines the peculiar velocities around the growing density fluctuations. Lahav et al. (1991) see this immunity to Λ reflecting the cosmic vacuum as a uniform background which does not have local

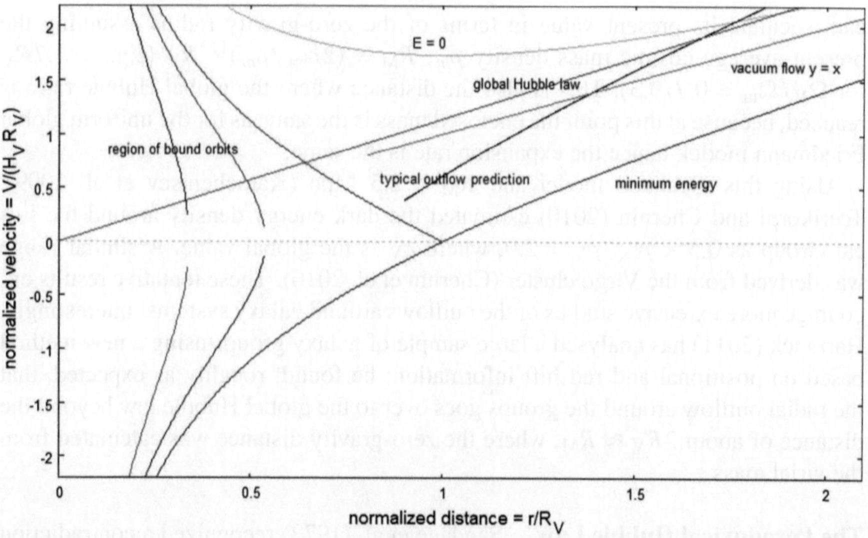

Fig. 12.5 The dynamical structure of a gravitating system within dark energy background, given as a normalized Hubble diagram. A typical outflow pattern for the present time is also shown (from Teerikorpi and Chernin 2010; reproduced with permission © ESO)

force effects. However, in our lumpy universe there are situations where local effects of dark energy are expected and where these might be used to measure its density (Chernin et al. 2006; Teerikorpi et al. 2005). The fact alone that one has to modify the virial theorem for a gravitationally bound system in order to account for the antigravity of dark energy shows that dark energy can have local effects (Chernin et al. 2009).

Dark Energy on Galaxy Group Scales In round numbers, the dark energy density may be written as

$$\rho_V/7 \times 10^{-30}\ \text{g/cm}^3 = (M/2 \times 10^{12} M_\odot) \Big/ \left(\frac{8\pi}{3} (R_V/1.3\ \text{Mpc})^3 \right). \qquad (12.8)$$

Thus one would be able to determine the dark energy density around a galaxy group if its mass M and the zero-gravity radius R_V were known from observations.

The expected decelerations and accelerations in the nearby velocity field are very small and impossible to measure directly. That is why Chernin and others have considered the possibility to see the signature of the zero-gravity distance in the observed distance-velocity field.

Chernin et al. (2006) noted that the size of the group or the zero-velocity distance is a strict lower limit to R_V, giving an upper limit to the local dark energy density ρ_{loc} (for a fixed group mass). An upper limit to R_V (and from it a lower limit to ρ_{loc}) is obtained from the distance where the local outflow approaches the global Hubble rate. This distance R_M is expected to give the size of the volume from which the mass M has been gathered during the formation of the group (so-called Einstein-Straus vacuole; Table 12.1 gives R_M for several examples). One can calculate its present value in terms of the zero-gravity radius assuming the present average cosmic mass density ρ_m: $R_M = (2\rho_{loc}/\rho_m)^{1/3} R_V$ (e.g., $= 1.7 R_V$ for $\Omega_V/\Omega_m = 0.7/0.3$). This is also the distance where the global Hubble ratio is reached, because at this point the enclosed mass is the same as for the uniform global Friedmann model, hence the expansion rate is the same.[4]

Using this idealized model and $R_M \approx 2.5$ Mpc (Karachentsev et al. 2009), Teerikorpi and Chernin (2010) estimated the dark energy density around the Local Group as $0.5 < \rho_{loc}/\rho_V < 2.0$, where ρ_V is the global value. A similar range was derived from the Virgo cluster (Chernin et al. 2010). These tentative results encourage more extensive studies of the outflows around galaxy systems. Interestingly, Hartwick (2011) has analysed a large sample of galaxy groups using a new method based on positional and redshift information; he found, roughly as expected, that the radial outflow around the groups goes over to the global Hubble law beyond the distance of about $2R_V \approx R_M$, where the zero-gravity distance was calculated from the virial mass.

The Paradoxical Hubble Law Sandage et al. (1972) recognized a contradiction between the observed linear Hubble law and a possible hierarchical galaxy distribution (or more generally, a very non-uniform distribution). They used the existence of the local Hubble law as an argument against de Vaucouleurs's hierarchical model.

A strong deflection from a linear distance–velocity relation expected within the hierarchical inhomogeneities has been confirmed by calculations (Haggerty and Wertz 1972; Fang et al. 1991; Gromov et al. 2001). Baryshev (1992b, 1994) and Baryshev et al. (1998) emphasized that the observed linear redshift-distance relation inside the power-law fractal density distribution creates the "Hubble-de Vaucouleurs paradox", i.e. the coexistence of both laws on the same scales.

Using asymptotically homogeneous LTB models (cf. Sect. 9.3.2), Gromov et al. (2001) found the necessary conditions for the linear Hubble law existing within a fractal structure with dimension $D = 2$. The larger the scale of homogeneity R_{hom} the smaller should be the density parameter Ω_m, e.g., for $R_{hom} = 100$ Mpc the linear Hubble law exists at distances $r > 1$ Mpc if the density parameter $\Omega_m < 0.01$ (which agrees with one possible explanation given by Sandage et al. (1972)—a very low cosmic density).[5]

Chernin (2001) suggested that the cold local Hubble flow is a signature of the dominance of the cosmological vacuum (the Λ-term). Baryshev et al. (2001) ex-

[4]This is the same 1.7 as in the global scale factor ratio leading to $z_V = 0.7$. Namely, the requirement that the global acceleration is zero when $\rho_V = 0.5\rho_m(z_V)$ leads in terms of the current mass density to the condition $2\rho_V = (1+z_V)^3\rho_m$ or $(2\rho_{loc}/\rho_m)^{1/3} = (1+z_V) = 1.7$.

[5]The LTB models, which are asymptotically Friedmann-like on large scales, have three possible initial conditions: the density distribution, the velocity field, and the bang time function, and any two of these determine the third one. Therefore, Gromov et al. (2001) pointed out that the Hubble law may be linear even inside a fractal distribution if the bang time function is suitably broad. This explanation may sound too artificial; however, a variant of it has been applied by Célérier et al. (2010) to explain the Hubble diagram of distant SNIa supernovae without the need to assume a big local hole (or dark energy).

tended this explanation to include dark energy which may vary in time. Cosmological N-body simulations by Macciò et al. (2005) confirmed that inclusion of cosmic vacuum in the calculations leads to significantly lower velocity dispersions in Local Volume–like regions than what happens without dark energy. However, the simulations of Hoffman et al. (2008) and Martinez-Vaquero et al. (2009) did not appear to find any such significant difference, and further investigations are clearly desirable.

12.4 Conceptual Problems of the Expanding Space Physics

The physics of space expansion is a relatively little studied subject, even though expansion is an essential ingredient of standard cosmology and the expanding space within the Friedmann model has properties which may sound paradoxical. These features, like violations of the ordinary speed limit and energy conservation, are usually viewed as new cosmological physics differing from our ordinary laboratory experience. They may also reflect real limitations of the model or may point ways to somewhere beyond the current theory. In any case, they underline the need of conceptual clarity in describing what the model means and what is observed. Even new crucial tests may be inspired.

Such unexpected phenomena were brought into light after an analysis of the Friedmann model by Harrison (1981, 1993, 1995). The non-Doppler nature of the cosmological redshift in the standard model has been discussed by Kiang (2003), Davis and Lineweaver (2004), Whiting (2004), and Francis et al. (2007) in attempts to clarify some "common big bang misconceptions" and the "expanding confusions" in the literature. We wish to emphasize here that the common cause of these interesting and problematic issues is the geometrical description of expanding space in the frame of general relativity, where there is no well-defined concept of the energy-momentum tensor and hence no energy-momentum conservation for matter plus gravity (cf. Chap. 5).

12.4.1 The Physical Meaning of Space Expansion

Nowadays, when the redshift is almost universally interpreted as due to the recession of distant galaxies, the concept of space expansion is commonly used in cosmology textbooks. In Chap. 2 we mentioned two opposite views on physical space as argued by Poincaré and Einstein. A mathematical theory of space is based on internally consistent abstract geometrical concepts of geometry, and its development does not depend on physical experiments. A theory of physical space is based on measurements of distances by means of real physical processes and units of length. Poincaré wrote that in the geometry-physics unity the Euclidean geometry is the simplest one and more convenient for physics, while Einstein considered that in more complex geometry the physical laws may have simpler forms and the geometry of space may be experimentally tested (Einstein's practical geometry).

Later Milne (1934) emphasized that space itself has no existence and it is physically preferable to use static space, as in ordinary physics, and consider the expansion of matter as motion in this space. The question raised by Poincaré and Milne about the physical meaning of space may be stated for expanding space as follows: is the expansion physically measurable? If "yes" then the cosmological redshift can be interpreted as a physical effect of expanding space. For instance, if the expansion of space would be necessarily accompanied by growing physical units then it could not be measured and the cosmological redshift would be caused by another physical phenomenon.

The theoretical analyses by Barnes et al. (2006) and Gron and Elgaroy (2007) have led them to conclude that the receding motion of galaxies within expanding Friedmann models has a physical meaning that differs from ordinary motion of a particle in static space. From this perspective, we find it useful to view the term *expansion* as meaning *continuous creation* of space, just because the volume of any comoving finite sphere is increasing. This concept of "space creation" serves to throw light on the various enigmas arising within the Friedmann model as will be considered below.

The Friedmann model gives an exact mathematical description of expanding space within general relativity. The increasing scale factor $S(t)$, along with the metric distance $r(t) = S(t)\chi$, physically corresponds to expanding space, a continuous adding of vacuum, which is controlled by the homogeneous matter distribution in the whole universe. Any comoving finite-sized box in the expanding universe increases its volume all the time, obtaining more and more cubic centimetres. This process may be visualized by a 2D analogy of an expanding sphere in 3D space, where the surface of the sphere increases with time. For 2D beings their universe grows with time (gets more square centimetres).

In the simple dust model, it is the classical empty vacuum which is created, while in the Λ model, space is created together with physical vacuum (presented by the constant Λ term). In comparison, in the steady state theory, space is created along with matter whose density keeps constant (Hoyle's C-field).

It is generally regarded that bounded physical objects, such as atoms, the Earth, stars, and galaxies, do not participate in cosmological expansion (Cooperstock et al. 1998), making it impossible to measure space expansion directly in the laboratory. One may say that in a Friedmann world there is no space creation inside these objects, whereas outside of them space is created.

12.4.2 Violation of the Limiting Velocity

Harrison (1981, 1993) emphasized that the cosmological redshift appearing in the Friedmann model is not the familiar Doppler effect, as often stated, but a new physical phenomenon, the Lemaître effect.

The exact formula for the Doppler effect caused by the relative motion of bodies in static space is given by the special relativistic expression:

$$\nu_{\text{obs}} = \gamma^{-1}(1 - \vec{\beta}\vec{n})^{-1}\nu_{\text{emit}}, \tag{12.9}$$

Table 12.2 Velocities vs. redshift for different models

z	v_{dop}/c	$v_{exp}/c, 1$	$v_{exp}/c, 2$	$v_{exp}/c, 3$
1	0.6	0.6	0.75	1
2	0.8	0.84	1.33	2
6	0.96	1.25	3.43	6
10	0.98	1.4	5.45	10
1000	$1–2 \cdot 10^{-6}$	1.94	500	1000
∞	1	2	∞	∞

where $\gamma = (1-\beta^2)^{-\frac{1}{2}}$ is the Lorentz factor, $\vec{\beta} = \frac{\vec{v}}{c}$ is the velocity vector of the moving body and $\vec{n}$ is the unit vector in the direction of the observer. For a purely radial motion the angle Θ between $\vec{n}$ and $\vec{\beta}$ equals 180°, then according to Eq. (12.9) we have the redshift

$$(1+z)_{\rm Dop} = \left(\frac{c+v}{c-v}\right)^{1/2}. \tag{12.10}$$

The Doppler redshift $z \to \infty$ for $v \to c$.

The exact relativistic Doppler velocity-redshift relation is

$$\frac{v_{\rm Dop}(z)}{c} = \frac{2z+z^2}{2+2z+z^2}. \tag{12.11}$$

For $z \to \infty$, the velocity $v_{\rm Dop}(z) \to c$ corresponding to the limit $v \le c$.

In the Friedmann model, the exact velocity-redshift relation is

$$\frac{v_{\rm exp}(z)}{c} = \frac{r(z)}{R_{H0}}. \tag{12.12}$$

This shows that for redshifts large enough ($r(z) > R_{H0}$) the velocity of the receding galaxy can exceed the speed of light, $v_{\rm exp} > c$.

Clearly we have two different mathematical formulae which correspond to two different physical phenomena—Doppler and Lemaître effects. Equations (12.11) and (12.12) give the same results only in the first order of v/c.

Expansion velocities. Table 12.2 compares the Doppler and Friedmann velocities calculated for three specific models. Their $v_{\rm exp}$ vs. z relations are

$$\text{For } \Omega_{\rm m} = 1;\ \Omega_\Lambda = 0 \quad v_{\rm exp}(z) = \frac{2(\sqrt{1+z}-1}{\sqrt{1+z}}c, \tag{12.13}$$

$$\text{For } \Omega_{\rm m} = 0;\ \Omega_\Lambda = 0 \quad v_{\rm exp}(z) = \frac{z(1+z/2)}{1+z}c, \tag{12.14}$$

$$\text{For } \Omega_{\rm m} = 0;\ \Omega_\Lambda = 1 \quad v_{\rm exp}(z) = zc. \tag{12.15}$$

The usual explanation of such intriguing results, which illustrate the unusual physics of expanding space, is that space expansion is not like ordinary motion in space, and it is not limited by the velocity of light (in this sense, "general relativity violates special relativity" (Guth 1992)). One may also say that the superluminally "moving" objects in the Friedmann model do not carry any information from place A to place B; the galaxies in the ideal case do not move away from their constant positions in the comoving coordinate system. In terms of space creation, at every point between two galaxies new space appears and for large enough distances the resulting rate of increase of distance can be arbitrarily high.

12.4.3 Newtonian Form of the Relativistic Friedmann Equation

The formal reason why the Friedmann model allows expansion velocities exceeding the speed of light is that the Friedmann equation is identical to the Newtonian equation of motion, and in Newtonian physics there is no limit on the velocity.

Friedmann's equation may be written in the form

$$\frac{d^2S}{dt^2} = -\frac{4\pi G}{3} S\left(\rho + \frac{3p}{c^2}\right). \tag{12.16}$$

Because the Lagrangian comoving coordinates do not depend on time, one may rewrite (12.16) using (7.19) as

$$\frac{d^2r}{dt^2} = -\frac{GM_g(r)}{r^2}. \tag{12.17}$$

This is identical to the Newtonian equation, where the gravitating mass $M_g(r)$ of a comoving sphere with radius r is given by

$$M_g(r) = \frac{4\pi}{3}\left(\rho + \frac{3p}{c^2}\right) r^3. \tag{12.18}$$

Therefore the relativistic equation describing the dynamical evolution of the universe is similar to the non-relativistic Newtonian equation of motion of a test particle in the gravity field of a finite sphere containing a mass M_g within the radius r. The second term in Eq. (12.18) does not essential change the Newtonian character of the solutions.

Such a similarity was first mentioned by Milne (1934) and McCrea and Milne (1934); they viewed the Newtonian model as an approximation to the Friedmann model. Later many authors asserted that the Newtonian model can be used only at short distances compared to the Hubble distance. Here, however, we see that the Newtonian form of the Friedmann equation is exact and true for all distances.

The root of the puzzle lies in the geometrical description of gravity in general relativity and in the derivation of Friedmann's equation from Einstein's gravity equations, using the comoving synchronous coordinates with the universal cosmic time t and homogeneous unbounded matter distribution. The Newtonian form of the Friedmann equation throws light on the question why the recession velocities of distant galaxies can be larger than the speed of light—in Newtonian theory there is no limiting velocity.

The Energy of the Hubbloid The Hubble distance $R_{\mathrm{H}} = c/H$ is often taken to be "the size of the universe"; it is also the radius of the Hubbloid which we considered in Sect. 9.4.3 in connection with the large cosmological numbers. In the Newtonian description of an expanding uniform matter system R_{H} has special significance. The total potential energy of a sphere of radius R is $E_{\mathrm{pot}} = -(3/5)GM(R)/R$. The total kinetic energy of this expanding sphere is $E_{\mathrm{kin}} = (3/5)M(R)V(R)^2/2$. Denoting $\rho = \alpha\rho_{\mathrm{crit}}$ (where $\rho_{\mathrm{crit}} = 3H^2/8\pi G$), the energies for the sphere of radius R_{H} may be written as

$$E_{\mathrm{pot}} = -(3/5)\alpha M(R_{\mathrm{H}})c^2/2, \qquad E_{\mathrm{kin}} = (3/5)M(R_{\mathrm{H}})c^2/2. \tag{12.19}$$

For $\alpha = 1$, i.e. $\rho = \rho_{\mathrm{crit}}$, the total energy $E_{\mathrm{tot}} = E_{\mathrm{pot}} + E_{\mathrm{kin}}$ is zero (in, fact for spheres of all sizes). Furthermore, if we calculate the total rest mass energy $M(R_{\mathrm{H}})c^2$ within the "size of the universe" R_{H}, this automatically will be rather near the corresponding potential energy, if the cosmic density is not far from the critical density. Sometimes one has put emphasis on the coincidence $E_{\mathrm{pot}} + Mc^2 \approx 0$, calling this the total (gravitational plus rest mass) energy of the universe (e.g., Feynman et al. 1995, p. 10). We see that for expanding systems with $\alpha \sim 1$ this happens only around the radius R_{H} where the expansion velocity $= c$. For much smaller or larger spheres the values of E_{pot} and $M(R)c^2$ are quite different, $E_{\mathrm{pot}}/M(R)c^2 \propto (R/R_{\mathrm{H}})^2$.

12.4.4 Problems of Energy Conservation

As we discussed in Chap. 5, the equation $T^i_{k;i} = 0$ "does not generally express a law of conservation" (Landau and Lifshitz 1971), because of the mathematical structure of the covariant divergence in Riemannian space. In order to have the total (all kinds of matter plus gravity) energy-momentum conserved, one should consider energy-momentum pseudotensor, which could describe the gravity itself. However, this violates the tensor character of the laws of conservation and does not solve the problem of the energy density of the gravitational field in a geometrical description of gravity. The root of the problem lies in the equivalence principle and in the absence of a true gravity force in general relativity, while all other fundamental fields present true forces, have true EMTs, and operate in Minkowski space. Noether's theorem relates the conserved energy-momentum tensor of material fields to the maximal symmetry of the Minkowski space; this is why the EMT of the gravity field can not be properly defined in curved Riemannian space.

Energy in Co-moving Sphere In cosmology, the absence of a true energy-momentum tensor for the gravity field appears in the fact that energy is not conserved during space expansion. Consider a comoving sphere with radius $r(t) = S(t)\chi$. The energy in this sphere is

$$e(r) = \int_0^r T_0^0 dV = \frac{4\pi}{3}\varepsilon(t)S^3(t)\chi^3\sigma_k(\chi), \qquad (12.20)$$

where $\sigma_k(\chi) = \int_0^\chi I_k^2(y)dy$ which is equal to 1 for $k = 0$, to $\frac{3}{\chi^3}(\frac{\chi^2}{2} - \frac{\sin 2\chi}{4})$ for $k = 1$, and to $\frac{3}{\chi^3}(\frac{\sinh 2\chi}{4} - \frac{\chi^2}{2})$ for $k = -1$. To calculate the time dependence of the energy density we use the continuity equations (5.10) and (5.17) in the form

$$\dot{\varepsilon} = -3(\varepsilon + p)\frac{\dot{S}}{S}. \qquad (12.21)$$

For an ideal equation of state $p = \gamma\rho c^2$ this equation has the simple solution

$$\rho \propto S^{-3(1+\gamma)}. \qquad (12.22)$$

In particular, we have for dust, radiation and vacuum

$$\rho_{\rm dust} \propto S^{-3}, \qquad \rho_{\rm rad} \propto S^{-4}, \qquad \rho_{\rm vac} \propto \text{const}. \qquad (12.23)$$

Hence the energy inside a comoving sphere is all the time changing as

$$e(r) = \frac{4\pi}{3}\rho c^2 r^3 \sigma_k(\chi) \propto S^{-3\gamma}(t). \qquad (12.24)$$

The energies of dust, radiation and vacuum behave as

$$e_{\rm dust}(r) \propto \text{const}, \qquad e_{\rm rad}(r) \propto S^{-1}, \qquad e_{\rm vac}(r) \propto S^{+3}. \qquad (12.25)$$

Intriguingly, the continuity equations (5.10) and (5.17) can also be written in the form

$$dE + pdV = 0, \qquad (12.26)$$

where $dE = d(\varepsilon V) = d(\rho c^2 V)$ is the change of energy within the comoving volume $V = \text{const} \cdot S^3$. Equation (12.26) looks like the ordinary law of conservation of energy in thermodynamics. However, the cosmological case is essentially different.

In laboratory conditions (12.26) means that if the energy decreases inside a finite box, it reappears outside the box as the work produced by the pressure increasing the volume of the box. The work performed by the pressure inside the box is the cause of the energy decrease in the box.

In cosmological conditions Eq. (12.26) allows one to calculate how much the energy increases or decreases inside a finite comoving volume but it does not tell where the energy comes from or where it goes. This is because the cosmological pressure does not produce work. It was noted by Harrison (1981, 1995) that in a uniform unbounded Friedmann model one may imagine the whole universe partitioned into macroscopic cells, each of comoving volume V, and all having their contents in identical states. The energy pdV lost from any one cell cannot reappear in neighbouring cells because all cells experience identical losses. So the usual idea of an expanding cell performing work on its surroundings does not apply in this case.

Edward Harrison emphasized: "The conclusion, whether we like it or not, is obvious: energy in the universe is not conserved" (Harrison 1981, p. 276). Peebles (1993, p. 139) concluded similarly, when he considered the energy loss inside a comoving sphere of the photon gas (see our Eq. (12.25)): "The resolution of this apparent paradox is that while energy conservation is a good local concept, ... there is not a general global energy conservation in general relativity."

In fact, only for the simplest dust model ($p = 0$) one may speak about the energy conservation in an expanding universe. For any matter with $p \neq 0$ within any local comoving volume, the energy is not conserved. This is because in general relativity there is no energy-momentum tensor of the gravity field and there is no gravity force in usual physical sense.

Continuous Creation of Gravitating Mass It is also interesting to consider the active gravitating mass of the cosmological fluid, which may be positive or negative and may change sign with the cosmic time t. For one fluid with the equation of state $p = \gamma \rho c^2$ the active gravitating mass (Eq. (12.18)) is

$$M_g(r) = +\frac{4\pi}{3}(1+3\gamma)\rho r^3 \propto S^{-3\gamma}(t). \tag{12.27}$$

So for dust, radiation and vacuum (or dark energy) we get

$$M_{\text{dust}}(r) = +\frac{4\pi}{3}\rho_{\text{dust}} r^3 \propto \text{const}, \tag{12.28}$$

$$M_{\text{rad}}(r) = +\frac{4\pi}{3}2\rho_{\text{rad}} r^3 \propto S^{-1}(t), \tag{12.29}$$

$$M_{\text{vac}}(r) = -\frac{4\pi}{3}2\rho_{\text{vac}} r^3 \propto -S^{+3}(t). \tag{12.30}$$

In the case of dust, the gravitating mass does not depend on time, while for radiation the gravitating mass continuously disappears in the expanding universe.

The strangest example is the vacuum, where the gravitating mass is negative: the vacuum antigravity between any two galaxies participating in cosmological expansion increases in time due to the continuous creation of gravitating (actually "antigravitating") vacuum mass. In this sense the continuous creation of matter in the classical Steady State model was just a particular case of the new physics of expanding space. All these processes are supposed to happen in our local environment of the galaxy universe, which makes Practical Cosmology a most exciting scientific adventure of our times.

Appendix A
Useful Astronomical and Physical Constants

Astronomical units:

Astronomical unit AU = $1.49597870 \times 10^{13}$ cm
Parsec pc = 206 265 AU = 3.0857 10^{18} cm = 3.26 ly
Light-year ly = 0.9461×10^{18} cm = 0.3066 pc
Hubble distance = $c/H_0 = 2997.9\, h^{-1}$ Mpc
Sidereal year 1 yr = 365.2564 d = 3.156×10^7 s
Hubble time $T_{\rm H} = H_0^{-1} = 9.7776 \times 10^9\, h^{-1}$ yr
Solar mass 1 $M_\odot = 1.989 \times 10^{33}$ g

Physical constants:

Velocity of light $c = 2.9979 \times 10^{10}$ cm/s
Gravitational constant $G = 6.67 \times 10^{-8}$ cm^3 g^{-1} s^{-2}
Planck's constant $h = 6.6256 \times 10^{-27}$ erg s
Boltzmann's constant $k = 1.3806 \times 10^{-16}$ erg K^{-1}
Mass of electron $m_{\rm e} = 9.1091 \times 10^{-28}$ g
Mass of proton $m_{\rm p} = 1.6725 \times 10^{-24}$ g
Mass of neutron $m_{\rm n} = 1.6725 \times 10^{-24}$ g

Y. Baryshev, P. Teerikorpi, *Fundamental Questions of Practical Cosmology*,
Astrophysics and Space Science Library 383,
DOI 10.1007/978-94-007-2379-5,

Appendix B
Why General Relativity Is Principally Different from Field Gravity

There have been many discussions about the derivation of Einstein's field equations from the spin 2 theory (another name for the field theory), and hence about the possible identity of general relativity and the field approach.

In his lectures on gravitation, Feynman tried to derive the full Einsteinian Lagrangian by iterating the Lagrangian of the spin 2 field. Misner et al. (1973, Chap. 7) wrote that "tensor theory in flat spacetime is internally inconsistent; when repaired, it becomes general relativity". They refered to the papers by Feynman (1963), Weinberg (1965), and Deser (1970) on a "field" derivation of Einstein's equations.

However, Straumann (2000) pointed out internal inconsistencies in such attempts to derive Einstein's equations from the spin 2 field theory: (1) general relativity having black hole solutions violates the simple topological structure of the Minkowski space of the field gravity, and (2) general relativity has lost the energy-momentum tensor of the gravity field together with the conservation laws (a direct consequence of the global symmetry of the Minkowski space). In his review, Padmanabhan (2008) showed that all derivations of general relativity from a spin 2 field are based on some additional assumptions that are equivalent to the geometrization of the gravitational interaction.

Indeed, general relativity and field gravity rest on incompatible physical principles (such as non-inertial frames and Riemann geometry of curved space versus inertial frames with Minkowski geometry of flat space). The geometrical approach eliminates the gravity force, as already de Sitter (1916) noted: "Gravitation is thus, properly speaking, not a 'force' in the new theory". This however leads to the problem of energy precisely because the work done by a force changes the energy. Within the field approach the gravity force is directly defined in an ordinary sense as the fourth interaction and has quantum nature (Feynman 1971).

Y. Baryshev, P. Teerikorpi, *Fundamental Questions of Practical Cosmology*,
Astrophysics and Space Science Library 383,
DOI 10.1007/978-94-007-2379-5,

Appendix C
The Gravitational Potential of a Fractal Matter Ball with Finite Radius

For a *homogeneous* matter distribution $\rho = \rho_0$ the solution of Eq. (9.5) inside the ball has the form (Baryshev and Kovalevski 1990):

$$\frac{\varphi(x)}{c^2} = -\frac{1}{2} + \frac{\mathrm{sh}(x)}{2x\,\mathrm{ch}(x_0)} \tag{C.1}$$

Here $x = r/R_{\mathrm{H}}$ is the dimensionless radius in units of the Hubble radius $R_{\mathrm{H}} = c/t_{\mathrm{H}} = c/(8\pi G\rho_0)^{1/2}$, $x_0 = r_0/R_{\mathrm{H}}$ and r_0 is the radius of the ball. The gravitating mass of this ball is

$$M(r) = M_{\mathrm{H}}x\left(1 - \frac{\mathrm{th}(x)}{x}\right), \tag{C.2}$$

where $M_{\mathrm{H}} = R_{\mathrm{H}}c^2/2G$ is the Hubble mass.

For sufficiently small distances ($r \ll R_{\mathrm{H}}$), the gravitational potential has Newtonian behavior, and for large distances ($r \gg R_{\mathrm{H}}$) the mass grows linearly so that the gravitational potential in the center of the ball asymptotically reaches the value $-c^2/2$.

In the case of the *fractal dark matter distribution* with $D = 2$ the rest mass density law is $\rho(r) = \rho_0 r_0/r$ and the solution of Eq. (9.5) inside the ball has the form (Nagirner 2006):

$$\frac{\varphi(x)}{c^2} = -\frac{1}{2} + \frac{1}{\sqrt{x}}\left[C_1 I_1(4\sqrt{x}) + C_2 K_1(4\sqrt{x})\right] \tag{C.3}$$

where I_1 and K_1 are the modified Bessel functions and x is the dimensionless distance. Using ordinary conditions for the gravitational potential of a finite ball with radius $x = x_0$, one finds that $C_2 = 0$ and $C_1 = 1/(4\,I_0(4\sqrt{x_0}))$, where $I_0(x)$ is the modified Bessel function.

The total gravitating mass inside the fractal ball of radius r is:

$$M(r) = M_{\mathrm{H}}x\left(1 - \frac{I_1(4\sqrt{x})}{2\sqrt{x}\,I_0(4\sqrt{x})}\right). \tag{C.4}$$

Here $x = r/R_{\mathrm{H}}$ is the dimensionless radius in units of the Hubble radius and $R_{\mathrm{H}} = c^2/(2\pi G\rho_0 r_0)$. ρ_0 and r_0 define the lower cutoff of the fractal structure and $M_{\mathrm{H}} = R_{\mathrm{H}}c^2/2G$ is the Hubble mass as above.

Y. Baryshev, P. Teerikorpi, *Fundamental Questions of Practical Cosmology*, Astrophysics and Space Science Library 383,
DOI 10.1007/978-94-007-2379-5, © Springer Science+Business Media B.V. 2012

References

Abell, G.O.: The distribution of rich clusters of galaxies. Astrophys. J. Suppl. Ser. **3**, 211 (1958)
Abell, G.O.: Evidence regarding second-order clustering of galaxies and interactions between clusters of galaxies. Astron. J. **66**, 607 (1961)
Abell, G.O., Corwin, H.G., Olowin, R.P.: A catalog of rich clusters of galaxies. Astrophys. J. Suppl. Ser. **70**, 1 (1989)
Abramowicz, M., Kluzniak, W., Lasota, J.-P.: No observational proof of the black hole event horizon. Astron. Astrophys. **396**, L31 (2002)
Abramowitz, M., Stegun, I.A.: Handbook of Mathematical Functions with Formulas, Graphs, and Mathematical Tables. Dover, New York (1964)
Adelberger, E.G., Heckel, B.R., Nelson, A.E.: Tests of the gravitational inverse-square law. Annu. Rev. Nucl. Part. Sci. **53**, 77 (2003)
Aglietta, M., Badino, G., Bologna, G., et al.: On the event observed in the Mont Blanc Underground Neutrino Observatory during the occurrence of supernova 1987a. Europhys. Lett. **3**, 1315 (1987)
Aguirre, A.N.: Dust versus cosmic acceleration. Astrophys. J. **512**, L19 (1999)
Alfvén, H.: Cosmology in the plasma universe. Laser Part. Beams **6**, 639 (1989)
Amaldi, E., Bonifazi, P., Castellano, M., et al.: Data recorded by the Rome room temperature gravitational wave antenna, during the supernova SN 1987a in the Large Magellanic Cloud. Europhys. Lett. **3**, 1325 (1987)
Amanullah, R., Lidman, C., Rubin, D., et al.: Spectra and HST light curves of six type Ia supernovea at $0.511 < z < 1.12$ and the Union2 compilation. Astrophys. J. **716**, 712 (2010)
Ambartzumian, V.A.: Fluctuations in numbers of extragalactic nebulae and galactic extinction. Bull. Abastuman Obs. **4**, 17 (1940) (in Russian)
Ambartzumian, V.A.: On the theory of fluctuations in the apparent distribution of stars on the sky. Commun. Byurakan Obs. **6** (1951) (in Russian)
Amelino-Camelia, G.: Quantum theory's last challenge. Nature **408**, 661 (2000)
Amelino-Camelia, G., Smolin, L.: Prospects for constraining quantum gravity dispersion with near term observations. Phys. Rev. D **80**, 084017 (2009)
Amelino-Camelia, G., Lammerzahl, C., Macias, A., Muller, H.: The search for quantum gravity signals. In: Gravitation and Cosmology: 2nd Mexican Meeting on Mathematical and Experimental Physics. AIP Conf. Proc., vol. 758, p. 30 (2005)
Anderson, J.D.: Is there something we do not know about gravity? Astronomy **37**, 22 (2009)
Andreev, A.Yu., Komberg, B.V.: Cosmological parameters and the large numbers of Eddington and Dirac. Astron. Rep. **44**, 139 (2000)
Argon, A.L., Greenhill, L.J., Moran, J.M., et al.: The IC 133 water vapor maser in the galaxy M33: A geometric distance. Astrophys. J. **615**, 702 (2004)
Arp, H.C.: A very small, condensed galaxy. Astrophys. J. **142**, 402 (1965)
Arp, H.C.: Redshifts of companion galaxies. Nature **225**, 1034 (1970)

Y. Baryshev, P. Teerikorpi, *Fundamental Questions of Practical Cosmology*,
Astrophysics and Space Science Library 383,
DOI 10.1007/978-94-007-2379-5,

Arp, H.C.: Arguments for a Hubble constant near $H_0 = 55$. Astrophys. J. **571**, 615 (2002)

Arrhenius, S.: Zur Frage nach der Unendlichkeit der Welt. Ark. Mat., Astron. Fys. **5**(12) (1908)

Astone, P., Babusci, D., Bassan, M., et al.: Study of the coincidences between the gravitational wave detectors EXPLORER and NAUTILUS in the year 2001. Class. Quantum Gravity **19**, 5449 (2002)

Astone, P., Babusci, D., Ballantini, R., et al.: The 2003 run of the Explorer-Nautilus gravitational wave experiment. Class. Quantum Gravity **23**, S169 (2006)

Aurela, A.M.: Ann. Univ. Turkuensis A **162**, 85 (1973)

Axenides, M., Perivolaropoulos, L.: Dark energy and the quietness of the local Hubble flow. Phys. Rev. D **65**, 127301 (2002)

Babadzhanyants, M.K., Belokon, E.T.: Optical manifestations of superluminal expansion of components belonging to the millisecond radio structure in the quasar 3C 345. Astrophysics **23**, 639 (1985)

Babadzhanyants, M.K., Belokon, E.T.: 3C 120: Connection between the optical variability and superluminal components of the millisecond radio structure. Astrophysics **27**, 588 (1987)

Babak, S.V., Grishchuk, L.P.: Finite-range gravity and its role in gravitational waves, black holes and cosmology. Phys. Rev. B **66**, 184507 (2002)

Baertschiger, T., Sylos Labini, F.: Growth of correlations in gravitational N-body simulations. Phys. Rev. D **69**, 123001 (2004)

Baertschiger, T., Joyce, M., Sylos Labini, F.: Power-law correlation and discreteness in cosmological N-body simulations. Astrophys. J. **581**, L63 (2002)

Bahcall, N.A.: Large-scale structure in the universe indicated by galaxy clusters. Annu. Rev. Astron. Astrophys. **26**, 631 (1988)

Bahcall, N.A., Burgett, W.S.: Are superclusters correlated on very large scale? Astrophys. J. **300**, L35 (1986)

Bahcall, N.A., Soneira, R.M.: The spatial correlation function of rich clusters of galaxies. Astrophys. J. **270**, 20 (1983)

Bahcall, J., Wolf, R.A.: Fine-structure transitions. Astrophys. J. **152**, 701 (1968)

Bahcall, N.A., Dong, F., Hao, L., et al.: The richness-dependent cluster correlation function: Early Sloan digital sky survey data. Astrophys. J. **599**, 814 (2003)

Bajan, K., Flin, P., Godlowski, W., Pervushin, V.P.: On the investigations of galaxy redshift periodicity. Phys. Part. Nucl. Lett. **4**, 5 (2007)

Balian, R., Schaeffer, R.: Scale-invariant matter distribution in the universe. II—Bifractal behaviour. Astron. Astrophys. **226**, 373 (1989)

Barnes, K.J.: Lagrangian theory for the second-rank-tensor field. J. Math. Phys. **6**, 788 (1965)

Barnes, L.A., Francis, M.J., James, J.B., Lewis, G.F.: Joining the Hubble flow: Implications for expanding space. Mon. Not. R. Astron. Soc. **373**, 382 (2006)

Barnothy, J.: Galaxy-quasar associations and globular clusters. Bull. Am. Astron. Soc. **6**, 212 (1974)

Barrow, J.D.: The mysterious lore of large numbers. In: Bertotti, B., et al. (eds.) Modern Cosmology in Retrospect, p. 67. Cambridge Univ. Press, Cambridge (1990)

Barrow, J.D., Tipler, F.J.: The Anthropic Cosmological Principle. Oxford Univ. Press, Oxford (1988)

Baryshev, Yu.V.: The hierarchical structure of metagalaxy—a review of problems. Rep. SAO Russ. Acad. Sci. **14**, 24 (1981) (in English: 1984, Allerton Press)

Baryshev, Yu.V.: On the gravitational radiation of the binary system with the pulsar PSR1913+16. Astrophysics **18**, 93 (1982)

Baryshev, Yu.V.: Equations of motion of test particles in Lorentz-covariant tensor theory of gravity. Vest. Leningr. Univ. Ser. 1 **4**, 113 (1986)

Baryshev, Yu.V.: Conservation laws and equations of motion in the field gravitation theory. Vest. Leningr. Univ. Ser. 1 **2**, 80 (1988)

Baryshev, Yu.V.: Pulsation of supermassive star in the tensor field gravitation theory. In: Variability of Blazars, p. 52. Cambridge Univ. Press, Cambridge (1992a)

Baryshev, Yu.: Modern state of observational cosmology. VINTI, Itogi Nauki i Techniki, Series Classical Field Theory and Gravity Theory, vol. 4: Gravitation and Cosmology, p. 89 (1992b) (in Russian)
Baryshev, Yu.V.: On the fractal nature of the large-scale structure of the universe. Astron. Astrophys. Trans. **5**, 15 (1994)
Baryshev, Yu.V.: On a possibility of scalar gravitational wave detection from the binary pulsar PSR1913+16. In: Coccia, E., Pizzella, G., Ronga, F. (eds.) Proc. of the First Amaldi Conference on Gravitational Wave Experiments, p. 251. World Sci. Publ. Co., Singapore (1995). gr-qc/9911081
Baryshev, Yu.V.: Field theory of gravitation: Desire and reality. Gravitation **2**, 69 (1996)
Baryshev, Yu.V.: Signals from SN1987A in Amaldi-Weber antennas as possible detection of scalar gravitational waves. Astrophysics **40**, 377 (1997)
Baryshev, Yu.V.: Translational motion of rotating bodies and tests of the equivalence principle. Gravit. Cosmol. **8**, 232 (2002)
Baryshev, Yu.V.: Spatial distribution of galaxies and tests of the relativistic cosmology. Dissertation on doctor of physical-mathematical sciences degree, St.-Petersburg University, St.-Petersburg (2003) (in Russian)
Baryshev, Yu.V.: Conceptual problems of the standard cosmological model. In: Proc. of the 1-st Int. Conf. Crisis in Cosmology. AIP Conf. Proc., vol. 822, p. 23 (2006)
Baryshev, Yu., Bukhmastova, Yu.: The method of a two-point conditional column density for estimating the fractal dimension of the distribution of galaxies. Astron. Lett. **30**, 444 (2004)
Baryshev, Yu., Ezova, Yu.: Gravitational mesolensing by king objects and quasar–galaxy associations. Astron. Rep. **41**, 436 (1997)
Baryshev, Yu.V., Kovalevski, M.A.: Homogeneous ball in the field gravitation theory. Vest. Leningr. Univ. Ser. 1 **1**, 86 (1990)
Baryshev, Yu.V., Paturel, G.: Statistics of the detection rates for tensor and scalar gravitational waves from the local galaxy universe. Astron. Astrophys. **371**, 378 (2001)
Baryshev, Yu.V., Raikov, A.A.: Note on characteristic quantities in cosmology. Astrofizika **28**, 689 (1988) (in Russian)
Baryshev, Yu.V., Raikov, A.A.: A quantum limitation on the gravitational interaction. In: Saamokhin, A.P., Rcheulishvili, G.L. (eds.) Proc. of the XVII Int. Workshop Problems on High Energy Physics and Field Theory, Protvino (1995)
Baryshev, Yu.V., Sokolov, V.V.: Relativistic tensor theory of gravitational field in flat space-time. Trans. Astron. Obs. Leningr. Univ. **38**, 36 (1983)
Baryshev, Yu.V., Sokolov, V.V.: Some astrophysical consequences of the dynamical treatment of gravity. Astrophysics **21**, 361 (1984)
Baryshev, Yu.V., Teerikorpi, P.: Kinematical models of double radio sources and the unified scheme I. Theoretical probability distributions of observable quantities. Astron. Astrophys. **295**, 11 (1995)
Baryshev, Yu., Teerikorpi, P.: Discovery of Cosmic Fractals. World Scientific, Singapore (2002), 408 pp (Polish translation: Poznawanie kosmicznego ladu Wszechswiat, WAM, Krakow, 2005; Italian translation: La scoperta dei frattali cosmici, Bollati Boringhieri, Torino, 2006)
Baryshev, Yu., Teerikorpi, P.: The fractal analysis of the large-scale galaxy distribution. Bull. Spec. Astrophys. Obs. **59**, 92 (2006)
Baryshev, Yu., Raikov, A.A., Yushchenco, A.V.: Quasar-galaxy associations as a result of gravitational lensing by objects of medium masses. In: Surdej, J., Fraipont-Caro, D., Gosset, E., Refsdal, S., Remy, M. (eds.) Gravitational Lenses in the Universe. Procs. of the 31st Liege International Astrophysical Colloquium, p. 307. Universite de Liege, Institut d'Astrophysique, Liege (1993)
Baryshev, Yu.V., Raikov, A.A., Sergeev, A.G., Tron, A.A.: A new approach to the large cosmological numbers coincidences. Astron. Astrophys. Trans. **5**, 27 (1994a)
Baryshev, Yu., Sylos-Labini, F., Montuori, M., Pietronero, L.: Facts and ideas in modern cosmology. Vistas Astron. **38**, 419 (1994b)
Baryshev, Yu.V., Gubanov, A.G., Raikov, A.A.: On possibility of observational testing of the frequency dependence of gravitational bending of light. Gravitation **2**, 72 (1996a)

Baryshev, Yu.V., Raikov, A.A., Tron, A.A.: Microwave background radiation and cosmological large numbers. Astron. Astrophys. Trans. **10**, 135 (1996b)

Baryshev, Yu., Sylos-Labini, F., Montuori, M., Pietronero, L., Teerikorpi, P.: On the fractal structure of galaxy distribution and its implications for cosmology. Fractals **6**, 231 (1998)

Baryshev, Yu., Chernin, A., Teerikorpi, P.: The cold local Hubble flow as a signature of dark energy. Astron. Astrophys. **378**, 729 (2001)

Basu, D.: Selection effects in the redshift distribution of gamma-ray bursts and associated quasi-stellar objects and active galaxies. Astrophys. J. **618**, L71 (2005)

Bauer, H.: On the energy components of the gravitational field. Phyz. Z. **19**, 163 (1918)

Beckwith, S.V.W., Stiavelli, M., Koekemoer, A.M., et al.: The Hubble ultra deep field. Astron. J. **132**, 1729 (2006)

Behr, A.: Zur Entfernungsskala der extragalaktischen Nebel. Astron. Nachr. **279**, 97 (1951)

Bekenstein, J.D.: The modified Newtonian dynamics—MOND—and its implications for new physics. Contemp. Phys. **47**, 387 (2006). astro-ph/0701848 (2007)

Bell, M.B., Comeau, S.P.: Selection effects in the SDSS quasar sample: the filter gap footprint. Astrophys. Space Sci. **326**, 11 (2010)

Belokon, E.T.: Optical variability of the quasar 3C273 and superluminal motion in its milliarcsecond radio jet. Sov. Astron. **35**, 1 (1991)

Benetti, S., Cappellaro, E., Mazzali, P.A., et al.: The diversity of Type Ia supernovae: Evidence for systematics? Astrophys. J. **623**, 1011 (2005)

Benzi, R., Paladin, G., Parisi, G., Vulpiani, A.: On the multifractal nature of fully developed turbulence and chaotic systems. J. Phys. A **17**, 3251 (1984)

Berdyugin, A., Teerikorpi, P.: Interstellar polarization at high galactic latitudes from distant stars VI. Extended polarization map and connection with the local spiral structure. Astron. Astrophys. **384**, 1050 (2002)

Berdyugin, A., Piirola, V., Teerikorpi, P.: Interstellar polarization at high galactic latitudes from distant stars VII. A complete map for south latitudes $-70° > b > -90°$. Astron. Astrophys. **424**, 837 (2004)

Berendzen, R., Hart, R., Seeley, D.: Man Discovers the Galaxies. Science History, New York (1976), 228 pp

Bernstein, R.A., Freedman, W.L., Madore, B.F.: The first detection of the extragalactic background light at 3000, 5500, and 8000 Å. Astrophys. J. **571**, 85 (2002)

Bernstein, R.A., Freedman, W.L., Madore, B.F.: Corrections of errors in "The first detection of the extragalactic background light at 3000, 5500, and 8000 Å". Astrophys. J. **632**, 713 (2005)

Bertolami, O., Paramos, J., Turyshev, S.: General theory of relativity: Will it survive the next decade. In: Dittus, H., Laemmerzahl, C., Turyshev, S. (eds.) Lasers, Clocks, and Drag-Free: Technologies for Future Exploration in Space and Tests of Gravity, p. 27. Springer, Berlin (2006a)

Bertolami, O., de Matos, C.J., Grenouilleau, J.C., Minster, O., Volonté, S.: Perspectives in fundamental physics in space. Acta Astronaut. **59**, 490 (2006b)

Best, J.S.: An examination of the large-scale clustering of the Las Campanas redshift survey. Astrophys. J. **541**, 519 (2000)

Beutler, F., Blake, C., Colless, M., et al.: The 6dF galaxy survey: Baryon acoustic oscillations and the local Hubble constant (2011). arXiv:1106.3366

Bharadwaj, S., Gupta, A.K., Seshadri, T.R.: Nature of clustering in the Las Campanas redshift survey. Astron. Astrophys. **351**, 405 (1999)

Binggeli, B., Tarenghi, M., Sandage, A.: The abundance and morphological segregation of dwarf galaxies in the field. Astron. Astrophys. **228**, 42 (1990)

Binney, J.: On the impossibility of advection dominated accretion (2003). astro-ph/0308171

Birkhoff, G.D.: Flat space-time and gravitation. Proc. Natl. Acad. Sci. USA **30**, 324 (1944)

Bisnovatyi-Kogan, G.S., Lovelace, R.V.E.: Magnetic field limitations on advection-dominated flows. Astrophys. J. **529**, 978 (2000)

Blondin, S., Davis, T.M., Krisciunas, K.: Time dilation in Type Ia supernova spectra at high redshift. Astrophys. J. **682**, 724 (2008)

Bogolubov, N., Shirkov, D.: Introduction to Quantum Field Theory. Nauka, Moscow (1976)
Bolzonella, M., Miralles, J.-M., Pelló, R.: Photometric redshifts based on standard SED fitting procedures. Astron. Astrophys. **363**, 476 (2000)
Bonamente, M., Joy, M.K., LaRoque, S.J., Carlstrom, J.E., Reese, E.D., Dawson, K.S.: Determination of the cosmic distance scale from Sunyaev-Zeldovich effect and Chandra X-ray measurements of high-redshift galaxy clusters. Astrophys. J. **647**, 25 (2006)
Bondi, H.: Spherically symmetrical models in general relativity. Mon. Not. R. Astron. Soc. **107**, 410 (1947)
Bondi, H.: Cosmology. Cambridge University Press, Cambridge (1952) (2nd edition 1968)
Bondi, H., Gold, T.: The steady-state theory of the expanding Universe. Mon. Not. R. Astron. Soc. **108**, 252 (1948)
Bonnor, W.: A non-uniform relativistic cosmological model. Mon. Not. R. Astron. Soc. **159**, 261 (1972)
Bottaccio, M., Amici, A., Miocchi, P., Capuzzo-Dolcetta, R., Montuori, M., Pietronero, L.: Clustering in gravitating N-body systems. Europhys. Lett. **57**, 315 (2002)
Bottinelli, L., Gouguenheim, L.: Redshifts of companion galaxies. Astron. Astrophys. **26**, 85 (1973)
Bottinelli, L., Gouguenheim, L., Paturel, G., Teerikorpi, P.: The Malmquist bias and the value of H_0 from the Tully-Fisher relation. Astron. Astrophys. **156**, 157 (1986)
Bottinelli, L., Fouqué, P., Gouguenheim, L., Paturel, G., Teerikorpi, P.: Cluster population incompleteness bias and the value of H_0 from the Tully-Fisher relation. Astron. Astrophys. **181**, 1 (1987)
Bottinelli, L., Gouguenheim, L., Paturel, G., Teerikorpi, P.: The Malmquist bias in the extragalactic distance scale: Controversies and misconceptions. Astrophys. J. **328**, 4 (1988)
Bottinelli, L., Gouguenheim, L., Paturel, G., Teerikorpi, P.: Extragalactic database: VI Inclination corrections for spiral galaxies and disk opaqueness in the B-band. Astron. Astrophys. **296**, 64 (1995)
Bouwens, R.J., Illingworth, G.D., Blakeslee, J.P., Broadhurst, T.J., Franx, M.: Galaxy size evolution at high redshift and surface brightness selection effects: Constraints from the Hubble Ultra Deep Field. Astrophys. J. **611**, L1 (2004)
Bouwens, R.J., Illingworth, G.D., Labbe, I., et al.: A candidate redshift $z \approx 10$ galaxy and rapid changes in that population at an age of 500 Myr. Nature **469**, 504 (2011)
Branch, D.: High-velocity matter in a classical Type I supernova: The demise of Type Ia homogeneity. Astrophys. J. **316**, L81 (1987)
Brémond, A.: Vesto Melvin Slipher (1875–1969) et la naissance de l'astrophysique extragalactique. Doctor's Thesis, l'Universite Lyon I-Claude Bernard (2008)
Bridgman, P.W.: Dimensional Analysis. Yale Univ. Press, New Haven (1920)
Bridgman, P.W.: The Nature of Physical Theory. Princeton University Press, Princeton (1936)
Bronstein, M.D.: On the possible theory of the world as a whole. In: Main Problems of Cosmic Physics, p. 186. ONTI, Kiev (1934)
Bronstein, M.D.: Quantization of gravitational waves. J. Exp. Theor. Phys. **6**, 195 (1936)
Brumberg, V.A.: Essential Relativistic Celestial Mechanics. Adam Hildger IOP Publ. Ltd, New York (1991)
Brunthaler, A., Reid, M.J., Falcke, H., Greenhill, L.J., Henkel, C.: The geometric distance and proper motion of the triangulum galaxy (M33). Science **307**, 1440 (2005)
Bruzual, G., Charlot, S.: Stellar population synthesis at the resolution of 2003. Mon. Not. R. Astron. Soc. **344**, 1000 (2003)
Brynjolfsson, A.: Magnitude-redshift relation for SNe Ia, time dilation, and plasma redshift (2006). arXiv:astro-ph/0602500
Buchalter, A., Helfand, D.J., Becker, R.H., White, R.L.: Constraining Ω_0 with the angular size–redshift relation of double-lobed quasars in the FIRST survey. Astrophys. J. **494**, 503 (1998)
Bukhmastova, Yu.L., Baryshev, Yu.V.: Constraining the nature of galaxy haloes with gravitational mesolensing of QSOs by halo substructure objects. In: Baryshev, Yu.V., Taganov, I.N., Teerikorpi, P. (eds.) Problems of Practical Cosmology. Proceedings of the International Conference

held at Russian Geographical Society, 23–27 June, 2008, St. Petersburg, vol. 1, p. 18. TIN, St.Petersburg (2008)

Burbidge, G.: The reality of anomalous redshifts in the spectra of some QSOs and its implications. Astron. Astrophys. **309**, 9 (1996)

Burbidge, G.: Noncosmological Redshifts. Publ. Astron. Soc. Pac. **113**, 899 (2001)

Burbidge, G.: The state of cosmology. In: Pecker, J.-C., Narlikar, J. (eds.) Current Issues in Cosmology, p. 3. Cambridge University Press, Cambridge (2006)

Burbidge, G., Napier, W.M.: The distribution of redshifts in new samples of quasi-stellar objects. Astron. J. **121**, 21 (2001)

Buta, R., de Vaucouleurs, G.: Inner ring structures in galaxies as distance indicators. IV. Distances to several groups, clusters, the Hercules supercluster, and the value of the Hubble constant. Astrophys. J. **266**, 1 (1983)

Butkevich, A., Berdyugin, A., Teerikorpi, P.: Statistical biases in stellar astronomy: The Malmquist bias revisited. Mon. Not. R. Astron. Soc. **362**, 321 (2005)

Byrd, G.G., Valtonen, M.J.: Origin of redshift differentials in galaxy groups. Astrophys. J. **289**, 535 (1985)

Byrd, G.G., Chernin, A.D., Valtonen, M.J.: Cosmology: Foundations and Frontiers. URSS, Moscow (2007)

Cabré, A., Gaztañaga, E.: Clustering of luminous red galaxies—I. Large-scale redshift-space distortions. Mon. Not. R. Astron. Soc. **393**, 1183 (2009)

Caldwell, R.R., Dave, R., Steinhardt, P.J.: Cosmological imprint of an energy component with general equation of state. Phys. Rev. Lett. **80**, 1582 (1998)

Caldwell, R.R., Kamionkowski, M., Weinberg, N.N.: Phantom energy and cosmic doomsday. Phys. Rev. Lett. **91**, 071301 (2003)

Calzetti, D., Einasto, J., Giavalisco, M., Ruffini, R., Saar, E.: The correlation function of galaxies in the direction of the Coma cluster. Astrophys. Space Sci. **137**, 101 (1987)

Calzetti, D., Giavalisco, M., Ruffini, R.: The normalization of the correlation functions for extragalactic structures. Astron. Astrophys. **198**, 1 (1988)

Calzetti, D., Giavalisco, M., Ruffini, R.: The angular two-point correlation function and the cellular fractal structure of the universe. Astron. Astrophys. **226**, 1 (1989)

Canizares, C.R.: Gravitational focusing and the association of distant quasars with foreground galaxies. Nature **291**, 620 (1981)

Carlstrom, J.E., Holder, G.P., Reese, E.D.: Cosmology with the Sunyaev-Zel'dovich effect. Annu. Rev. Astron. Astrophys. **40**, 643 (2002)

Carpenter, E.F.: Some characteristics of associated galaxies I. A density restriction in the metagalaxy. Astrophys. J. **88**, 344 (1938)

Carroll, S.M., Press, W.H., Turner, E.L.: The cosmological constant. Annu. Rev. Astron. Astrophys. **30**, 499 (1992)

Carroll, S.M., De Felice, A., Duvvuri, V., et al.: The cosmology of generalized modified gravity models. Phys. Rev. D **71**(2005), 063513 (2005)

Carter, B.: Large number coincidences and the anthropic principle in cosmology. In: Confrontation of Cosmological Theories with Observational Data; Proceedings of the Symposium, Krakow, Poland, September 10–12, 1973, p. 291. Reidel, Dordrecht (1974)

Célérier, M.-N., Bolejko, K., Krasiński, A.: A (giant) void is not mandatory to explain away dark energy with a Lemaître-Tolman model. Astron. Astrophys. **518**, A21 (2010)

Chand, H., Srianand, R., Petitjean, P., Aracil, B.: Probing the cosmological variation of the fine-structure constant: Results based on VLT-UVES sample. Astron. Astrophys. **417**, 853 (2004)

Chandrasekhar, S.: Cosmological constants. Nature **139**, 757 (1937)

Chapline, G.: Dark energy stars (2005). astro-ph/0503200

Charlier, C.: Wie ein Unendliche Welt aufgebaut sein kann. Ark. Mat. Astron. Fys. **4**(24), 1 (1908)

Charlier, C.: How an infinite world may be built up. Ark. Mat. Astron. Fys. **16**(22), 1 (1922)

Chernin, A.: The rest mass of primordial neutrinos, and gravitational instability in the hot universe. Sov. Astron. **25**, 14 (1981)

Chernin, A.: Cosmic vacuum. Phys. Usp. **44**, 1153 (2001)

Chernin, A.: Physical vacuum and cosmic coincidence problem. New Astron. **7**, 113 (2002)
Chernin, A., Teerikorpi, P., Baryshev, Yu.: Non-Friedmann cosmology for the Local Universe, significance of the universal Hubble constant and short-distance indicators of dark energy. Astron. Astrophys. **456**, 13 (2006)
Chernin, A.D., Teerikorpi, P., Valtonen, M.J., Dolgachev, V.P., Domozhilova, L.M. Byrd G.G.: Local dark matter and dark energy as estimated on a scale of ~ 1 Mpc in a self-consistent way. Astron. Astrophys. **507**, 1271 (2009)
Chernin, A.D., Karachentsev, I.D., Nasonova, O.G., et al.: Dark energy domination in the virgocentric flow. Astron. Astrophys. **520**, A104 (2010)
Chiao, R.: Conceptual tensions between quantum mechanics and general relativity: are there experimental consequences? (2003). gr-qc/0303100
Clifton, T., Ferreira, P.G., Padilla, A., Skordis, C.: Modified gravity and cosmology (2011). arXiv:1106.2476 [astro-ph]
Coccia, E., Dubath, F., Maggiore, M.: On the possible sources of gravitational wave bursts detectable today. Phys. Rev. D **70**, 084010 (2004)
Cohen, S.E., Windhorst, R.A., Odewahn, S.C., Chiarenza, C.A., Driver, S.P.: The Hubble Space Telescope WFPC2 B-band parallel survey: A study of galaxy morphology for magnitudes $18 \leq B \leq 27$. Astron. J. **125**, 1762 (2003)
Coleman, P.H., Pietronero, L.: The fractal structure of the Universe. Phys. Rep. **213**, 311 (1992)
Coleman, P.H., Pietronero, L., Sanders, R.H.: Absence of any characteristic correlation length in the CfA galaxy catalogue. Astron. Astrophys. **200**, L32 (1988)
Coles, P.: An unprincipled universe? Nature **391**, 120 (1998)
Colless, M., Peterson, B., Jackson, C., et al. (the 2dFGRS team): The 2dF galaxy redshift survey: Final data release (2003). astro-ph/0306581
Combes, F.: Fractal structure driven by self-gravity: molecular clouds and the universe. Celest. Mech. Dyn. Astron. **72**, 91 (1999)
Conklin, E.K., Bracewell, R.N.: Limits on small scale variations in the cosmic background radiation. Nature **216**, 777 (1967)
Cooperstock, F.I., Faraoni, V., Vollick, D.N.: The influence of the cosmological expansion on local systems. Astrophys. J. **503**, 61 (1998)
Courtois, H., Paturel, G., Sousbie, T., Sylos Labini, F.: The LEDA galaxy distribution: I. Maps of the Local Universe. Astron. Astrophys. **423**, 27 (2004)
Crawford, D.F.: Angular size in a static universe. Astrophys. J. **440**, 446 (1995)
Da Costa, L.N., Pellegrini, P.S., Davis, M., Meiksin, A., Sargent, W.L.W., Tonry, J.L.: Southern sky redshift survey: The catalog. Astrophys. J. Suppl. Ser. **75**, 935 (1991)
Da Costa, L.N., Geller, M.J., Pellegrini, P.S., et al.: A complete southern sky redshift survey. Astrophys. J. Lett. **424**, L1 (1994)
Damour, T., Taylor, J.: On the orbital period change of the binary pulsar PSR 1913+16. Astrophys. J. **366**, 501 (1991)
Davis, M.: Is the universe homogeneous on large scales? In: Turok, N. (ed.) Proc. of the Conference Critical Dialogues in Cosmology, p. 13. World Scientific, Singapore (1997)
Davis, T.M., Lineweaver, C.H.: Expanding confusion: Common misconceptions of cosmological horizons and the superluminal expansion of the universe. Publ. Astron. Soc. Aust. **21**, 97 (2004)
Davis, M., Peebles, P.J.E.: A survey of galaxy redshifts. V. The two-point position and velocity correlations. Astrophys. J. **267**, 465 (1983)
Davis, M., Meiksin, A., Strauss, M.A., da Costa, L.N., Yahil, A.: On the universality of the two-point galaxy correlation function. Astrophys. J. **333**, L9 (1988)
de Bernardis, P., Ade, P., Bock, J., et al.: A flat Universe from high-resolution maps of the cosmic microwave background radiation. Nature **404**, 955 (2000)
de Lapparent, V., Geller, M., Huchra, J.: A slice of the universe. Astrophys. J. **302**, L1 (1986)
de Sitter, W.: On Einstein's theory of gravitation, and its astronomical consequences. I. Mon. Not. R. Astron. Soc. **LXXVI**(9), 699 (1916)
de Sitter, W.: On Einstein's theory of gravitation, and its astronomical consequences. III. Mon. Not. R. Astron. Soc. **LXXVIII**(1), 3 (1917)

de Vaucouleurs, G.: Evidence for a local supergalaxy. Astron. J. **58**, 30 (1953)
de Vaucouleurs, G.: Further evidence for a local super-cluster of galaxies: Rotation and expansion. Astron. J. **63**, 253 (1958)
de Vaucouleurs, G.: Recent studies of clusters and superclusters. Astron. J. **66**, 629 (1961)
de Vaucouleurs, G.: The case for a hierarchical cosmology. Science **167**, 1203 (1970)
de Vaucouleurs, G.: The large-scale distribution of galaxies and clusters of galaxies. Publ. Astron. Soc. Pac. **83**, 113 (1971)
de Vaucouleurs, G.: The velocity-distance relation and the Hubble constant for nearby groups of galaxies. In: Evans, D.S. (ed.) External Galaxies and Quasi-Stellar Objects. IAU Symp., vol. 44, p. 353. Reidel, Dordrecht (1972)
de Vaucouleurs, G.: Extragalactic distance scale, Malmquist bias and Hubble constant. Mon. Not. R. Astron. Soc. **202**, 367 (1983a)
de Vaucouleurs, G.: The distance scale of the universe. In: Sky & Telescope, December 1983, p. 511 (1983b)
de Vaucouleurs, G., de Vaucouleurs, A.: Reference Catalogue of Bright Galaxies. Univ. of Texas Press, Austin (1964)
De Vaucouleurs, G., de Vaucouleurs, A., Corwin, H.G. Jr.: Second Reference Catalogue of Bright Galaxies. Univ. of Texas Press, Austin (1976)
De Vaucouleurs, G., de Vaucouleurs, A., Corwin, H.G. Jr., Buta, R.J., Paturel, G., Fouqu'e, P.: Third Reference Catalogue of Bright Galaxies. Springer, New York (1991)
De Vega, H., S'anches, N., Combes, F.: Self-gravity as an explanation of the fractal structure of the interstellar medium. Nature **383**, 56 (1996)
De Vega, H., S'anches, N., Combes, F.: The fractal structure of the universe: A new field theory approach. Astrophys. J. **500**, 8 (1998)
Deller, A.T., Bailes, M., Tingay, S.J.: Implications of a VLBI distance to the double pulsar J0737-3039A/B. Science **323**, 132 (2009)
Demianski, M., de Ritis, R., Marino, A.A., Piedipalumbo, E.: Approximate angular diameter distance in a locally inhomogeneous universe with nonzero cosmological constant. Astron. Astrophys. **411**, 33 (2003)
Deser, S.: Self-interaction and gauge invariance. Gen. Relativ. Gravit. **1**, 9 (1970)
Dicke, R.H.: Dirac's cosmology and Mach's principle. Nature **192**, 440 (1961)
Dirac, P.A.M.: Cosmological constants. Nature **139**, 323 (1937)
Disney, M.J.: Visibility of galaxies. Nature **263**, 573 (1976)
Djorgovski, S., Spinrad, H.: Towards the application of a metric size function in galactic evolution and cosmology. Astrophys. J. **251**, 417 (1981)
Dogterom, M., Pietronero, L.: The angular projection of fractals and its relevance for the galaxy distributions. Physica A **171**, 239 (1991)
Donato, F., Gentile, G., Salucci, P.: A constant dark matter halo surface density in galaxies. Mon. Not. R. Astron. Soc. **397**, 1169 (2009)
Dressler, A., Faber, S.M., Burstein, D., Davies, R.L., Lynden-Bell, D.: Spectroscopy and photometry of elliptical galaxies. I. A new distance estimator. Astrophys. J. **313**, 42 (1987)
Durrer, R., Sylos Labini, F.: A fractal galaxy distribution in a homogeneous universe? Astron. Astrophys. **339**, L85 (1998)
Durrer, R., Eckmann, J.-P., Sylos Labini, F., Montuori, M., Pietronero, L.: Angular projections of fractal sets. Europhys. Lett. **40**, 491 (1997)
Dymnikova, I.: The cosmological term as a source of mass. Class. Quantum Gravity **19**, 725 (2002)
Eckmann, J.-P., Järvenpää, E., Järvenpää, M., Procaccia, I.: On the fractal dimension of the visible universe. In: Klonowski, W. (ed.) Simplicity Behind Complexity, Euroattractor, 2002. Pabst Science, Lengerich (2004)
Eddington, A.S.: On a formula for correcting statistics for the effects of a known error of observation. Mon. Not. R. Astron. Soc. **73**, 359 (1913)
Eddington, A.S.: Stellar Movements and the Structure of the Universe. Macmillan, London (1914)
Eddington, A.S.: Space, Time and Gravitation. Cambridge University Press, Cambridge (1923a)
Eddington, A.S.: The Mathematical Theory of Relativity, p. 161. Cambridge University Press, Cambridge (1923b)

Eddington, A.S.: The correction for statistics for accidental error. Mon. Not. R. Astron. Soc. **100**, 354 (1940)

Einasto, J., Klypin, A., Saar, E.: Structure of superclusters and supercluster formation—V. Spatial correlation and voids. Mon. Not. R. Astron. Soc. **219**, 457 (1986)

Einasto, J., Einasto, M., Saar, E., et al.: Superclusters of galaxies from the 2dF redshift survey. II. Comparison with simulations. Astron. Astrophys. **462**, 397 (2007)

Einstein, A.: Die Feldgleichungen der Gravitation. Preuss. Akad. Wiss, Berlin (1915), Sitzber., 844

Einstein, A.: Die Grundlagen der allgemeinen Relativitätstheorie. Ann. Phys. **49**, 769 (1916)

Einstein, A.: Kosmologiche Betrachtungen zur allgemeinen Relativitätstheorie. Sitzungsber. Berl. Akad. **1**, 142 (1917)

Einstein, A.: Notiz zu E. Schrödingers Arbeit "Die Energiekomponenten des Gravitationsfeldes". Phys. Z. **19**, 115 (1918)

Einstein, A.: Geometry and experience. Lecture before the Preussian Academy of Sciences Jan. 27, 1927 (1921). Printed in Einstein: Ideas and Opinions (1973, Souvenir Press, London)

Einstein, A.: Bemergung zur Seletyschen Arbeit: Beiträge zum kosmologischen Problem. Ann. Phys., Ser. 4 **69**, 436 (1922)

Einstein, A., Grossmann, M.: Entwurf einer verallgemeinerten Relativitätstheorie und einer Theorie der Gravitation. Z. Math. Phys. **62**, 225 (1913)

Eisenhauer, F., Schödel, R., Gentzel, R., et al.: A geometric determination of the distance to the galactic center. Astrophys. J. **597**, L121 (2003)

Eisenstein, D.J., Hu, W.: Baryonic features in the matter transfer function. Astrophys. J. **496**, 605 (1998)

Eisenstein, D.J., Zehavi, I., Hogg, D.W.: Detection of the baryon acoustic peak in the large-scale correlation function of SDSS luminous red galaxies. Astrophys. J. **633**, 560 (2005)

Ekholm, T., Teerikorpi, P., Theureau, G., et al.: Kinematics of the local Universe. X. H_0 from the inverse B-band Tully-Fisher relation using diameter and magnitude limited samples. Astron. Astrophys. **347**, 99 (1999a)

Ekholm, T., Lanoix, P., Paturel, G., Teerikorpi, P., Fouqué, P.: Investigations of the local supercluster velocity field: II. A study using Tolman-Bondi solution and galaxies with accurate distances from the Cepheid PL-relation. Astron. Astrophys. **351**, 827 (1999b)

Ekholm, T., Baryshev, Yu., Teerikorpi, P., Hanski, M., Paturel, G.: On the quiescence of the Hubble flow in the vicinity of the local group: A study using galaxies with distances from the Cepheid PL-relation. Astron. Astrophys. **368**, L17 (2001)

Fabbri, R., Melchiorri, F., Natale, V.: The Sunyaev-Zeldovich effect in the millimetric region. Astrophys. Space Sci. **59**, 223 (1978)

Falcke, H., Melia, F., Agol, E.: Viewing the shadow of the black hole in the galactic center. Astrophys. J. **528**, L13 (2000)

Falconer, K.: Fractal Geometry. Wiley, New York (1990)

Fall, S.M.: Galaxy correlations and cosmology. Rev. Mod. Phys. **51**, 21 (1979)

Fall, S.M., Efstathiou, G.: Formation and rotation of disc galaxies with haloes. Mon. Not. R. Astron. Soc. **193**, 189 (1980)

Fang, L.L., Mo, H.J., Ruffini, R.: The cellular structure of the universe and cosmological tests. Astron. Astrophys. **243**, 283 (1991)

Feast, M.W.: A problem in distance-determination for Mira variables with an appendix on OB-star distances. Vistas Astron. **13**, 207 (1972)

Feast, M.W.: A diameter effect in the Tully–Fisher relation and its consequences. Mon. Not. R. Astron. Soc. **266**, 255 (1994)

Feng, J.L.: Dark matter candidates from particle physics and methods of detection. Annu. Rev. Astron. Astrophys. **48**, 495 (2010)

Ferguson, H.C., Dickinson, M., Giavalisco, M., et al.: The size evolution of high-redshift galaxies. Astrophys. J. **600**, L107 (2004)

Fesenko, B.I.: The galactic extinction from the apparent distribution of galaxies. Sov. Astron. **19**, 177 (1975)

Feynman, R.: Quantum theory of gravitation. Acta Phys. Pol. **XXIV**, 697 (1963)
Feynman, R.: Lectures on Gravitation. California Institute of Technology, Pasadena (1971)
Feynman, R., Morinigo, F., Wagner, W.: Feynman Lectures on Gravitation. Addison-Wesley, Reading (1995)
Fierz, M., Pauli, W.: On relativistic wave equations for particles of arbitrary spin in an electromagnetic field. Proc. R. Soc. Lond. A **173**, 211 (1939)
Fischer, K.B., Huchra, J.P., Strauss, M.A., Davis, M., Yahil, A., Schlegel, D.: The IRAS 1.2 Jy survey: Redshift data. Astrophys. J. Suppl. Ser. **100**, 69 (1995)
Fouqué, P., Bottinelli, L., Gouguenheim, L., Paturel, G.: The extragalactic distance scale. II. The unbiased distance to the Virgo cluster from the B-band Tully-Fisher relation. Astrophys. J. **349**, 1 (1990)
Fouqué, P., Arriagada, P., Storm, J., et al.: A new calibration of Galactic Cepheid period-luminosity relations from B to K bands, and a comparison to LMC relations. Astron. Astrophys. **476**, 73 (2007)
Fournier d'Albe, E.: Two New Worlds. Longmans Green, London (1907)
Fowler, W.: Massive stars, relativistic polytrops and gravitational radiation. Rev. Mod. Phys. **36**, 545 (1964)
Fowler, W.: The stability of supermassive stars. Astrophys. J. **144**, 180 (1966)
Francis, M.J., Barnes, L.A., James, J.B., Lewis, G.F.: Expanding space: The root of all evil? Publ. Astron. Soc. Aust. **24**, 95 (2007)
Freedman, W.L., Madore, B.F.: The Hubble constant. Annu. Rev. Astron. Astrophys. **48**, 673 (2010)
Freedman, W.L., Madore, B.F., Gibson, B.K., et al.: Final results from the Hubble Space Telescope Key Project to measure the Hubble constant. Astrophys. J. **553**, 47 (2001)
Friedmann, A.: On the curvature of space. Z. Phys. **10**, 377 (1922)
Friedmann, A.: The World as Space and Time. Academia, Petrograd (1923) (in Russian)
Friedmann, A.: On a possibility of world with constant negative curvature of space. Z. Phys. **21**, 326 (1924)
Gabrielli, A.: Notes on the average "bi-conditional" density in fractals (2005) (private communication)
Gabrielli, A., Sylos Labini, F.: Fluctuations in galaxy counts: A new test for homogeneity vs. fractality. Europhys. Lett. **54**, 286 (2001)
Gabrielli, A., Joyce, M., Sylos Labini, F.: Glass-like universe: Real-space correlation properties of standard cosmological models. Phys. Rev. D **65**(8), 083523 (2002)
Gabrielli, A., Sylos Labini, F., Durrer, R.: Biasing in Gaussian random fields and galaxy correlations. Astrophys. J. **531**, L1 (2004)
Gabrielli, A., Sylos Labini, F., Joyce, M., Pietronero, L.: Statistical Physics for Cosmic Structures. Springer, Berlin (2005)
Gamow, G.: Variability of elementary charge and quasistellar objects. Phys. Rev. Lett. **19**, 913 (1967)
Gamow, G., Ivanenko, D.D., Landau, L.D.: World constants and limiting transitions. J. Russ. Phys.-Chem. Soc. **60**, 13 (1928)
Gardner, J.P., Cowie, L.L., Wainscoat, R.J.: Galaxy number counts from K=10 to K=23. Astrophys. J. **415**, L9 (1993)
Gaztañaga, E.: Correlation between galaxies and quasi-stellar objects in the Sloan digital sky survey: A signal from gravitational lensing magnification? Astrophys. J. **589**, 82 (2003)
Gentile, G., Famaey, B., Zhao, H., Salucci, P.: Universality of galactic surface densities within one dark halo scale-length. Nature **461**, 672 (2009)
Gill, D.: The determination of the solar parallax. Part I. Observatory **1**, 7 (1877)
Giovanelli, R., Haynes, M.P., Herter, T., et al.: The I band Tully-Fisher relation for cluster galaxies: A template relation, its scatter and bias corrections. Astron. J. **113**, 53 (1997)
Godlowski, W., Bajan, K., Flin, P.: Weak redshift discretisation in the local group of galaxies? Astron. Nachr. **327**, 104 (2006)
Goldhaber, G., Groom, D.E., Kim, A., et al.: Timescale stretch parameterization of Type Ia supernova B-band light curves. Astrophys. J. **558**, 359 (2001)

Gott, J.R. III, Juric, M., Schlegel, D., et al.: A map of the universe. Astrophys. J. **624**, 463 (2005)
Gouguenheim, L.: Neutral hydrogen content of small galaxies. Astron. Astrophys. **3**, 281 (1969)
Gould, A.: Selection, covariance, and Malmquist bias. Astrophys. J. **412**, L55 (1993)
Governato, F., Moore, B., Cen, R., Stadel, J., Lake, G., Quinn, T.: The local group as a test of cosmological models. New Astron. **2**, 91 (1997)
Grene, M.: A Portrait of Aristotle. The University of Chigaco Press, Chicago (1963)
Grischuk, L.P., Petrov, A.N., Popova, A.D.: Exact theory of the (Einstein) gravitational field in an arbitrary background space-time. Commun. Math. Phys. **94**, 379 (1984)
Gromov, A., Baryshev, Yu., Suson, D., Teerikorpi, P.: Lemaître-Tolman-Bondi model: Fractality, non-simultaneus bang time and the Hubble law. Gravit. Cosmol. **7**, 140 (2001)
Gromov, A., Baryshev, Yu., Teerikorpi, P.: Two-fluid matter-quintessence FLRW models: Energy transfer and the equation of state of the universe. Astron. Astrophys. **415**, 813 (2004)
Gron, O., Elgaroy, O.: Is space expanding in the Friedmann universe models? Am. J. Phys. **75**, 151 (2007)
Gudmundsson, E.H., Björnsson, G.: Dark energy and the observable universe. Astrophys. J. **565**, 1 (2002)
Guth, A.H.: Inflationary universe: A possible solution to the horizon and flatness problems. Phys. Rev. D **23**, 347 (1981)
Guth, A.H.: The big bang and cosmic inflation. CTP Preprint 2144, 1-47 (1992)
Guthrie, B.N.G., Napier, W.M.: Redshift periodicity in the local supercluster. Astron. Astrophys. **310**, 353 (1996)
Gutiérrez, C.M., López-Corredoira, M.: The value of the fine structure constant over cosmological times. Astrophys. J. **713**, 46 (2010)
Haggerty, M.J.: A Newtonian model for the creation of a hierarchical cosmology. Astrophys. J. **166**, 257 (1971)
Haggerty, M.J., Wertz, J.R.: On the redshift–magnitude relation in hierarchical cosmologies. Mon. Not. R. Astron. Soc. **155**, 495 (1972)
Hamuy, M., Phillips, M.M., Maza, J., Suntzeff, N.B., Schommer, R.A., Aviles, R.: Hubble diagram of distant type IA supernovae. Astron. J. **109**, 1 (1995)
Hanski, M.O., Theureau, G., Ekholm, T., Teerikorpi, P.: Kinematics of the local universe IX. The Perseus-Pisces supercluster and the Tolman-Bondi model. Astron. Astrophys. **378**, 345 (2001)
Harrison, E.R.: Cosmology–The Science of the Universe, 1st edn. Cambridge University Press, Cambridge (1981), 2nd edition 2000
Harrison, E.R.: Darkness at Night. Harvard University Press, Cambridge (1987)
Harrison, E.R.: The redshift-distance and velocity-distance laws. Astrophys. J. **403**, 28 (1993)
Harrison, E.R.: Mining energy in an expanding universe. Astrophys. J. **446**, 63 (1995)
Harrison, E.R., Noonan, T.W.: Interpretation of extragalactic redshifts. Astrophys. J. **232**, 18 (1979)
Hartwick, F.D.A.: The velocity field around groups of galaxies. Astron. J. **141**, 198 (2011)
Haugan, M.P., Lämmerzahl, C.: Principles of equivalence: Their role in gravitation physics and experiments that test them. Lect. Notes Phys. **562**, 195 (2001)
Hawkins, G.S.: Expansion of the universe. Nature **194**, 563 (1962)
Hawkins, M.R.S.: On time dilation in quasar light curves. Mon. Not. R. Astron. Soc. **405**, 1940 (2010)
Hawkins, E., Maddox, S.J., Merrifield, M.R.: No periodicities in 2dF redshift survey data. Mon. Not. R. Astron. Soc. **336**, L13 (2002)
Hawkins, E., Maddox, S., Cole, S., et al.: The 2dF galaxy redshift survey: correlation functions, peculiar velocities and the matter density of the universe. Mon. Not. R. Astron. Soc. **346**, 78 (2003)
Haynes, M., Giovanelli, R.: Large-scale structure in the local universe: The Pisces-Perseus supercluster. In: Rubin, V.C., Coyne, G. (eds.) Large-Scale Motion in the Universe. Princeton University Press, Princeton (1988)
Heinämäki, P., Suhhonenko, I., Saar, E., Einasto, M., Einasto, J., Virtanen, H.: Light-cone simulations: Evolution of dark matter haloes (2005). astro-ph/0507197

Hendry, M.A., Simmons, J.F.L.: Optimal galaxy distance estimators. Astrophys. J. **435**, 515 (1994)

Hendry, M.A., Simmons, J.F.L.: Distance estimation in cosmology. Vistas Astron. **39**, 297 (1995)

Herrnstein, J.R., Moran, J.M., Greenhill, L.J., et al.: A geometric distance to the galaxy NGC4258 from orbital motions in a nuclear gas disk. Nature **400**, 539 (1999)

Hilbert, D.: Die Grundlagen der Physik. Göttingen Nachr. **4**, 21 (1917)

Hodge, P.W.: The extragalactic distance scale. Annu. Rev. Astron. Astrophys. **19**, 357 (1981)

Hoffman, Y., Martinez-Vaquero, L.A., Yepes, G., Gottlöber, S.: The local Hubble flow: Is it a manifestation of dark energy? Mon. Not. R. Astron. Soc. **386**, 390 (2008)

Hogg, D., Eisenstein, D., Blanton, M., et al.: Cosmic homogeneity demonstrated with luminous red galaxies. Astrophys. J. **624**, 54 (2005)

Holden, D.J.: An investigation of the clustering of radio sources. Mon. Not. R. Astron. Soc. **133**, 225 (1966)

Holmberg, E.: Distribution of clusters of galaxies as related to galactic absorption. Astron. Astrophys. **35**, 121 (1974)

Holtsmark, J.: Über die Verbreiterung von Spektrallinien. Ann. Phys. **58**, 577 (1919)

Horellou, C., Nord, M., Johansson, D., Lévy, A.: Probing the cosmic microwave background temperature using the Sunyaev-Zeldovich effect. Astron. Astrophys. **441**, 435 (2005)

Howell, A., Sullivan, M., Conley, A., Carlberg, R.: Predicted and observed evolution in the mean properties of Type Ia supernovae with redshift. Astrophys. J. **667**, L37 (2007)

Hoyle, F.: A new model for the expanding universe. Mon. Not. R. Astron. Soc. **108**, 372 (1948)

Hoyle, F.: The relation of radio astronomy to cosmology. In: Bracewell, R.N. (ed.) Radio Astronomy. IAU Symp., vol. 9, p. 529 (1959)

Hoyle, F.: The universe: Past and present reflections. Annu. Rev. Astron. Astrophys. **20**, 1 (1982)

Hoyle, F., Fowler, W.: Nucleosynthesis in supernovae. Astrophys. J. **132**, 565 (1960)

Hoyle, F., Fowler, W.: On the nature of strong radio sources. Mon. Not. R. Astron. Soc. **125**, 169 (1963)

Hoyle, F., Sandage, A.: The second-order term in the redshift-magnitude relation. Publ. Astron. Soc. Pac. **68**, 301 (1956)

Hoyle, F., Burbidge, G., Narlikar, J.: A Different Approach to Cosmology. Cambridge University Press, Cambridge (2000)

Hu, W., Dodelson, S.: Cosmic microwave background anisotropies. Annu. Rev. Astron. Astrophys. **40**, 171 (2002)

Hubble, E.: Extra-galactic nebulae. Astrophys. J. **64**, 321 (1926)

Hubble, E.: A relation between distance and radial velocity among extra-galactic nebulae. Proc. Natl. Acad. Sci. USA **15**, 168 (1929)

Hubble, E.: The distribution of extra-galactic nebulae. Astrophys. J. **79**, 8 (1934)

Hubble, E.: The Realm of the Nebulae. Yale University Press, New Haven (1936a)

Hubble, E.: The luminosity function of nebulae. I. The luminosity function of resolved nebulae as indicated by their brightest stars. Astrophys. J. **84**, 158 (1936b)

Hubble, E.: The Observational Approach to Cosmology. Clarendon, Oxford (1937) (68 pp)

Hubble, E., Humason, M.L.: The velocity–distance relation among extra-galactic nebulae. Astrophys. J. **74**, 43 (1931)

Hubble, E., Tolman, R.: Two methods of investigating the nature of the nebular red-shift. Astrophys. J. **82**, 302 (1935)

Huber, D., Pfenniger, D.: Long-range correlations in self-gravitating N-body systems. Astron. Astrophys. **386**, 359 (2002)

Huchra, J., Davis, M., Latham, D., Tonry, J.: A survey of galaxy redshifts. IV. The data. Astrophys. J. Suppl. Ser. **52**, 89 (1983)

Huchra, J., Geller, M., de Lapparent, V., Burg, R.: The center for astrophysics redshift survey. In: Audouze, J., Pelletan, M.-C., Szalay, A. (eds.) Large Scale Structures of the Universe. IAU Symposium, vol. 130, p. 105 (1988)

Hughes, D.W.: The introduction of absolute magnitude (1902–1929). J. Astron. Hist. Herit. **9**, 173 (2006)

Humason, M.L., Mayall, N.U., Sandage, A.: Redshifts and magnitudes of extragalactic nebulae. Astron. J. **61**, 97 (1956)

Impey, C., Bothun, G.: Low surface brightness galaxies. Annu. Rev. Astron. Astrophys. **35**, 267 (1997)

Ivanenko, D.D., Sokolov, A.: Quantum gravitation theory. Trans. Mosc. Univ. **8**, 103 (1947)

Jaakkola, T.: On the redshifts of galaxies. Nature **234**, 534 (1971)

Jaakkola, T.: Equilibrium cosmology. In: Arp, H.C., Keys, C.R., Rudnicki, K. (eds.) Progress in New Cosmologies: Beyond the Big Bang, p. 111. Plenum, New York (1993)

Jaakkola, T., Moles, M.: The type-redshift relation in the Vaucouleurs' groups of galaxies. Astron. Astrophys. **53**, 389 (1976)

Jaakkola, T., Moles, M., Vigier, J.P.: Empirical status in cosmology and the problem of the nature of the redshifts. Astron. Nachr. **300**, 229 (1979)

Jackson, N.: The Hubble constant. Living Rev. Relativ. **10**, 4 (2007). www.livingreviews.org/lrr-2007-4

Jackson, J.C., Dogdson, M.: Deceleration without dark matter. Mon. Not. R. Astron. Soc. **285**, 806 (1997)

Jackson, J.C., Jannetta, A.L.: Legacy data and cosmological constraints from the angular-size/redshift relation for ultra-compact radio sources. J. Cosmol. Astropart. Phys. **0611**, 002 (2006)

Järvenpää, E., Järvenpää, M., MacManus, P., O'Neil, T.C.: Visible parts and dimensions. Nonlinearity **16**, 803 (2003)

Jõeveer, M., Einasto, J.: Has the universe the cell structure? In: Longair, M., Einasto, J. (eds.) The Large-Scale Structure of the Universe. IAU Symposium, vol. 79, p. 241 (1978)

Johnson, M.: Knut Lundmark and Man's March into Space. A memorial volume. Värld och Vetande, Göteborg (1961), 134 pp

Jones, B., Martinez, V., Saar, E., Einasto, J.: Multifractal description of the large-scale structure of the universe. Astrophys. J. **332**, L1 (1988)

Jones, B., Coles, P., Martinez: Heterotopic clustering. Mon. Not. R. Astron. Soc. **259**, 146 (1992)

Jones, B., Martinez, V., Saar, E., Trimble, V.: Scaling laws in the distribution of galaxies. Rev. Mod. Phys. **76**, 1211 (2004)

Joyce, M., Sylos Labini, F.: Luminosity density estimation from redshift surveys and the mass density of the universe. Astrophys. J. **554**, L1 (2001)

Joyce, M., Montuori, M., Sylos Labini, F.: Fractal correlations in the CfA2-south redshift survey. Astrophys. J. **514**, L5 (1999)

Joyce, M., Anderson, P.W., Montuori, M., Pietronero, L., Sylos Labini, F.: Fractal cosmology in an open universe. Europhys. Lett. **50**, 416 (2000)

Joyce, M., Sylos Labini, F., Gabrielli, A., Montuori, M., Pietronero, L.: Basic properties of galaxy clustering in the light of recent results from the Sloan Digital Sky Survey. Astron. Astrophys. **443**, 11 (2005)

Kaiser, N.: On the spatial correlations of Abell clusters. Astrophys. J. **284**, L9 (1984)

Kalman, G.: Lagrangian formalism in relativistic dynamics. Phys. Rev. **123**, 384 (1961)

Kankare, E., Hanski, M., Theureau, G., Teerikorpi, P.: Revisiting the optical depth of spiral galaxies using the Tully-Fisher B relation. Astron. Astrophys. **493**, 23 (2009)

Kapahi, V.K.: The angular size-redshift relation as a cosmological tool. In: Observational Cosmology. IAU Symp, vol. 124, p. 251. Riedel, Dordrecht (1987)

Kapner, D.J., Cook, T.S., Adelberger, E.G., et al.: Tests of the gravitational inverse-square law below the dark-energy length scale. Phys. Rev. Lett. **98**(2), 021101 (2007)

Karachentsev, I.D.: Some statistical characteristics of superclusters of galaxies. Astrofizika **2**, 307 (1966) (English translation in Astrophysics **2**, 159)

Karachentsev, I.D.: Average statistical characteristics of systems of galaxies and the problem of the existence of hidden virial mass. Commun. Byurakan Obs. **39**, 76 (1968) (in Russian)

Karachentsev, I.D.: On the formulation of the Copernican cosmological principle. Acta Cosmol. **2**, 43 (1974)

Karachentsev, I., Kopylov, A.: A cut of large-scale structure across the Coma supercluster. In: Audouze, J., Pelletan, M.-C., Szalay, A. (eds.) Large Scale Structures of the Universe. IAU Symposium, vol. 130, p. 139 (1988)

Karachentsev, I.D., Lipovetskii, V.A.: Absorbing material in clusters of galaxies. Sov. Astron. **12**, 909 (1969)

Karachentsev, I., Makarov, D.: The galaxy motion relative to nearby galaxies and the local velocity field. Astron. J. **111**, 794 (1996)

Karachentsev, I., Makarov, D.I., Sharina, M.E., et al.: Local galaxy flows within 5 Mpc. Astron. Astrophys. **398**, 493 (2003)

Karachentsev, I.D., Tully, R.B., Dolphin, A., et al.: The Hubble flow around the Centaurus A/M83 galaxy complex. Astron. J. **133**, 504 (2007)

Karachentsev, I.D., Kashibadze, O.G., Makarov, D.I., Tully, R.B.: The Hubble flow around the local group. Mon. Not. R. Astron. Soc. **393**, 1265 (2009)

Karlsson, K.G.: On the existence of significant peaks in the quasar redshift distribution. Astrophys. J. **58**, 237 (1977)

Karttunen, H., Kröger, P., Oja, H., Poutanen, M., Donner, K.J. (eds.): Fundamental Astronomy. Springer, Berlin (2006)

Kashlinsky, A., Atrio-Barandela, F., Kocevski, D., Ebeling, H.: A measurement of large-scale peculiar velocities of clusters of galaxies: Results and cosmological implications. Astrophys. J. **686**, L49 (2008)

Kaspi, S., Smith, P.S., Netzer, H., Maoz, D., Jannuzi, B.T., Giveon, U.: Reverberation measurements for 17 quasars and the size-mass-luminosity relations in activc galactic nuclei. Astrophys. J. **533**, 631 (2000)

Kawahara, H., Kitayama, T., Sasaki, S., Suto, Y.: Systematic error in the Hubble constant measurement from the Sunyaev-Zeldovich effect. Astrophys. J. **674**, 11 (2008)

Kellermann, K.I.: The cosmological deceleration parameter estimated from the angular-size/redshift relation for compact radio sources. Nature **361**, 134 (1993)

Kerscher, M., Szapudi, I., Szalay, A.S.: A comparison of estimators for the two-point correlation function. Astrophys. J. **535**, L13 (2000)

Kiang, T.: On the clustering of rich clusters of galaxies. Mon. Not. R. Astron. Soc. **135**, 1 (1967)

Kiang, T.: Time, distance, velocity, redshift: A personal guided tour (2003). astro-ph/0308010

Kiang, T., Saslaw, W.C.: The distribution in space of clusters of galaxies. Mon. Not. R. Astron. Soc. **143**, 129 (1969)

Kim, J., Park, Ch., Gott, J.R., Dubinski, J.: The horizon run N-body simulation: Baryon acoustic oscillations and topology of large-scale structure of the universe. Astrophys. J. **701**, 1547 (2009)

King, I.R.: The structure of star clusters. III. Some simple dynamical models. Astron. J. **71**, 64 (1966)

King, J.A., Webb, J.K., Murphy, M.T., Carswell, R.F.: Stringent null constraint on cosmological evolution of the proton-to-electron mass ratio. Phys. Rev. Lett. **101**, 251304 (2008)

Klypin, A.A., Kopylov, A.I.: The spatial covariance function for rich clusters of galaxies. Sov. Astron. Lett. **9**, 41 (1983)

Klypin, A., Einasto, J., Einasto, M., Saar, E.: Structure and formation of superclusters—X: Fractal properties of superclusters. Mon. Not. R. Astron. Soc. **237**, 929 (1989)

Klypin, A., Kravtsov, A.V., Valenzuela, O., Prada, F.: Where are the missing galactic satellites? Astrophys. J. **522**, 82 (1999)

Klypin, A., Hoffman, Y., Kravtsov, A.V., Gottlöber, S.: Constrained simulations of the real universe: The local supercluster. Astrophys. J. **596**, 19 (2003)

Komberg, B.: A binary system as a quasar model. Sov. Astron. **11**, 727 (1968)

Kopylov, A., Kuznetsov, D., Fetisova, T., Shvartsman, V.: Possible inhomogeneities in the Universe on scales of 100–300 Mpc from observations with the 6-meter telescope. In: Audouze, J., Pelletan, M.-C., Szalay, A. (eds.) Large Scale Structures of the Universe. IAU Symp., vol. 130, p. 129 (1988)

Koranyi, D.M., Strauss, M.A.: Testing the Hubble law with the IRAS 1.2 Jy redshift survey. Astrophys. J. **477**, 36 (1997)

Koratkar, A.P., Gaskell, C.M.: Radius-luminosity and mass-luminosity relationships for active galactic nuclei. Astrophys. J. **370**, L61 (1991)

Kowalski, M., Rubin, D., Aldering, G., et al.: Improved cosmological constraints from new, old, and combined supernova data sets. Astrophys. J. **686**, 749 (2008)

Kraan-Korteweg, R.C., Tammann, G.: A catalogue of galaxies within 10 Mpc. Astron. Nachr. **300**, 181 (1979)

Kraan-Korteweg, R.C., Cameron, L.M., Tammann, G.A.: 21 centimeter line width distances of cluster galaxies and the value of H_0. Astrophys. J. **331**, 620 (1988)

Kron, R.G.: Evolution in the galaxy population. In: Binggeli, B., Buser, R. (eds.) The Deep Universe, pp. 233–316. Springer, Berlin (1995)

Lahav, O., Lilje, P.B., Primack, J.R., Rees, M.J.: Dynamical effects of the cosmological constant. Mon. Not. R. Astron. Soc. **251**, 128 (1991)

Landau, L.D., Lifshitz, E.M.: Statistical Physics. Pergamon, London (1958)

Landau, L.D., Lifshitz, E.M.: The Classical Theory of Fields. Pergamon, Oxford (1971)

Landy, S.D., Szalay, A.S.: A general analytical solution to the problem of Malmquist bias due to lognormal distance errors. Astrophys. J. **391**, 494 (1992)

Lanoix, P., Paturel, G., Garnier, R.: Bias in the Cepheid period-luminosity relation. Astrophys. J. **517**, 188 (1999)

LaViolette, P.A.: Is the universe really expanding? Astrophys. J. **301**, 544 (1986)

Leavitt, H.S., Pickering, E.C.: Periods of 25 variable stars in the small Magellanic cloud. Harvard Coll. Obs. Circ. **173**, 1 (1912)

Lebedev, V., Lebedeva, I.: The spatial correlation functions of superclusters of galaxies. Sov. Atron. Lett. **14**, 1822 (1988)

Lee, M.G., Freedman, W.L., Madore, B.F.: The tip of the red giant branch as a distance indicator for resolved galaxies. Astrophys. J. **417**, 553 (1993)

Lehti, R.: Time in relativity theory and cosmology. In: Pihlström, S., Siitonen, A., Vilkko, R. (eds.) Aika, p. 82 (2000) (in Finnish)

Lehti, R.: The concept of space in relativity theory and cosmology. In: Martikainen, E. (ed.) Infinity, Causality and Determinism. Peter Lang, Oxford (2002)

Lehto, A.: Periodic time and the stationary properties of matter. Chin. J. Phys. **28**, 215 (1990)

Lehto, A.: 3-D period doubling and magnetic moments of particles. Astrophys. Space Sci. **244**, 321 (1996)

Lehto, A.: On the Planck scale and properties of matter. Nonlinear Dyn. **55**, 279 (2009)

Leibundgut, B.: Cosmological implications from observations of Type Ia supernovae. Annu. Rev. Astron. Astrophys. **39**, 67 (2001)

Lemaître, G.: Un Univers homogène de masse constante et de rayon croissant rendant compte de la vitesse radiale des nébuleuses extra-galactiques. Ann. Soc. Sci. Brux. **47**, 49 (1927)

Lemaître, G.: A homogeneous universe of constant mass and increasing radius accounting for the radial velocity of extra-galactic nebulae. Mon. Not. R. Astron. Soc. **91**, 483 (1931) (translation from the original 1927 paper)

Lemson, G., Sanders, R.H.: On the use of the conditional density as a description of galaxy clustering. Mon. Not. R. Astron. Soc. **252**, 319 (1991)

Lerner, E.: Magnetic vortex filaments, universal invariants, and the fundamental constants. IEEE Trans. Plasma Sci.; Special Issue on Cosmic Plasma **PS-14**, 690 (1986)

Levshakov, S.A., Molaro, P., Lopez, S., et al.: A new measure of $\Delta\alpha/\alpha$ at redshift $z = 1.84$ from very high resolution spectra of Q1101-264. Astron. Astrophys. **466**, 1077 (2007)

Lilly, S.J., Cowie, L.L., Gardner, J.P.: A deep imaging and spectroscopic survey of faint galaxies. Astrophys. J. **369**, 79 (1991)

Limber, D.N.: The analysis of counts of the extragalactic nebulae in terms of a fluctuating density field. Astrophys. J. **117**, 134 (1953)

Liske, J., Grazian, A., Vanzella, E., et al.: Cosmic dynamics in the era of extremely large telescopes. Mon. Not. R. Astron. Soc. **386**, 1192 (2008)

Loeb, A.: Direct measurement of cosmological parameters from the cosmic deceleration of extragalactic objects. Astrophys. J. **499**, L111 (1998)

Logunov, A.A., Folomeshkin, V.N.: Problem of energy-momentum and gravity theory. Theor. Math. Phys. **32**, 291 (1977)

Logunov, A.A., Mestvirishvili, M.A.: The Relativistic Theory of Gravitation. Mir, Moscow (1989)

Longair, M.S.: Personal view—the large scale structure of the universe. In: Longair, M., Einasto, J. (eds.) The Large-Scale Structure of the Universe. IAU Symposium, vol. 79, p. 451 (1978)

López-Corredoira, M.: Observational cosmology: Caveats and open questions in the standard model. Recent Res. Dev. Astron. Astrophys. **1**, 561 (2003). astro-ph/0310214

López-Corredoira, M.: Angular size test on the expansion of the universe. Int. J. Mod. Phys. D **19**(3), 245 (2010). arXiv:1002.0525

López-Corredoira, M., Gutiérrez, C.M.: Research on candidates for non-cosmological redshifts. AIP Conf. Proc. **822**, 75 (2006). astro-ph/0509630

LoSecco, J., Mathews, G., Wang, Y.: Prospects for constraining cosmology with the extragalactic cosmic microwave background temperature. Phys. Rev. D **64**, 123002 (2001)

Lovyagin, N.Yu.: Statistical properties of the spatial distribution of galaxies. Astron .Bull. SAO RAN **64**, 213 (2009)

Lubin, L., Sandage, A.: The Tolman surface brightness test for the reality of the expansion. IV. Astron. J. **122**, 1084 (2001)

Luminet, J.-P., Weeks, J.R., Riazuelo, A., Lehoucq, R., Uzan, J.-P.: Dodecahedral space topology as an explanation for weak wide-angle temperature correlations in the cosmic microwave background. Nature **425**, 593 (2003)

Lundmark, K.: Die Stellung der kugelförmigen Sternhaufen und Spiralnebel zu unserem Sternsystem. Astron. Nachr. **209**, 369 (1919)

Lundmark, K.: The Relations of the Globular Clusters and Spiral Nebulae to the Stellar System (1920) (Kungl. Svenska Vetenskapsakademiens Handlingar, vol. 60, No. 8, Stockholm, 1920)

Lundmark, K.: The motions and the distances of spiral nebulae. Mon. Not. R. Astron. Soc. **85**, 865 (1925)

Lundmark, K.: A preliminary classification of nebulae. Ark. Mat. Astron. Fys. **19**(8) (1926)

Lundmark, K.: Studies of anagalactic nebulae. Medd. f. Astron. Obs. Upsala **30** (1927)

Lundmark, K.: The distance indicators of astronomy. In: Dixième congrès des mathèmaticiens Scandinaves, Copenhague, 1946. Medd. Lunds Obs. Ser. I, No. 163 (1946)

Luo, X., Schramm, D.N.: Fractals and cosmological large-scale structure. Science **256**, 513 (1992)

Lutz, T.E., Kelker, D.H.: On the use of trigonometric parallaxes for the calibration of luminosity systems: Theory. Publ. Astron. Soc. Pac. **85**, 573 (1973)

Luzzi, G., Shimon, M., Lamagna, L., et al.: Redshift dependence of the cosmic microwave background temperature from Sunyaev-Zeldovich measurements. Astrophys. J. **705**, 1122 (2009)

Lynden-Bell, D.: The dynamical age of the local group of galaxies. Observatory **101**, 111 (1981)

Lynden-Bell, D., Faber, S.M., Burstein, D., et al.: Spectroscopy and photometry of elliptical galaxies. V—Galaxy streaming toward the new supergalactic center. Astrophys. J. **326**, 19 (1988)

Macciò, A.V., Governato, F., Horellou, C.: The signature of dark energy on the local Hubble flow. Mon. Not. R. Astron. Soc. **359**, 941 (2005)

Maddox, J.: The universe as a fractal structure. Nature **329**, 195 (1987)

Mädler, J.H.v.: Geschichte der Himmelskunde, vol. 2. Druck und Verlag von George Westermann, Brauschweig (1873)

Madore, B.F., Freedman, W.L.: The Cepheid distance scale. Publ. Astron. Soc. Pac. **103**, 933 (1991)

Maeda, K., Benetti, S., Stritzinger, M., et al.: An asymmetric explosion as the origin of spectral evolution diversity in type Ia supernovae. Nature **466**, 82 (2010)

Magueijo, J., Barrow, J., Sandvik, H.B.: Is it e or is it c? Experimental tests of varying α. Phys. Lett. B **549**, 284 (2002)

Malec, A.L., Buning, R., Murphy, M.T., et al.: Keck telescope constraints on cosmological variation of the proton-to-electron mass ratio. Mon. Not. R. Astron. Soc. **403**, 1541 (2010)

Malmquist, K.G.: A study of the stars of spectral type A. Lund Medd. Ser. II, No. 22 (1920)

Malmquist, K.G.: On some relations in stellar statistics. Lund Medd. Ser. I, No. 100 (1922)

Mandelbrot, B.B.: Les objects fractals: forme, hasard et dimension. Flammarion, Paris (1975)

Mandelbrot, B.B.: Fractals: Form, Chance and Dimension. W.H. Freeman, New York (1977)

Mandelbrot, B.B.: The Fractal Geometry of Nature. W.H. Freeman, New York (1982)

Mandelbrot, B.B.: The fractal range of the distribution of galaxies; crossover to homogeneity, and multifractals. In: Mezzetti, M., et al. (eds.) Large Scale Structure and Motions in the Universe, p. 259. Kluwer Academic, Dordrecht (1989)
Marscher, A., Jorstad, S.G., Larionov, V.M., Aller, M.F., Lähteenmäki, A.: Multi-frequency observations of gamma-ray blazar 1633+382. J. Astrophys. Astron. **41J** (2011)
Martinez, V.J., Jones, B.J.T.: Why the universe is not a fractal. Mon. Not. R. Astron. Soc. **242**, 517 (1990)
Martinez, V.J., Saar, E.: Statistics of the Galaxy Distribution. Chapman & Hall/CRC, New York (2002)
Martinez, V.J., Arnalte-Mur, P., Saar, E., et al.: Reliability of the detection of the baryon acoustic peak. Astrophys. J. Lett. **696**, L93 (2009)
Martinez-Vaquero, L.A., Yepes, G., Hoffman, Y., Gottlöber, S., Sivan, M.: Constrained simulations of the local universe—II. The nature of the local Hubble flow. Mon. Not. R. Astron. Soc. **397**, 2070 (2009)
Massey, R., Rhodes, J., Ellis, R., et al.: Dark matter maps reveal cosmic scaffolding. Nature **445**, 286 (2007a)
Mather, J.C., Fixsen, D.J., Shafer, R.A., Mosier, C., Wilkinson, D.T.: Calibrator design for the COBE far infrared absolute spectrophotometer (FIRAS). Astrophys. J. **512**, 511 (1999)
Mattig, W.: Über den Zusammenhang zwischen Rotverschiebung und scheinbarer Helligkeit. Astron. Nachr. **284**, 109 (1958)
Mattila, K.: Photometry of the intergalactic optical surface brightness in the Coma cluster. Astron. Astrophys. **60**, 425 (1977)
Mattila, K.: Observations of the extragalactic background light. In: The Galactic and Extragalactic Background Radiation, p. 257. Kluwer Academic, Dordrecht (1990)
Mattila, K.: Has the optical extragalactic background light been detected? Astrophys. J. **591**, 119 (2003)
Mattsson, T.: Dark energy as a mirage. Gen. Relativ. Gravit. **42**, 567 (2010)
Mazur, P., Mottola, E.: Gravitational vacuum condensate stars. Proc. Natl. Acad. Sci. USA **111**, 9545 (2004)
McCrea, W., Milne, E.: Newtonian universes and the curvature of space. Q. J. Math. **5**, 73 (1934)
McGaugh, S.: A possible local counterpart to the excess population of faint galaxies. Nature **367**, 538 (1994)
McKellar, A.: Molecular lines from the lowest states of diatomic molecules composed of atoms probably present in interstellar space. Publ. Dom. Astrophys. Obs. **7**, 251 (1941)
McVittie, G.C.: Distance and large redshifts. Q. J. R. Astron. Soc. **15**, 246 (1974)
Ménard, B., Kilbinger, M., Scranton, R.: On the impact of intergalactic dust on cosmology with Type Ia supernovae. Mon. Not. R. Astron. Soc. **406**, 1815 (2010)
Menger, K.: Mensuration and mathematical connections of observable material. In: Measurement, Definitions, and Theories, New York (1959)
Milgrom, M.: A modification of the Newtonian dynamics as a possible alternative to the hidden mass hypothesis. Astrophys. J. **270**, 365 (1983)
Milne, E.A.: A Newtonian expanding universe. Q. J. Math. **5**, 64 (1934)
Milne, E.A.: Relativity, Gravitation and World Structure. Clarendon, Oxford (1935)
Minchin, R.F., Disney, M.J., Parker, Q.A., et al.: The cosmological significance of low surface brightness galaxies found in a deep blind neutral hydrogen survey. Mon. Not. R. Astron. Soc. **355**, 1303 (2004)
Misner, C., Thorne, K., Wheeler, J.: Gravitation. Freeman, San Francisco (1973)
Mitra, A.: Non-occurence of trapped surfaces and black holes in spherical gravitational collapse. Found. Phys. Lett. **13**, 543 (2000)
Mitra, A.: On the final state of spherical gravitational collapse. Found. Phys. Lett. **15**, 439 (2002)
Mitra, A.: Radiation pressure supported stars in Einstein gravity: Eternally collapsing objects. Mon. Not. R. Astron. Soc. **369**, 492 (2006)
Miyoshi, M., Moran, J., Herrnstein, J., et al.: Evidence for a black hole from high rotation velocities in a sub-parsec region of NGC4258. Nature **373**, 127 (1995)

Molaro, P., Levshakov, S., et al.: The cosmic microwave background radiation temperature at $z = 3.025$ toward QSO 0347-3819. Astron. Astrophys. **381**, L64 (2002)
Moles, M., Campos, A., Kjaergaard, P., Fasano, G., Bettoni, D.: On the use of scaling relations for the Tolman test. Astrophys. J. **495**, L31 (1998)
Moles, M., Benitez, N., Aguerri, J.A.L., et al.: The Alhambra survey: A large area multimedium-band optical and near-infrared photometric survey. Astrophys. J. **136**, 1325 (2008)
Montuori, M., Sylos Labini, F.: Angular correlations of galaxy distribution. Astrophys. J. **487**, L21 (1997)
Moore, B., Ghigna, S., Governato, F.: Dark matter substructure within galactic halos. Astrophys. J. **524**, L19 (1999)
Moore, B., Calcaneo-Roldan, C., Stadel, J., et al.: Dark matter in Draco and the local group: Implications for direct detection experiments. Phys. Rev. D **64**, 063508 (2001)
Moshinsky, M.: On the interacting Birkhoff's gravitational field with the electromagnetic and pair fields. Phys. Rev. **80**, 514 (1950)
Multamäki, T., Vilja, I.: Constraining Newtonian stellar configurations in f(R) theories of gravity. Phys. Lett. B **659**, 843 (2008)
Multamäki, T., Putaja, A., Vagenas, E., Vilja, I.: Energy-momentum complexes in f(R) theories of gravity. Class. Quantum Gravity **25**, 075017 (2008)
Mureika, J.R.: Fractal holography: A geometric re-interpretation of cosmological large scale structure. J. Cosmol. Astropart. Phys. **5**, 21 (2007)
Murphy, M.T., Webb, J.K., Flambaum, V.V.: Possible evidence for a variable fine-structure constant from QSO absorption lines: motivations, analysis and results. Mon. Not. R. Astron. Soc. **327**, 1208 (2001)
Nabokov, N.V., Baryshev, Yu.V.: Classical cosmological tests for galaxies of the Hubble ultra deep field. Astrophys. Bull. **63**, 244 (2008a)
Nabokov, N.V., Baryshev, Yu.V.: A search for super-large structures in deep galaxy surveys. In: Baryshev, Yu.V., Taganov, I.N., Teerikorpi, P. (eds.) Problems of Practical Cosmology. Proceedings of the International Conference held at Russian Geographical Society, 23–27 June, 2008, St. Petersburg, vol. 1, p. 69. TIN, St.Petersburg (2008b)
Nabokov, N.V., Baryshev, Yu.V.: Method for analyzing the spatial distribution of galaxies on gigaparsec scales. I. Initial principles. Astrophysics **53**, 91 (2010a)
Nabokov, N.V., Baryshev, Yu.V.: Method for analyzing the spatial distribution of galaxies on gigaparsec scales. II. Application to a grid of the HUDF-FDF-COSMOS-HDF surveys. Astrophysics **53**, 101 (2010b)
Nagirner, D.: 2006 (private communication)
Napier, W.M.: A statistical evaluation of anomalous redshift claims. Astrophys. Space Sci. **285**, 419 (2003)
Narayan, R., Quataert, E.: Black hole accretion. Science **307**, 77 (2005)
Narayan, R., Garcia, M.R., McClintock, J.E.: Advection-dominated accretion and black hole event horizons. Astrophys. J. **478**, L79 (1997)
Narlikar, J.: Introduction to Cosmology, 2nd edn. Cambridge University Press, Cambridge (1993)
Nesvizhevsky, V.V., Protasov, K.V.: Constrains on non-Newtonian gravity from the experiment on neutron quantum states in the Earth's gravitational field. Class. Quantum Gravity **21**, 4557 (2004)
Nesvizhevsky, V.V., Borner, H.G., Petukhov, A.K., et al.: Quantum states of neutrons in the Earth's gravitational field. Nature **415**, 297 (2002)
Neumann, C.: Allgemeine Untersuchungen über das Newtonsche Prinzip der Fernwirkungen. Teubner, Leipzig (1896)
Neyman, J., Scott, E.L.: A theory of the spatial distribution of galaxies. Astrophys. J. **116**, 144 (1952)
Nicoll, J.F., Segal, I.E.: Correction of a criticism of the phenomenological quadratic redshift-distance law. Astrophys. J. **258**, 457 (1982)
Niemi, S.-M., Valtonen, M.: The origin of redshift asymmetries: How ΛCDM explains anomalous redshift. Astron. Astrophys. **494**, 857 (2009)

Nikiforov, I.I.: The distance to the center of the galaxy: The current state-of-the-art in measuring R_0. In: Byrd, G.G., Kholshevnikov, K.V., Mylläri, A.A., Nikiforov, I.I., Orlov, V.V. (eds.) Order and Chaos in Stellar and Planetary Systems, San Francisco, CA. ASP Conference Series, vol. 316, p. 199 (2004)

Nilsson, K., Valtonen, M., Kotilainen, J., Jaakkola, T.: On the redshift–apparent size diagram of double radio sources. Astrophys. J. **413**, 453 (1993)

Noether, E.: Invariante Variationsprobleme. In: Nachr. v.d. Ges. d. Wiss. zu Göttingen, p. 235 (1918)

Norberg, P., Baugh, C.M., Hawkins, E., et al.: The 2dF galaxy redshift survey: Luminosity dependence of galaxy clustering. Mon. Not. R. Astron. Soc. **328**, 64 (2001)

Norberg, P., Baugh, C.M., Hawkins, E., et al.: The 2dF galaxy redshift survey: The dependence of galaxy clustering on luminosity and spectral type. Mon. Not. R. Astron. Soc. **332**, 828 (2002)

North, J.D.: The Measure of the Universe. Clarendon, Oxford (1965)

Ogievetsky, V.I., Polubarinov, I.V.: Interacting field of spin 2 and the Einstein equations. Ann. Phys. **35**, 167 (1965)

Okun, L.B.: Fundamental constants of physics. Usp. Phys. Nauk **161**(9), 177 (1991)

Okun, L.B., Selivanov, K.G., Telegdi, V.L.: On the interpretation of the redshift in a static gravitational field. Am. J. Phys. **68**, 15 (2000)

Olson, D.W., Silk, J.: Primordial inhomogeneities in the expanding universe. II—General features of spherical models at late times. Astrophys. J. **233**, 395 (1979)

Öpik, E.: An estimate of the distance of the Andromeda Nebula. Astrophys. J. **55**, 406 (1922)

Oschepkov, S.A., Raikov, A.A.: Post-Newtonian polytropes in alternative gravity theories. Gravitation **1**, 44 (1995)

Ostriker, J.P., Cowie, L.L.: Galaxy formation in an intergalactic medium dominated by explosions. Astrophys. J. **243**, L127 (1981)

Paczynski, B.: Gamma-ray burst—supernova relation. In: Livio, M., Panagia, N., Sahu, K. (eds.) Proc. of the STSI 1999 May Symposium (13): "Supernovae and Gamma Ray Bursts; The Largest Explosions Since the Big Bang", p. 1. Cambridge University Press, Cambridge (2001)

Padmanabhan, T.: Understanding our universe: Current status and open issues (2005). gr-qc/0503107

Padmanabhan, T.: From gravitons to gravity: myth and reality. Int. J. Mod. Phys. D **17**, 367 (2008). gr-qc/0409089

Padmanabhan, N., Schlegel, D.J., Seljak, U.: The clustering of luminous red galaxies in the Sloan Digital Sky Survey imaging data. Mon. Not. R. Astron. Soc. **378**, 852 (2007)

Padovani, P., Urry, C.M.: Luminosity functions, relativistic beaming and unified theories of high-luminosity radio sources. Astrophys. J. **387**, 449 (1992)

Paladin, G., Vulpiani, A.: Anomalous scaling laws in multifractal objects. Phys. Rep. **156**, 147 (1987)

Park, C., Vogeley, M.S., Geller, M.J., Huchra, J.P.: Power spectrum, correlation function, and tests for luminosity bias in the CfA redshift survey. Astrophys. J. **431**, 569 (1994)

Paturel, G.: Direct determination of extragalactic distances from the 21 centimeter line width and the method of "sosies". Astrophys. J. **282**, 382 (1984)

Paturel, G., Baryshev, Yu.V.: Prediction of the sidereal time distribution of gravitational wave events for different detectors. Astrophys. J. **592**, L99 (2003a)

Paturel, G., Baryshev, Yu.V.: Sidereal time analysis as a tool for the study of the space distribution of sources of gravitational waves. Astron. Astrophys. **398**, 377 (2003b)

Paturel, G., Teerikorpi, P.: The extragalactic Cepheid distance bias: Numerical simulations. Astron. Astrophys. **413**, L31 (2004)

Paturel, G., Teerikorpi, P.: The extragalactic Cepheid bias: Significant influence on the cosmic distance scale. Astron. Astrophys. **443**, 883 (2005)

Paturel, G., Teerikorpi, P.: The extragalactic Cepheid bias: A new test using the period-luminosity-color relation. Astron. Astrophys. **425**, 423 (2006)

Paturel, G., Bottinelli, L., Gouguenheim, L., Fouqué, P.: New determination of the pole of a "hypergalactic" large scale system. Astron. Astrophys. **189**, 1 (1988)

Paturel, G., Bottinelli, L., Di Nella, H., Gouguenheim, L., Teerikorpi, P.: Kinematics of the local universe: Completeness of the sample. Astron. Astrophys. **289**, 711 (1994)

Paturel, G., Lanoix, P., Garnier, R., et al.: Distances for 17 nearby galaxies based on the HIPPARCOS calibration of the P-L relation. In: Proc. of the ESA Symposium 'Hipparcos—Venice 97', July 1997. ESA SP, vol. 402, p. 629 (1997a)

Paturel, G., Bottinelli, L., Di Nella, H., et al.: The extragalactic data base: VII—Reduction of astrophysical parameters. Astron. Astrophys. Suppl. Ser. **124**, 109 (1997b)

Paturel, G., Teerikorpi, P., Theureau, G., Fouqué, P., Musella, I., Terry, J.N.: Calibration of the distance scale from galactic Cepheids II. Use of the HIPPARCOS calibration. Astron. Astrophys. **389**, 19 (2002)

Peacock, J.A.: Cosmological Physics. Cambridge Univ. Press, Cambridge (1999)

Peebles, P.J.E.: Significance of the first brightest galaxies in rich clusters. Astrophys. J. **153**, 13 (1968)

Peebles, P.J.E.: The gravitational instability picture and the nature of the distribution of galaxies. Astrophys. J. **189**, L51 (1974)

Peebles, P.J.E.: Large scale clustering in the Universe. In: Longair, M.S., Einasto, J. (eds.) The Large Scale Structure of the universe. Proc. of IAU Symp., vol. 79, p. 217. Reidel, Dordrecht (1978)

Peebles, P.J.E.: The Large-Scale Structure of the Universe. Princeton Univ. Press, Princeton (1980)

Peebles, P.J.E.: Summary. In: Audouze, J., Pelletan, M.-C., Szalay, A. (eds.) Large Scale Structures of the Universe. IAU Symposium, vol. 130, p. 495 (1988)

Peebles, P.J.E.: Principles of Physical Cosmology. Princeton Univ. Press, Princeton (1993)

Peebles, P.J.E.: From precision cosmology to accurate cosmology (2002). astro-ph/0208037

Peebles, P.J.E., Ratra, B.: The cosmological constant and dark energy. Rev. Mod. Phys. **75**, 599 (2003)

Peirani, S., de Freitas Pacheco, J.A.: Dynamics of nearby groups of galaxies: The role of the cosmological constant. Astron. Astrophys. **488**, 845 (2008)

Penzias, A., Wilson, R.: A measurement of excess antenna temperature at 4080 Mc/s. Astrophys. J. **142**, 419 (1965)

Penzias, A.A., Schraml, J., Wilson, R.W.: Observational constraints on a discrete-source model to explain the microwave background. Astrophys. J. Lett. **157**, L49 (1969)

Perdang, J.: Self-gravitational fractal configuration. Vistas Astron. **33**, 371 (1990)

Perlmutter, S., Aldering, G., Goldhaber, G., et al.: Measurements of Ω and Λ from 42 high–redshift supernovae. Astrophys. J. **517**, 565 (1999)

Petrosian, V.: Surface brightness and evolution of galaxies. Astrophys. J. **210**, L53 (1976)

Petrovskaya, I.V.: The distribution function of the force acting on a star for powerful interaction. Pisma Astron. Zh. **12**, 562 (1986)

Pfleiderer, J.: Data-set description and bias. In: Statistical Methods in Astronomy. ESA SP, vol. 201 (1983)

Pietronero, L.: The fractal structure of the Universe: correlations of galaxies and clusters and the average mass density. Physica A **144**, 257 (1987)

Pietronero, L., Kuper, R.: Stochastic approach to large scale clustering of matter in the universe. In: Pietronero, L., Tosatti, E. (eds.) Fractals in Physics, p. 319. Elsevier, Amsterdam (1986)

Pietronero, L., Sylos Labini, F.: Cosmological principle and the debate about large-scale structures distribution. Lect. Notes Phys. **455**, 17 (1995)

Pietronero, L., Montuori, M., Sylos Labini, F.: On the fractal structure of the visible universe. In: Turok, N. (ed.) Proc. of the Conference "Critical Dialogues in Cosmology", p. 24. World Scientific, Singapore (1997)

Planck, M.: Über irreversible Strahlungsvorgänge. Sitz. Preuss.-Akad. Wiss. **5**, 479 (1899)

Poggianti, B.: K and evolutionary corrections from UV to IR. Astron. Astrophys. Suppl. Ser. **122**, 399 (1997)

Poincaré, H.: La science et l'hypothèse (1902) (Flammarion, Paris, 1968)

Poincaré, H.: Sur la dynamique de l'électron. C. R. Acad. Sci. **140**, 1504 (1905)

Poincaré, H.: Sur la dynamique de l'électron. Rend. Circ. Mat. Palermo **21**, 129 (1906)

Pskovskii, Yu.P.: Light curves, color curves, and expansion velocity of type I supernovae as functions of the rate of brightness decline. Sov. Astron. **21**, 675 (1977)

Pynzar, A.V.: Correlation between the scattering parameters for pulsars and the emission measure of the galactic background. Astron. Rep. **39**, 406 (1995)

Ragazzoni, R., Turatto, M., Gaessler, W.: Lack of observational evidence for quantum structure of space-time at Planck scales. Astrophys. J. **587**, L1 (2003)

Raikov, A.A., Orlov, V.V., Beketov, O.B.: Inhomogeneities in the spatial distribution of gamma-ray bursts. Astrophysics **53**, 396 (2010)

Rauzy, S., Triay, R.: The Tully-Fisher relation: Correspondence between the inverse and direct approaches. Astron. Astrophys. **307**, 726 (1996)

Refsdal, S.: On the possibility of determining Hubble's parameter and the masses of galaxies from the gravitational lens effect. Mon. Not. R. Astron. Soc. **307** (1964)

Refsdal, S., Surdej, J.: Gravitational lenses. Rep. Prog. Phys. **56**, 117 (1994)

Rephaeli, Y.: On the determination of the degree of cosmological Compton distortions and the temperature of the cosmic blackbody radiation. Astrophys. J. **241**, 858 (1980)

Reshetnikov, V.P.: Sky surveys and deep fields of ground-based and space telescopes. Phys. Usp. **48**, 1109 (2005)

Ribeiro, M.B.: On modeling a relativistic hierarchical (fractal) cosmology by Tolman's spacetime. III. Numerical results. Astrophys. J. **415**, 469 (1993)

Riess, A.G., Filippenko, A.V., Challis, P., et al.: Observational evidence from supernovae for an accelerating universe and a cosmological constant. Astron. J. **116**, 1009 (1998)

Riess, A.G., Strolger, L.-G., Tonry, J., et al.: Type Ia supernova discoveries at $z > 1$ from the HST: Evidence for past deceleration and constraints on dark energy evolution. Astrophys. J. **607**, 665 (2004)

Robaina, A.R., Cepa, J.: Redshift-distance relations from Type Ia supernova observations. New constraints on grey dust models. Astron. Astrophys. **464**, 465 (2007)

Roberts, M.S.: The gaseous content of galaxies (survey lecture). In: Evans, D.S. (ed.) External Galaxies and Quasi-Stellar Objects. IAU Symp., vol. 44, p. 12. Reidel, Dordrecht (1972)

Robertson, S.L., Leiter, D.J.: Evidence for intrinsic magnetic moments in black hole candidates. Astrophys. J. **565**, 447 (2002)

Robertson, S.L., Leiter, D.J.: On intrinsic magnetic moments in black hole candidates. Astrophys. J. Lett. **596**, L203 (2003)

Robertson, S.L., Leiter, D.J.: On the origin of the universal radio-X-ray luminocity correlation in black hole candidates. Astrophys. J. Lett. **596**, L203 (2004)

Robertson, S.L., Leiter, D.J.: The magnetospheric eternally collapsing object (MECO) model of galactic black hole candidates and active galactic nuclei. In: Kreitler, P.V. (ed.) New Developments in Black Hole Reseach. Nova Science, New York (2005). astro-ph/0602453

Rosenbaum, S.D., Bomans, D.J.: The environment of low surface brightness galaxies. Astron. Astrophys. **422**, L5 (2004)

Rost, F.: Estimates of the fractal dimension and density of the universe from the empirical mass-radius relation in cluster systems. Preprint (2004)

Rudnick, L., Brown, S., Williams, L.R.: Extragalactic radio sources and the WMAP cold spot. Astrophys. J. **671**, 40 (2007)

Rudnicki, K.: The Cosmological Principles. Jagiellonian University, Cracow (1995)

Ruffini, R., Song, D.J., Taraglio, S.: The "ino" mass and the cellular large-scale structure of the universe. Astron. Astrophys. **190**, 1 (1988)

Rust, B.W.: Use of supernovae light curves for testing the expansion hypothesis and other cosmological relations. Ph.D. Thesis, Oak Ridge National Lab., TN (1974)

Sakai, S., Madore, B.F., Freedman, W.L., et al.: Detection of the tip of the red giant branch in NGC 3379 (M105) in the Leo I group using the Hubble space telescope. Astrophys. J. **478**, 49 (1997)

Sakai, S., Mould, J.R., Hughes, S.M.G., et al.: The Hubble space telescope key project on the extragalactic distance scale. XXIV. The calibration of Tully-Fisher relations and the value of the Hubble constant. Astrophys. J. **529**, 698 (2000)

Samuel, S.: Comments on an expanding universe (2005). astro-ph/0512282

Sandage, A.: Current problems in the extragalactic distance scale. Astrophys. J. **127**, 513 (1958)

Sandage, A.: The ability of the 200-inch telescope to discriminate between selected world models. Astrophys. J. **133**, 355 (1961)

Sandage, A.: The change of redshift and apparent luminosity of galaxies due to the deceleration of the expanding universes. Astrophys. J. **136**, 319 (1962)

Sandage, A.: The age of the galaxies and globular clusters: Problems of finding the Hubble constant and deceleration parameter. In: O'Connel, D.J.K. (ed.) Nuclei of Galaxies, p. 61. North-Holland, Amsterdam (1971)

Sandage, A.: The redshift–distance relation. I. Angular diameter of first-ranked cluster galaxies as a function of redshift: The aperture correction to magnitudes. Astrophys. J. **173**, 485 (1972)

Sandage, A.: The brightest stars in nearby galaxies. IV—The color-magnitude diagram for the brightest red and blue stars in NGC 2403. Astron. J. **89**, 621 (1984)

Sandage, A.: The redshift-distance relation. IX. Astrophys. J. **307**, 1 (1986)

Sandage, A.: Observational tests of world models. Annu. Rev. Astron. Astrophys. **26**, 561 (1988a)

Sandage, A.: Cepheids as distance indicators when used near their detection limit. Publ. Astron. Soc. Pac. **100**, 935 (1988b)

Sandage, A.: The case for $H_0 \approx 55$ from the 21 centimeter linewidth absolute magnitude relation for field galaxies. Astrophys. J. **331**, 606 (1988c)

Sandage, A.: The search for the curvature of space. Phys. Scr. **T43**, 7 (1992)

Sandage, A.: Bias properties of extragalactic distance indicators. I. The Hubble constant does not increase outward. Astrophys. J. **430**, 1 (1994)

Sandage, A.: Astronomical problems for the next three decades. In: Mamaso, A., Munch, G. (eds.) Key Problems in Astronomy and Astrophysics. Cambridge University Press, Cambridge (1995a)

Sandage, A.: Practical cosmology: Inventing the past. In: Binggeli, Buser, R. (eds.) The Deep Universe, pp. 1–232. Springer, Berlin (1995b)

Sandage, A.: Bias properties of extragalactic distance indicators. VIII, H_0 from distance-limited luminosity class and morphological type-specific luminosity functions for Sb, Sbc, and Sc galaxies calibrated using Cepheids. Astrophys. J. **527**, 479 (1999)

Sandage, A.: Breaking the Code of Cosmic Evolution: An Informal History of the Mount Wilson Observatory (1904–1950). Cambridge Univ. Press, Cambridge (2004)

Sandage, A.: The classification of galaxies: Early history and ongoing developments. Annu. Rev. Astron. Astrophys. **43**, 581 (2005)

Sandage, A.: The Tolman surface brightness test for the reality of the expansion. V. Provenance of the test and a new representation of the data for three remote Hubble space telescope galaxy clusters. Astron. J. **139**, 728 (2010)

Sandage, A., Bedke, J.: The Carnegie Atlas of Galaxies, vols. I, II. Carnegie Institution, Washington (1994)

Sandage, A., Tammann, G.A.: Steps toward the Hubble constant. II. The brightest stars in late-type spiral galaxies. Astrophys. J. **191**, 603 (1974)

Sandage, A., Tammann, G.A.: Steps toward the Hubble constant. V. The Hubble constant from nearby galaxies and the regularity of the local velocity field. Astrophys. J. **196**, 313 (1975)

Sandage, A., Tammann, G.A.: The Hubble constant as derived from 21 cm linewidths. Nature **307**, 326 (1984)

Sandage, A., Tammann, G.A.: Absolute magnitude calibrations of population I and II Cepheids and other pulsating variables in the instability strip of the HR diagram. Annu. Rev. Astron. Astrophys. **44**, 93 (2006a)

Sandage, A., Tammann, G.A.: The distance to the Virgo cluster from a recalibrated Tully-Fisher relation based on HST Cepheids and a demonstrated Teerikorpi cluster incompleteness bias (2006b). astro-ph/0608677

Sandage, A., Tammann, G.A., Hardy, E.: Limits on the local deviation of the universe for a homogeneous model. Astrophys. J. **172**, 253 (1972)

Sandage, A., Tammann, G.A., Federspiel, M.: Bias properties of extragalactic distance indicators. IV. Demonstration of the population incompleteness bias inherent in the Tully-Fisher method applied to clusters. Astrophys. J. **452**, 1 (1995)

Sandage, A., Tammann, G.A., Saha, A., Reindl, B., Macchetto, F.D., Panagia, N.: The Hubble constant: A summary of the HST program for the luminosity calibration of Type Ia supernovae by means of Cepheids. Astrophys. J. **253**, 843 (2006)

Sarkar, D., Amblard, A., Holz, D.E., Cooray, A.: Lensing and supernovae: Quantifying the bias on the dark energy equation of state. Astrophys. J. **678**, 1 (2008)

Schechter, P.: An analytic expression of the luminosity function of galaxies. Astrophys. J. **203**, 297 (1976)

Schechter, P.: Mass-to-light ratios for elliptical galaxies. Astron. J. **85**, 801 (1980)

Schild, R.E., Leiter, D.J., Robertson, S.L.: Observations supporting the existence of an intrinsic magnetic moment inside the central compact object within the quasar Q0957+561. Astron. J. **132**, 420 (2006)

Schlegel, D.J., Finkbeiner, D.P., Davis, M.: Maps of dust infrared emission for use in estimation of reddening and cosmic microwave background radiation foregrounds. Astrophys. J. **500**, 525 (1998)

Schneider, P., Ehlers, J., Falco, E.E.: Gravitational Lensing. Springer, Berlin (1992)

Schrödinger, E.: The energy components of the gravitational field. Phys. Z. **19**, 4 (1918)

Schulman, L.S., Seiden, P.E.: Hierarchical structure in the distribution of galaxies. Astrophys. J. **311**, 1 (1986)

Scott, E.L.: The brightest galaxy in a cluster as a distance indicator. Astron. J. **62**, 248 (1957)

Scranton, R., Ménard, B., Richards, G.T., Nichol, R.C.: Detection of cosmic magnification with the Sloan digital sky survey. Astrophys. J. **633**, 589 (2005)

Segal, I.E.: Covariant chronogeometry and extreme distances. Astron. Astrophys. **18**, 143 (1972)

Segal, I.E.: Mathematical Cosmology and Extragalactic Astronomy. Academic Press, New York (1976)

Segal, I.E., Nicoll, J.F.: Tests of local redshift–distance laws on the basis of a complete optical sample. Astrophys. J. **465**, 578 (1996)

Selety, F.J.: Beiträgen zum kosmologischen Problem. Ann. Phys. **68**(12), 281 (1922)

Selety, F.J.: Erwiderung auf die Bemergungen Einsteins über meine Arbeit "Beiträgen zum kosmologischen Problem". Ann. Phys. **72**(17), 58 (1923)

Shane, C.D.: Distribution of galaxies. In: Sandage, A., Sandage, M., Kristian, J. (eds.) Galaxies and the Universe. The University of Chicago Press, Chigaco (1975)

Shane, C.D., Wirtanen, C.A.: The distribution of galaxies. Publ. Lick Obs. **22**(1) (1967)

Shapley, H.: The Inner Metagalaxy. Yale University Press, New Haven (1957), 204 pp

Shapley, H., Ames, A.: A survey of the external galaxies brighter than the thirteenth magnitude. Ann. Harvard Coll. Obs. **88**(2) (1932)

Shapley, H., Shapley, M.B.: Studies based on the colors and magnitudes in stellar clusters. Fourteenth paper: Further remarks on the structure of the galactic system. Astrophys. J. **50**, 107 (1919)

Shectman, S.A., Landy, S.D., Oemler, A., et al.: The Las Campanas redshift survey. Astrophys. J. **470**, 172 (1996)

Shmaonov, T.: Prib. Teh. Eksp. **1**, 83 (1957) (in Russian)

Silk, J., White, S.D.M.: The determination of q_0 using X-ray and microwave observations of galaxy clusters. Astrophys. J. **226**, L103 (1978)

Sillanpää, A., Takalo, L.O., Pursimo, T., et al.: Confirmation of the 12-year optical outburst cycle in blazar OJ 287. Astron. Astrophys. **305**, L17 (1996)

Slipher, V.M.: Spectrographic observations of nebulae. Pop. Astron. **23**, 21 (1915)

Slipher, V.M.: Nebulae. Proc. Am. Philos. Soc. **56**, 403 (1917)

Smith, H. Jr.: Is there really a Lutz-Kelker bias? Reconsidering calibration with trigonometric parallaxes. Mon. Not. R. Astron. Soc. **338**, 891 (2003)

Smith, R.W.: The Expanding Universe: Astronomy's 'Great Debate' 1900–1931. Cambridge Univ. Press, Cambridge (1982), 220 pp

Sokolov, V.V.: Linear and nonlinear gravidynamics: static field of a collapsar. Astrophys. Space Sci. **191**, 231 (1992a)

Sokolov, V.V.: Nonlinear gravidynamics: energy-momentum tensor of collapsar field. Astrophys. Space Sci. **197**, 87 (1992b)

Sokolov, V.V.: The properties of the strong static field of a collapsar in gravidynamics. Astrophys. Space Sci. **197**, 179 (1992c)

Sokolov, V.V.: Scalar gravitational waves and observational limitations for the energy-momentum tensor of a gravitational field. Astrophys. Space Sci. **198**, 53 (1992d)

Sokolov, V.V.: Identifications of the Gamma-bursts: Optical transients and host galaxies. Dissertation on doctor of physical-mathematical sciences degree, SAO RAS (2002) (in Russian)

Sokolov, V.V., Baryshev, Yu.V.: Field theory approach to gravity. Energy momentum tensor of the field. Gravit. Relativ. Theory **17**, 34 (1980)

Sokolov, V.V., Bisnovatyi-Kogan, G.S., Kurt, V.G., Gnedin, Yu.N., Baryshev, Yu.V.: Observational constraints on the angular and spectral distributions of photons in gamma-ray burst sources. Astron. Rep. **50**, 612 (2006)

Soneira, R.M.: A cosmological redshift-distance square law? Astrophys. J. **230**, L63 (1979)

Soneira, R.M., Peebles, P.J.E.: Is there evidence for a spatially homogeneous population of field galaxies? Astrophys. J. **211**, 1 (1977)

Soneira, R.M., Peebles, P.J.E.: A computer model universe: simulation of the nature of the galaxy distribution in the Lick catalog. Astron. J. **83**, 845 (1978)

Spaenhauer, A.M.: A systematic comparison of four methods to derive stellar space densities. Astron. Astrophys. **65**, 313 (1978)

Spergel, D.N., Steinhardt, P.J.: Observational evidence for self-interacting cold dark matter. Phys. Rev. Lett. **84**, 3760 (2000)

Spergel, D.N., Verde, L., Peiris, H.V., et al.: First year Wilkinson microwave anisotropy probe (WMAP) observations: Determination of cosmological parameters. Astrophys. J. Suppl. **148**, 175 (2003)

Spergel, D.N., Bean, R., Dore, O., et al.: Three-year Wilkinson microwave anisotropy probe (WMAP) observations: Implications for cosmology. Astrophys. J. Suppl. **170**, 377 (2007)

Springel, V., White, S.D., Jenkins, A., et al.: Simulations of the formation, evolution and clustering of galaxies and quasars. Nature **435**, 629 (2005)

Starkman, G.D., Trotta, R.: Why anthropic reasoning cannot predict Λ. Phys. Rev. Lett. **97**(20), 201301 (2006)

Stickel, M., Lemke, D., Mattila, K., Haikala, L.K., Haas, M.: Far-infrared emission of intracluster dust in the Coma galaxy cluster. Astron. Astrophys. **329**, 55 (1998)

Straumann, N.: Reflections on gravity (2000). astro-ph/0006423

Strauss, M.A., Huchra, J.P., Davis, M., Yahil, A., Fisher, K.B., Tonry, J.: A redshift survey of IRAS galaxies. VII. The infrared and redshift data for the 1.936 Jansky sample. Astrophys. J. Suppl. Ser. **83**, 29 (1992)

Sulentic, J.W.: Redshift differentials in a complete sample of galaxy groups. Astrophys. J. **286**, 442 (1984)

Suntola, T.: The Dynamic Universe, Toward a Unified Picture of Physical Reality. Physics Foundations Society/CreateSpace, Seattle (2011)

Sunyaev, R.A., Zeldovich, Ya.B.: The observations of relic radiation as a test of the nature of X-ray radiation from the clusters of galaxies. Comm. Astrophys. Space Phys. **4**, 173 (1972)

Surdej, J., Claeskens, J.F., Crampton, D., et al.: Gravitational lensing statistics based on a large sample of highly luminous quasars. Astron. J. **105**, 2064 (1993)

Sylos Labini, F.: Isotropy, homogeneity, and dipole saturation. Astrophys. J. **433**, 464 (1994)

Sylos Labini, F.: Inhomogeneities in the universe (2011). arXiv:1103.5974, contribution to Classical and Quantum Gravity special issue "Inhomogeneous Cosmological Models and Averaging in Cosmology"

Sylos Labini, F., Amendola, L.: The power spectrum in a strongly inhomogeneous universe. Astrophys. J. **468**, L1 (1996)

Sylos Labini, F., Pietronero, L.: Multifractality as a link between luminosity and space distribution of visible matter. Astrophys. J. **469**, 26 (1996)

Sylos Labini, F., Vasilyev, N.L.: Extension and estimation of correlations in cold dark matter models. Astron. Astrophys. **477**, 381 (2008)

Sylos Labini, F., Montuori, M., Pietronero, L.: Scale-invariance of galaxy clustering. Phys. Rep. **293**, 61 (1998)

Sylos Labini, F., Vasilyev, N.L., Baryshev, Yu.V.: Power law correlations in galaxy distribution and finite volume effects from the Sloan digital sky survey data release four. Astron. Astrophys. **465**, 23 (2007)

Sylos Labini, F., Vasilyev, N.L., Baryshev, Yu.V.: Large-scale fluctuations in the distribution of galaxies from the two-degree galaxy redshift survey. Astron. Astrophys. **496**, 7 (2009a)

Sylos Labini, F., Vasilyev, N.L., Baryshev, Yu.V.: Breaking the self-averaging properties of spatial galaxy fluctuations in the Sloan digital sky survey data release six. Astron. Astrophys. **508**, 17 (2009b)

Sylos Labini, F., Vasilyev, N.L., Baryshev, Yu.V., Lopez-Corredoira, M.: Absence of anti-correlations and of baryon acoustic oscillations in the galaxy correlation function from the Sloan digital sky survey DR7. Astron. Astrophys. **505**, 981 (2009c)

Szalay, A.S., Schramm, D.N.: Are galaxies more strongly correlated than clusters? Nature **314**, 718 (1985)

Tabur, V., Kiss, L.L., Bedding, T.R.: Hipparcos calibration of the tip of the red giant branch. Astrophys. J. **703**, L72 (2009)

Taganov, I.N.: Conception of quantum cosmology. In: Baryshev, Yu., Taganov, I.N., Teerikorpi, P. (eds.) Practical Cosmology II, p. 68. Russian Geographical Society, St.Petersburg (2008)

Tammann, G.A.: Precise determination of the distances of galaxies. In: Scientific Research with the Space Telescope. IAU Colloq., vol. 54, p. 263 (1979)

Tammann, G.A.: The ups and downs of the Hubble constant. In: Röser, S. (ed.) Reviews of Modern Astronomy, vol. 19 (2006); Karl Schwarzschild Lecture (in collaboration with B. Reindl)

Tammann, G.A., Sandage, A., Yahil, A.: The determination of cosmological parameters. In: Balian, R., Adouze, J., Schramm, D.N. (eds.) Physical Cosmology, p. 53. North-Holland, Amsterdam (1980)

Tammann, G.A., Sandage, A., Reindl, B.: New period-luminosity and period-color relations for classical Cepheids: I. Cepheids in the galaxy. Astron. Astrophys. **404**, 423 (2003)

Tanyukhin, I.M.: Energy of gravitational field and models of neutron stars. Diplom work, St.-Petersburg University (1995) (in Russian)

Teerikorpi, P.: A note on the velocity–distance relationship for nearby galaxies and galaxy groups. Observatory **95**, 105 (1975a)

Teerikorpi, P.: On the effect of the luminosity selection on the redshift–distance relationship. Astron. Astrophys. **45**, 117 (1975b)

Teerikorpi, P.: Local Hubble constant from nearby Sc galaxies. Astron. Astrophys. **50**, 455 (1976)

Teerikorpi, P.: A study of galactic absorption as revealed by the Rubin et al. sample of Sc galaxies. Astron. Astrophys. **64**, 379 (1978)

Teerikorpi, P.: A study of galactic absorption as revealed by the reddenings of quasars. Astron. Astrophys. **98**, 300 (1981a)

Teerikorpi, P.: On the Hubble diagram for quasars as corrected for galactic absorption: Evidence for a separate class of the most luminous quasars. Astron. Astrophys. **98**, 309 (1981b)

Teerikorpi, P.: Note on the use of Type I supernovae as cosmic clocks. Acta Cosmol. **10**, 21 (1981c)

Teerikorpi, P.: Studies on the Hubble law with emphasis on selection effects and galactic absorption. Report Series No. D4, Dep. Phys. Sci. Univ. Turku (1981d)

Teerikorpi, P.: Malmquist bias in a relation of the form $M = a \cdot p + b$. Astron. Astrophys. **141**, 407 (1984)

Teerikorpi, P.: Inner rings in galaxies and the value of H_0. Vestn. Leningr. State Univ. **2**, 121 (1986)

Teerikorpi, P.: Cluster population incompleteness bias and distances from the Tully-Fisher relation: Theory and numerical examples. Astron. Astrophys. **173**, 39 (1987)

Teerikorpi, P.: Lundmark's unpublished 1922 nebula classification. J. Hist. Astron. **20**, 165 (1989)

Teerikorpi, P.: A statistical effect in stellar reddenings with implications for extinction at high galactic latitudes. Astron. Astrophys. **235**, 362 (1990a)

Teerikorpi, P.: Theoretical aspects in the use of the inverse Tully-Fisher relation for distance determinations. Astron. Astrophys. **234**, 1 (1990b)

Teerikorpi, P.: On general Malmquist corrections to direct and inverse Tully–Fisher distance moduli. Astron. Astrophys. **280**, 443 (1993)

Teerikorpi, P.: Observational selection bias affecting the determination of the extragalactic distance scale. Annu. Rev. Astron. Astrophys. **35**, 101 (1997)

Teerikorpi, P.: Cosmological Malmquist bias in the Hubble diagram at high redshifts. Astron. Astrophys. **339**, 647 (1998)

Teerikorpi, P.: Evidence for the class of the most luminous quasars II. Variability, polarization, and the gap in the M_V distribution. Astron. Astrophys. **353**, 77 (2000)

Teerikorpi, P.: Evidence for the class of the most luminous quasars III. Linear sizes and the core-size relation of double radio sources. Astron. Astrophys. **375**, 752 (2001)

Teerikorpi, P.: Dusty haloes of galaxies at intermediate redshifts. Astron. Astrophys. **386**, 865 (2002)

Teerikorpi, P.: Evidence for the class of the most luminous quasars IV. Cosmological Malmquist bias and the Λ term. Astron. Astrophys. **399**, 829 (2003)

Teerikorpi, P.: Influence of a generalized Eddington effect on galaxy counts. Astron. Astrophys. **424**, 73 (2004)

Teerikorpi, P.: On Öpik's distance evaluation method in a cosmological context. Astron. Astrophys. **531**, A10 (2011)

Teerikorpi, P., Chernin, A.D.: The Hubble diagram for a system within dark energy: The location of the zero-gravity radius and the global Hubble rate. Astron. Astrophys. **516**, A93 (2010)

Teerikorpi, P., Kotilainen, J.: Reddenings of quasars vs. PSA counts at high galactic latitudes. Astrophys. Space Sci. **155**, 19 (1989)

Teerikorpi, P., Paturel, G.: Evidence for the extragalactic Cepheid distance bias from the kinematical distance scale. Astron. Astrophys. **381**, L37 (2002)

Teerikorpi, P., Bottinelli, L., Gouguenheim, L., Paturel, G.: Investigations of the local supercluster velocity field: I. Observations close to Virgo, using Tully-Fisher distances, and the Tolman-Bondi expanding sphere. Astron. Astrophys. **260**, 17 (1992)

Teerikorpi, P., Hanski, M., Theureau, G., et al.: The radial space distribution of KLUN-galaxies up to 200 Mpc: Incompleteness or evidence for the behaviour predicted by fractal dimension $\approx$2?. Astron. Astrophys. **334**, 395 (1998)

Teerikorpi, P., Ekholm, T., Hanski, M., Theureau, G.: Theoretical aspects of the inverse Tully-Fisher relation as a distance indicator: incompleteness in $\log V_{\rm max}$, the relevant slope, and the calibrator sample bias 1999. Astron. Astrophys. **343**, 713 (1999)

Teerikorpi, P., Gromov, A., Baryshev, Yu.: Limits on dark energy–matter interaction from the Hubble relation for two-fluid FLRW models. Astron. Astrophys. **407**, L9 (2003)

Teerikorpi, P., Chernin, A.D., Baryshev, Yu.V.: The quiescent Hubble flow, local dark energy tests, and pairwise velocity dispersion in a $\Omega = 1$ universe. Astron. Astrophys. **440**, 791 (2005)

Teerikorpi, P., Chernin, A.D., Karachentsev, I.D., Valtonen, M.J.: Dark energy in the environments of the local group, the M81 group and the CenA group: The normalized Hubble diagram. Astron. Astrophys. **483**, 383 (2008)

Teerikorpi, P., Valtonen, M., Lehto, K., Lehto, H., Byrd, G., Chernin, A.: The Evolving Universe and the Origin of Life—The Search for Our Cosmic Roots. Springer, New York (2009)

Tegmark, M., Hamilton, A., Xu, Y.: The power spectrum of galaxies in the 2dF 100k redshift survey. Mon. Not. R. Astron. Soc. **335**, 887 (2002)

Tegmark, M., Blanton, M., Strauss, M.: The three-dimensional power spectrum of galaxies from the Sloan digital sky survey. Astrophys. J. **606**, 702 (2004)

Theureau, G., Hanski, M., Teerikorpi, P., et al.: Kinematics of the local universe IV Type dependence in the diameter Tully-Fisher relation and implications on the mass-luminosity structure. Astron. Astrophys. **319**, 435 (1997a)

Theureau, G., Hanski, M., Ekholm, T., et al.: Kinematics of the local universe. V. The value of H_0 from the Tully–Fisher B and D_{25} relations for field galaxies. Astron. Astrophys. **322**, 730 (1997b)

Theureau, G., Coudreau, G., Hallet, N., et al.: Kinematics of the local universe XII. 21-cm line measurements of 586 galaxies with the new Nancay receiver. Astron. Astrophys. **430**, 373 (2005)

Thirring, W.E.: An alternative approach to the theory of gravitation. Ann. Phys. **16**, 96 (1961)

Tifft, W.G.: Periodicity in the redshift intervals for double galaxies. Astrophys. J. **236**, 70 (1980)

Tifft, W.G.: Global redshift periodicities and periodicity structure. Astrophys. J. **468**, 491 (1996)

Tifft, W.G., Cocke, W.J.: Global redshift quantization. Astrophys. J. **287**, 492 (1984)

Tikhonov, A.V., Karachentsev, I.D.: Minivoids in the local volume. Astrophys. J. **653**, 969 (2006)

Tikhonov, A., Makarov, D.: Correlation properties of galaxies in the local supercluster. Astron. Lett. **29**, 289 (2003)

Tikhonov, A., Makarov, D., Kopylov, A.: Investigation of the clustering of galaxies, clusters of galaxies and superclusters with the correlation gamma function. Bull. Spec. Astrophys. Obs. **50**, 39 (2000)

Tolman, R.C.: On the estimation of distances in a curved universe with a non-static line element. Proc. Natl. Acad. Sci. USA **16**, 511 (1930)

Totsuji, H., Kihara, T.: The correlation function for the distribution of galaxies. Publ. Astron. Soc. Jpn. **21**, 221 (1969)

Triay, R., Lachieze-Rey, M., Rauzy, S.: On the Malmquist bias in the determination of H_0 and of distances of galaxies. Astron. Astrophys. **289**, 19 (1994)

Trimble, V.: The distance to the Crab nebula and NP 0532. Publ. Astron. Soc. Pac. **85**, 579 (1973)

Troitskii, V.S.: Physical constants and evolution of the universe. Astrophys. Space Sci. **139**, 389 (1987)

Tully, R.B.: Alignment of clusters and galaxies on scales up to 0.1c. Astrophys. J. **303**, 25 (1986)

Tully, R.B.: More about clustering on scale of 0.1c. Astrophys. J. **323**, 1 (1987)

Tully, R.B.: Origin of the Hubble constant controversy. Nature **334**, 209 (1988)

Tully, B., Fisher, R.: A new method of determining distances to galaxies. Astron. Astrophys. **54**, 661 (1977)

Tully, R.B., Pierce, M.J.: Distances to galaxies from the correlation between luminosities and line widths. III. Cluster template and global measurement of H_0. Astrophys. J. **533**, 744 (2000)

Tully, R.B., Shaya, E.J., Karachentsev, I.D., et al.: Our peculiar motion away from the local void. Astrophys. J. **676**, 184 (2008)

Turner, M.: Making sense of the new cosmology. Int. J. Mod. Phys. A **17**, 180 (2002)

Tyson, J.A.: Deep CD survey: Galaxy luminosity and color evolution. Astron. J. **96**, 1 (1988)

Unzicker, A.: Why do we still believe in Newton's Law? Facts, myths and methods in gravitational physics (2007). gr-qc/0702009

Uzan, J.-P.: The fundamental constants and their variation: Observational status and theoretical motivations. Rev. Mod. Phys. **75**, 403 (2003)

Valtonen, M.J., Byrd, G.G.: Redshift asymmetries in systems of galaxies and the missing mass. Astrophys. J. **303**, 523 (1986)

Valtonen, M.J., Lehto, H.J.: Outbursts in OJ287: A new test for the general theory of relativity. Astrophys. J. **481**, L5 (1997)

Valtonen, M.J., Lehto, H.J., Nilsson, K., et al.: A massive binary black-hole system in OJ287 and a test of general relativity. Nature **452**, 851 (2008a)

Valtonen, M.J., Niemi, S., Teerikorpi, P., et al.: Redshift asymmetry as an indication of dark energy around groups of galaxies. American Astronomical Society, DDA meeting 39.0103V (2008b)

Van den Bergh, S.: A preliminary luminosity classification of late-type galaxies. Astrophys. J. **131**, 215 (1960) (also: Astrophys. J. **131**, 558)

Vasilyev, N.: Investigation of the methods of correlation analysis of large scale structures and their application to the 2dF survey. Graduate work, Astronomy Department, St.Petersburg University (2004)

Vasilyev, N., Baryshev, Yu., Sylos Labini, F.: Large scale correlations in galaxy clustering from the VL samples of 2dFGRS. Astron. Astrophys. **447**, 431 (2006)

Veblen, O., Whitehead, J.H.C.: The Foundations of Differential Geometry. Cambridge University Press, Cambridge (1932)

Vettolani, G., Zucca, E., Zamorani, G., et al.: The ESO slice project (ESP) galaxy redshift survey. I. Description and first results. Astron. Astrophys. **325**, 954 (1997)

von der Pahlen, E.: Lehrbuch der Stellarstatistik. Johann Ambrosius Barth, Leipzig (1937)

von Seeliger, H.: Über das Newton'sche Gravitationsgesetz. Astron. Nachr. **137**, 129 (1895)

Wagner, R.: Exploring quantum gravity with very-high-energy gamma-ray instruments—Prospects and limitations. AIP Conf. Proc. **1112**, 187 (2009)

Walker, A.G.: Complete symmetric spaces. J. Lond. Math. Soc. **19**, 219 (1944)

Warwick, C.: The effect of irregular absorption on galaxy distribution. Proc. Natl. Acad. Sci. USA **36**, 415 (1950)

Watkins, R., Feldman, H.A., Hudson, M.J.: Consistently large cosmic flows on scales of 100h-1Mpc: a challenge for the standard ΛCDM cosmology. Mon. Not. R. Astron. Soc. **392**, 743 (2009)

Webb, S.: Measuring the Universe. The Cosmological Distance Ladder. Springer, Berlin (1999)

Webster, A.: The clustering of radio sources–II The Parkes 2700-MHz and Bologna B2 surveys. Mon. Not. R. Astron. Soc. **179**, 511 (1977)

Weinberg, S.: Photons and gravitons in perturbation theory: Derivation of Maxwell's and Einstein's equations. Phys. Rev. B **138**, 988 (1965)

Weinberg, S.: The First Three Minutes. Basic Books, New York (1977)

Weinberg, S.: The cosmological constant problem. Rev. Mod. Phys. **61**, 1 (1989)

Weisberg, J.M., Taylor, J.H.: General relativistic geodetic spin precession in binary pulsar B1913+16: Mapping the emission beam in two dimensions. Astrophys. J. **576**, 942 (2002)

Weisberg, J.M., Nice, D.J., Taylor, J.H.: Timing measurements of the relativistic binary pulsar PSR B1913+16. Astrophys. J. **722**, 1030 (2010)

Wertz, J.R.: A Newtonian big-bang hierarchical cosmological model. Astrophys. J. **164**, 227 (1971)

Wesson, P.S.: General-relativistic hierarchical cosmology: An exact model. Astrophys. Space Sci. **54**, 489 (1978)

Westphal, A., Abele, H., Baebler, S., et al.: A quantum mechanical description of the experiment on the observation of gravitationally bound states. Eur. Phys. J. C **51**, 367 (2007)

Weyl, H.: Eine neue Erweiterung der Relativitätstheorie. Ann. Phys. **59**, 129 (1919)

Whiting, A.B.: The expansion of space: Free-particle motion and the cosmological redshift. Observatory **124**, 173 (2004). astro-ph/0404095

Whitney, H.: The mathematics of physical quantities. Am. Math. Mon. **75**, 115 (1968) p. 227

Wiik, K., Valtaoja, E.: The geometry of the universe from high resolution VLBI data of AGN shocks. Astron. Astrophys. **366**, 1061 (2001)

Will, C.M.: Theory and Experiment in Gravitational Physics. Cambridge University Press, Cambridge (1993)

Will, C.M.: The confrontation between general relativity and experiment. Living Rev. Relativ. **9**, 3 (2005)

Wilms, J., Reynolds, C., Begelman, M., et al.: XMM-EPIC observation of MCG-6-30-15: Direct evidence for the extraction of energy from a spinning black hole? Mon. Not. R. Astron. Soc. **328**, L27 (2001)

Wilson, O.C.: Possible applications of supernovae to the study of the nebular redshifts. Astrophys. J. **90**, 634 (1939)

Wu, K., Lahav, O., Rees, M.: The large-scale smoothness of the universe. Nature **397**, 225 (1999)

Xulu, S.S.: The energy-momentum problem in general relativity. PhD Thesis (2003). arXiv: hep-th/0308070

Yahil, A., Tammann, G.A., Sandage, A.: The local group—the solar motion relative to its centroid. Astrophys. J. **217**, 903 (1977)

Yilmaz, H.: Toward a field theory of gravitation. Nuovo Cimento B **107**, 941 (1992)

Yilmaz, H.: Toward a comprehensible physical theory: gravity and quantum mechanics, a modern synthesis. In: Jeffers, S., et al. (eds.) The Present Status of the Quantum Theory of Light, p. 503. Kluwer Academic, Dordrecht (1997)

Yonehara, A., Umemura, M., Susa, H.: Quasar mesolensing—Direct probe to substructures around galaxies. Publ. Astron. Soc. Jpn. **55**, 1059 (2003)

Yoshii, Yu.: Detection and selection effects in observations of faint galaxies. Astrophys. J. **403**, 552 (1993)

Yoshii, Yu., Takara, F.: Galactic evolution and cosmology: probing the cosmological deceleration parameter. Astrophys. J. **326**, 1 (1988)

Zakharov, V.I.: Spin of virtual gravitons. Zh. Eksp. Teor. Fiz. **48**, 303 (1965)
Zehavi, I., Blanton, M.R., Frieman, J.A., et al.: Galaxy clustering in early Sloan digital sky survey redshift data. Astrophys. J. **571**, 172 (2002)
Zehavi, I., Zheng, Zh., Weinberg, D.H., et al.: The luminosity and color dependence of the galaxy correlation function. Astrophys. J. **630**, 1 (2005)
Zeldovich, Ya.B., Novikov, I.D.: Relativistic Astrophysics, vol. 2. The University of Chicago Press, Chicago (1984)
Zeldovich, Ya.B., Sunyaev, R.A.: The interaction of matter and radiation in a hot-model universe. Astrophys. Space Sci. **4**, 301 (1969a)
Zeldovich, Ya.B., Sunyaev, R.A.: The interaction of matter and radiation in a hot-model universe. Astrophys. Space Sci. **7**, 20 (1969b)
Zwicky, F.: On the red shift of spectral lines through interstellar space. Proc. Natl. Acad. Sci. USA **15**, 773 (1929)
Zwicky, F.: Die Rotverschiebung von extragalaktischen Nebeln. Helv. Phys. Acta **6**, 110 (1933)
Zwicky, F.: On the masses of nebulae and of clusters of nebulae. Astrophys. J. **86**, 217 (1937)
Zwicky, F.: Statistics of clusters of galaxies. In: Proceedings of the Third Berkeley Symposium on Mathematical Statistics and Probability, p. 113. University of California Press, Los Angeles (1955)
Zwicky, F.: Morphological Astronomy. Springer, Berlin (1957)
Zwicky, F., Herzog, E., Wild, P., Karpowicz, M., Kowal, C.T.: Catalogue of Galaxies and Clusters of Galaxies. California Institute of Technology, Pasadena (1961–68)

Index

Y. Baryshev, P. Teerikorpi, *Fundamental Questions of Practical Cosmology*, Astrophysics and Space Science Library 383,
DOI 10.1007/978-94-007-2379-5, © Springer Science+Business Media B.V. 2012

MIX
Papier aus verantwortungsvollen Quellen
Paper from responsible sources
FSC® C105338

If you have any concerns about our products,
you can contact us on
ProductSafety@springernature.com

In case Publisher is established outside the EU,
the EU authorized representative is:
Springer Nature Customer Service Center GmbH
Europaplatz 3, 69115 Heidelberg, Germany

Printed by Libri Plureos GmbH
in Hamburg, Germany